195
Advances in Polymer Science

Editorial Board:
A. Abe · A.-C. Albertsson · R. Duncan · K. Dušek · W. H. de Jeu
J.-F. Joanny · H.-H. Kausch · S. Kobayashi · K.-S. Lee · L. Leibler
T. E. Long · I. Manners · M. Möller · O. Nuyken · E. M. Terentjev
B. Voit · G. Wegner · U. Wiesner

Advances in Polymer Science

Recently Published and Forthcoming Volumes

**Supramolecular Polymers/Oligomers/
Polymeric Betains**
Vol. 201, 2006

Ordered Polymeric Nanostructures at Surfaces
Materials, Chemistry, Formation, Properties,
and Applications
Volume Editor: Vansco, G. J.
Vol. 200, 2006

Emissive Materials/Nanomaterials
Vol. 199, 2006

Surface-Initiated Polymerization II
Volume Editor: Jordan, R.
Vol. 198, 2006

Surface-Initiated Polymerization I
Volume Editor: Jordan, R.
Vol. 197, 2006

**Conformation-Dependent Design of Sequences
in Copolymers II**
Volume Editor: Khokhlov, A. R.
Vol. 196, 2006

**Conformation-Dependent Design of Sequences
in Copolymers I**
Volume Editor: Khokhlov, A. R.
Vol. 195, 2006

Enzyme-Catalyzed Synthesis of Polymers
Volume Editors: Kobayashi, S., Ritter, H.,
Kaplan, D.
Vol. 194, 2006

Polymer Therapeutics II
Polymers as Drugs, Conjugates and Gene
Delivery Systems
Volume Editors: Satchi-Fainaro, R., Duncan, R.
Vol. 193, 2006

Polymer Therapeutics I
Polymers as Drugs, Conjugates and Gene
Delivery Systems
Volume Editors: Satchi-Fainaro, R., Duncan, R.
Vol. 192, 2006

**Interphases and Mesophases in Polymer
Crystallization III**
Volume Editor: Allegra, G.
Vol. 191, 2005

Block Copolymers II
Volume Editor: Abetz, V.
Vol. 190, 2005

Block Copolymers I
Volume Editor: Abetz, V.
Vol. 189, 2005

**Intrinsic Molecular Mobility and Toughness
of Polymers II**
Volume Editor: Kausch, H.-H.
Vol. 188, 2005

**Intrinsic Molecular Mobility and Toughness
of Polymers I**
Volume Editor: Kausch, H.-H.
Vol. 187, 2005

Polysaccharides I
Structure, Characterization and Use
Volume Editor: Heinze, T.
Vol. 186, 2005

**Advanced Computer Simulation
Approaches for Soft Matter Sciences II**
Volume Editors: Holm, C., Kremer, K.
Vol. 185, 2005

Crosslinking in Materials Science
Vol. 184, 2005

Conformation-Dependent Design of Sequences in Copolymers I

Volume Editor: Alexei R. Khokhlov

With contributions by
P. G. Khalatur · A. R. Khokhlov · E. E. Makhaeva
I. M. Okhapkin · C. Wu · G. Zhang

The series *Advances in Polymer Science* presents critical reviews of the present and future trends in polymer and biopolymer science including chemistry, physical chemistry, physics and material science. It is adressed to all scientists at universities and in industry who wish to keep abreast of advances in the topics covered.
As a rule, contributions are specially commissioned. The editors and publishers will, however, always be pleased to receive suggestions and supplementary information. Papers are accepted for *Advances in Polymer Science* in English.
In references *Advances in Polymer Science* is abbreviated *Adv Polym Sci* and is cited as a journal.

Springer WWW home page: http://www.springer.com
Visit the APS content at http://www.springerlink.com/

Library of Congress Control Number: 2005934719

ISSN 0065-3195
ISBN-10 3-540-29513-5 Springer Berlin Heidelberg New York
ISBN-13 978-3-540-29513-6 Springer Berlin Heidelberg New York
DOI 10.1007/11569572

This work is subject to copyright. All rights are reserved, whether the whole or part of the material is concerned, specifically the rights of translation, reprinting, reuse of illustrations, recitation, broadcasting, reproduction on microfilm or in any other way, and storage in data banks. Duplication of this publication or parts thereof is permitted only under the provisions of the German Copyright Law of September 9, 1965, in its current version, and permission for use must always be obtained from Springer. Violations are liable for prosecution under the German Copyright Law.

Springer is a part of Springer Science+Business Media

springer.com

© Springer-Verlag Berlin Heidelberg 2006
Printed in Germany

The use of registered names, trademarks, etc. in this publication does not imply, even in the absence of a specific statement, that such names are exempt from the relevant protective laws and regulations and therefore free for general use.

Cover design: *Design & Production* GmbH, Heidelberg
Typesetting and Production: LE-TEX Jelonek, Schmidt & Vöckler GbR, Leipzig

Printed on acid-free paper   02/3141 YL - 5 4 3 2 1 0

Volume Editor

Prof. Alexei R. Khokhlov
Physics Department
Moscow State University
119992 Moscow, Russia
khokhlov@polly.phys.msu.ru

Editorial Board

Prof. Akihiro Abe
Department of Industrial Chemistry
Tokyo Institute of Polytechnics
1583 Iiyama, Atsugi-shi 243-02, Japan
aabe@chem.t-kougei.ac.jp

Prof. A.-C. Albertsson
Department of Polymer Technology
The Royal Institute of Technology
10044 Stockholm, Sweden
aila@polymer.kth.se

Prof. Ruth Duncan
Welsh School of Pharmacy
Cardiff University
Redwood Building
King Edward VII Avenue
Cardiff CF 10 3XF, UK
DuncanR@cf.ac.uk

Prof. Karel Dušek
Institute of Macromolecular Chemistry,
Czech
Academy of Sciences of the Czech Republic
Heyrovský Sq. 2
16206 Prague 6, Czech Republic
dusek@imc.cas.cz

Prof. W. H. de Jeu
FOM-Institute AMOLF
Kruislaan 407
1098 SJ Amsterdam, The Netherlands
dejeu@amolf.nl
and Dutch Polymer Institute
Eindhoven University of Technology
PO Box 513
5600 MB Eindhoven, The Netherlands

Prof. Jean-François Joanny
Physicochimie Curie
Institut Curie section recherche
26 rue d'Ulm
75248 Paris cedex 05, France
jean-francois.joanny@curie.fr

Prof. Hans-Henning Kausch
Ecole Polytechnique Fédérale de Lausanne
Science de Base
Station 6
1015 Lausanne, Switzerland
kausch.cully@bluewin.ch

Prof. Shiro Kobayashi
R & D Center for Bio-based Materials
Kyoto Institute of Technology
Matsugasaki, Sakyo-ku
Kyoto 606-8585, Japan
kobayash@kit.ac.jp

Prof. Kwang-Sup Lee
Department of Polymer Science &
Engineering
Hannam University
133 Ojung-Dong
Daejeon 306-791, Korea
kslee@hannam.ac.kr

Prof. L. Leibler
Matière Molle et Chimie
Ecole Supérieure de Physique
et Chimie Industrielles (ESPCI)
10 rue Vauquelin
75231 Paris Cedex 05, France
ludwik.leibler@espci.fr

Prof. Timothy E. Long
Department of Chemistry
and Research Institute
Virginia Tech
2110 Hahn Hall (0344)
Blacksburg, VA 24061, USA
telong@vt.edu

Prof. Ian Manners
School of Chemistry
University of Bristol
Cantock's Close
BS8 1TS Bristol, UK
ian.manners@bristol.ac.uk

Prof. Martin Möller
Deutsches Wollforschungsinstitut
an der RWTH Aachen e.V.
Pauwelsstraße 8
52056 Aachen, Germany
moeller@dwi.rwth-aachen.de

Prof. Oskar Nuyken
Lehrstuhl für Makromolekulare Stoffe
TU München
Lichtenbergstr. 4
85747 Garching, Germany
oskar.nuyken@ch.tum.de

Prof. E. M. Terentjev
Cavendish Laboratory
Madingley Road
Cambridge CB 3 OHE, UK
emt1000@cam.ac.uk

Prof. Brigitte Voit
Institut für Polymerforschung Dresden
Hohe Straße 6
01069 Dresden, Germany
voit@ipfdd.de

Prof. Gerhard Wegner
Max-Planck-Institut
für Polymerforschung
Ackermannweg 10
Postfach 3148
55128 Mainz, Germany
wegner@mpip-mainz.mpg.de

Prof. Ulrich Wiesner
Materials Science & Engineering
Cornell University
329 Bard Hall
Ithaca, NY 14853, USA
ubw1@cornell.edu

Advances in Polymer Science
Also Available Electronically

For all customers who have a standing order to Advances in Polymer Science, we offer the electronic version via SpringerLink free of charge. Please contact your librarian who can receive a password or free access to the full articles by registering at:

springerlink.com

If you do not have a subscription, you can still view the tables of contents of the volumes and the abstract of each article by going to the SpringerLink Homepage, clicking on "Browse by Online Libraries", then "Chemical Sciences", and finally choose Advances in Polymer Science.

You will find information about the

- Editorial Board
- Aims and Scope
- Instructions for Authors
- Sample Contribution

at springeronline.com using the search function.

Preface

For a long time, chemical industry was focused on polymers mainly from the viewpoint of obtaining advanced construction materials, such as plastics, rubbers, fibers, polymer composites. These materials provided a number of important benefits, including improved strength and long-term durability, light weight, environmental resistance, and design flexibility. Starting from about the early 1980s, the main focus of interest shifted to functional polymers. Among these are superabsorbents, nanoporous rate-controlling membranes, reversible adhesives, electro-conductive polymers and nanowires. In the 1990s, the scientific and industrial polymer community started to discuss "smart" or "intellectual" polymer systems (e.g., soft manipulators, polymer systems for controlled drug release, field-responsive polymers, shape memory networks, and self-healing coatings); the meaning behind these terms is that simply the functions performed by polymers become more sophisticated and diverse. The line of research concentrating on polymer systems with more and more complex functions will certainly be in the mainstream of polymer science in the 21st century.

One of the ways to obtain new polymers for sophisticated functions is connected to the synthesis of novel building blocks – monomer species – where the required function is linked to the chemical structure of these blocks. However, the potential of this approach is rather limited because complicated and diverse functions of polymeric materials would then require a very complex structure of monomers, which normally means that the organic synthesis is more expensive and less robust. An alternative approach is to use known building blocks and to try to design a copolymer with a given sequence of these units. At the present time, there are many synthetic and theoretical strategies directed towards varying the chemical sequences of copolymers: from the variation of their composition and blocky structure to more sophisticated features like "tapered" and gradient structures. In a broad sense, most conventional chemical syntheses, especially polymer syntheses, can be regarded as *bottom-up nanotechnology* leading to the assembly of building blocks into the final macromolecules. Unfortunately, in most conventional polymerization processes, the physical control of assembly during the reactions is practically impossible.

With these difficulties in mind, it is instructive to look at main biological macromolecules – proteins, DNA, and RNA – that have precise and specific structures. These polymers in living systems are responsible for functions, which are incomparably more complex and diverse than the functions that we normally discuss for synthetic copolymers. The molecular basis for this ability to perform sophisticated functions is associated with the primary sequences of biopolymers. In particular, the unique functions of proteins reflect their molecular structure, so that the sequence of amino acids in a protein defines its secondary and tertiary structure (fibrous versus globular, for example) as well as its function. Therefore, the study of biopolymers at the molecular level may point to new directions in materials design and construction – not just actually using biomolecules themselves to construct novel materials, but mimicking the specific primary sequences of biological polymers. Indeed, by taking this biomimetic route, significant and path-breaking achievements may be made by materials scientists and engineers. Thus, a promising path related to the design of new synthetic copolymers is to learn the rules of biologically driven controlled synthesis and directed assembly in vivo and eventually to apply these rules to the creation of engineered synthetic polymer systems from cheap and commonly available building blocks.

Original ideas connected with the biomimetic design of sequences in synthetic functional copolymers were formulated by us in 1998. They were based on the simple and well-known fact that the function of all globular proteins depends on two main factors: (1) they are globular and (2) they are soluble in aqueous medium. The combination of these two factors is nontrivial, e.g., for homopolymers and random copolymers the transition to globular conformation is usually accompanied by the precipitation of globules from the solution. Protein globules are soluble in water because of the special primary sequence: in the native conformation most of hydrophobic monomer units are in the core of the globule while hydrophilic monomer units form the envelope of this core. Keeping the biomimetic approach described above in mind, one can formulate the following problem: is it possible to design such a sequence of synthetic HP copolymer (copolymer consisting of monomer units of two types, H and P) that, in the most dense globular conformation, all hydrophobic H-units are in the core of this globule while the hydrophilic (polar) P-units form the envelope of this core? The corresponding bio-inspired two-letter HP copolymers generated on a computer were called *protein-like* copolymers. Of course, the degree of function complexity that we hope to achieve for designed copolymers, is much less than current biopolymers, but the behavior of copolymers with designed sequences can exhibit many useful features distinguishing them from the "scratch" (e.g., statistically random) sequences.

The approach formulated in 1998 is based on the assumption that a copolymer obtained under some preparation conditions is able to "remember" features of its original conformation from which it was built and to store the corresponding information in the resulting sequence. Therefore, this approach may be called *conformation-dependent sequence design*.

Most of the contributions collected in volumes 195 and 196 are dedicated to the review of the results obtained recently in this direction, i.e., dealing with the conformation-dependent sequence design of copolymers and the study of their properties. The contributions collected present the up-to-date research results on most of the topics related to this approach. Surface and interfacial phenomena are also major topics of these volumes. The individual chapters are diverse in purpose and in style. Some of them paint the field in broad, conceptual strokes, others in fine methodical detail. Some present information, others arguments or interpretation. Some summarize past activities, others point to future potential.

Volume 195 begins with a chapter by Khalatur and Khokhlov that gives a survey of the simulation methods as applied to the design of nontrivial sequences in synthetic copolymers aimed to achieve desired functional properties. Several new synthetic strategies allowing for the synthesis of copolymers with a broad variation of their sequence distributions are reported. Synthetic copolymers exhibiting long-range statistical correlations, large-scale compositional heterogeneities, and physical complexity are the focal point of this review. Roughly speaking, the physical complexity of a sequence is understood as the amount of information that is stored in that sequence about a particular environment.

Zhang and Wu review the experimental results on the folding of different hydrophilically or hydrophobically modified copolymers in extremely dilute solutions. The focus is on the formation of stable and soluble mesoglobules made of several amphiphilic copolymers and on the folding of single copolymers – both linear and grafted – into core-shell nanostructures that have various applications. All these features are directly related to the unique chemical sequence of the copolymer chains under discussion. The authors discuss the insights that can be obtained by the analysis of the data of laser light scattering and the use of this method as a potential probe of fine molecular structures, including the so-called "molten" globule state. Emphasis is put on the most recent achievements, although important historical contributions are also mentioned.

A concept of amphiphilicity, as applied to single monomer units of designed water-soluble polymers, is presented in the third chapter by Okhapkin, Makhaeva, and Khokhlov. The concept is relevant to biomolecular structures and assemblies in aqueous solution. The authors consider the substantial body of information obtained experimentally and theoretically on surface molecular chemical structures, including those that are prospective for surface catalysis. Unusual conformational behaviors of single amphiphilic polymers recently observed in simulations are also discussed in detail.

The problems related to the colloidal stability of amphiphilic polymers in water are reviewed by Aseyev, Tenhu, and Winnik in the first chapter of volume 196. The focus is on the derivatives of thermally responsive smart macromolecules – both on copolymers and homopolymers – which are present in a solution as stable micelles potentially having various applications.

One of the promising synthetic strategies of conformation-dependent sequence design is based on direct copolymerization under unusual conditions.

This strategy was first realized by Lozinsky et al., who studied the redox-initiated free-radical copolymerization of thermosensitive N-vinylcaprolactam with hydrophilic N-vinylimidazole at different temperatures, as well as by Chi Wu and coworkers. Lozinsky presents an extensive review of the experimental approaches, both already described in the literature and potential new ones, to chemical synthesis of protein-like copolymers capable of forming core-shell nanostructures in a solution.

Continuing along the chemical theme, Kuchanov and Khokhlov review comprehensively the diffusion-controlled polymer-analogous reactions and free-radical copolymerization. The field of polymer chemistry today is distinguished by its depth of mathematical and quantitative rigor in the pursuit of a wide array of challenging new subjects. These ingredients are present in full in this chapter, where the quantitative theory of solution and interphase free-radical copolymerization is discussed from the viewpoint of statistical chemistry and statistical physics of polymers. It is shown that the interaction of these two disciplines has significant impact on the development of new synthetic strategies and technologies in polymer chemistry, including those related to the conformation-dependent design of nontrivial copolymer sequences.

The final short historical review by Grosberg and Khokhlov is devoted exclusively to the discussion of fundamental ideas initially formulated in polymer physics by the outstanding Russian scientist I.M. Lifshitz and especially to today's development of his ideas. Major attention is focused on the statistical theory of heteropolymers covering such areas as protein folding and the sequence design both of protein macromolecules and of synthetic copolymers mimicking some protein properties. In particular, the ideas of sequence design in functional copolymers were originated within the school of Lifshitz; thus, in volume 196, it seems natural to give a review of the achievements along the lines of all major ideas initially formulated by Lifshitz in the field of polymer science.

All of the selected contributions that are present in these special volumes are good representatives for manifesting the importance of the concepts based on conformation-dependent sequence design. It has been our intention to provide the scientific and industrial polymer community with a comprehensive view of the current state of knowledge on designed polymers. Both volumes attempt to review what is currently known about these polymers in terms of their synthesis, chemical and physical properties, and applications. We will feel the volumes have been successful if some of the chapters presented here stimulate readers to become interested in and solve specific problems in this rapidly developing field of research.

Moscow, November 2005 *Alexei R. Khokhlov*

Contents

Computer-Aided Conformation-Dependent Design of Copolymer Sequences
P. G. Khalatur · A. R. Khokhlov 1

Folding and Formation of Mesoglobules in Dilute Copolymer Solutions
G. Zhang · C. Wu 101

Water Solutions of Amphiphilic Polymers:
Nanostructure Formation and Possibilities for Catalysis
I. M. Okhapkin · E. E. Makhaeva · A. R. Khokhlov 177

Author Index Volumes 101–195 211

Subject Index 237

Contents of Volume 196

Conformation-Dependent Design of Sequences in Copolymers II

Volume Editor: Alexei R. Khokhlov
ISBN: 3-540-29515-1

Temperature Dependence of the Colloidal Stability
of Neutral Amphiphilic Polymers in Water
V. O. Aseyev · H. Tenhu · F. M. Winnik

Approaches to Chemical Synthesis of Protein-Like Copolymers
V. I. Lozinsky

Role of Physical Factors in the Process of Obtaining Copolymers
S. I. Kuchanov · A. R. Khokhlov

After-Action of the Ideas of I. M. Lifshitz in Polymer
and Biopolymer Physics
A. Y. Grosberg · A. R. Khokhlov

…

Computer-Aided Conformation-Dependent Design of Copolymer Sequences

Pavel G. Khalatur[1,2] (✉) · Alexei R. Khokhlov[1,2,3]

[1]Institute of Organoelement Compounds, Russian Academy of Sciences, 117823 Moscow, Russia
khalatur@germany.ru

[2]Department of Polymer Science, University of Ulm, 89069 Ulm, Germany
khalatur@germany.ru

[3]Physics Department, Moscow State University, 119899 Moscow, Russia

1	Introduction: Two Paradigms in Sequence Design	5
2	New Synthetic Strategies in Sequence Design	8
2.1	Preliminary Remarks	8
2.2	CDSD Via Polymer-Analogous Modification	10
2.2.1	Protein-like Copolymers: Structure Dictates Sequence	10
2.2.2	Long-Range Correlations and Their Measure	14
2.2.3	Hydrophobic Modification of Hydrophilic Polymers	19
2.2.4	Adsorption-Tuned Copolymers	23
2.2.5	Design as a Simulation of Evolutionary Process	25
2.3	CDSD Via Copolymerization	31
2.3.1	Conditions for CDSD	31
2.3.2	Copolymerization with Simultaneous Globule Formation	32
2.3.3	Emulsion Copolymerization	36
2.3.4	Copolymerization Near a Selectively Adsorbing Surface	39
2.3.5	Copolymerization Near a Patterned Surface	43
2.4	Design of Monomeric Units	48
2.4.1	Amphiphilic Polymers	48
2.4.2	HA Model	49
3	Properties of Designed Copolymers	51
3.1	Single Chains	51
3.1.1	Coil-to-Globule Transition	51
3.1.2	Kinetics of the Collapse Transition	54
3.2	Phase Behavior: The Sequence-Assembly Problem	57
3.2.1	The Polymer RISM Theory	58
3.2.2	Field-Theoretic Calculation	63
3.2.3	Molecular Dynamics Simulation	64
3.2.4	Evolutionary Approach	67
3.3	Charged Hydrophobic Copolymers	70
3.3.1	Solution Properties	71
3.3.2	Designed Copolymers in the Presence of Monovalent Counterions	72
3.3.3	Effect of Multivalent Counterions	74
3.3.4	Stabilization Mechanism	76
3.3.5	Experimental Results	78

3.4	Hydrophobic-Amphiphilic Copolymers	79
3.4.1	Single Amphiphilic Chains	81
3.4.2	Coil-Globule Transition Versus Aggregation	86
3.5	Adsorption Selectivity	90
3.5.1	Adsorption-Tuned Copolymers	90
3.5.2	Molecular Dispenser	91
4	Conclusion	93
References		95

Abstract A survey is given of the simulation methods as applied to the design of nontrivial sequences in synthetic copolymers aimed at achieving desired functional properties. We consider a recently developed approach, called conformation-dependent sequence design (CDSD), which is based on the assumption that a copolymer obtained under certain preparation conditions is able to "remember" features of the original conformation in which it was built and to store the corresponding information in the resulting sequence. The emphasis is on copolymer sequences exhibiting large-scale compositional heterogeneities and long-range statistical correlations between monomer units. Several new synthetic strategies and polymerization processes that allow synthesis of copolymers with a broad variation of their sequence distributions are reported. We demonstrate that the CDSD polymer-analogous transformation is a versatile approach allowing various functional copolymers to be obtained. Another synthetic strategy is the CDSD step growth copolymerization which is carried out under special conditions. It includes the intrinsic possibilities of exploiting the heterogeneities of the reaction system to control the chemical microstructure of the synthesized copolymers, making possible new paradigms for synthesis and production of polymeric materials. In both cases, we try to show how the preparation conditions dictate copolymer sequences. Also, we discuss advances that have recently been achieved in the computer simulation and theoretical understanding of designed copolymers in solution and in bulk. The focus is on amphiphilic protein-like copolymers and on hydrophobic polyelectrolytes. Here, we demonstrate how copolymer sequence dictates structure and properties.

Keywords Charged heteropolymers · Copolymers · Phase behavior · Polyamphiphiles · Sequence design · Simulation · Solution properties

Abbreviations
A	amphiphilic group (monomer, monomeric unit)
α^2	chain expansion factor
$[A_M]$	mole fraction of intermolecular aggregates of size M
ATC	adsorption-tuned copolymer
ATRP	atom transfer radical polymerization
b	bond length
C	symmetric matrix of direct correlation functions
χ	Flory–Huggins interaction parameter
$\widetilde{\chi}$	effective interaction parameter
C_α^0	bulk concentration of monomer species α
$C_\alpha(r)$	instant local concentration of monomer species α
$C_\alpha(z)$	equilibrium concentration profile of monomer species α
$c(r)$	direct site-site pair correlation function

CRP	controlled radical polymerization
DCTT	degenerative chain transfer technique
CDSD	conformation-dependent sequence design
DF	density functional
DFA	detrended fluctuation analysis
ΔG	change in association free energy
D_L	block length dispersion
D_λ	dispersion within a sliding window of length λ
$\Delta \mu_s$	solvation free energy
$\Delta(q)$	determinant of matrix integral equation
ΔT^*	change in transition temperature
ε	energy parameter
E	unit diagonal matrix
ε^*	critical adsorption energy
ε_{PP}	attraction energy between hydrophilic (polar) segments
f	fraction of charged monomers
Φ	volume fraction of macromolecules
φ_α	average fraction of monomer species α in copolymer chain
ϕ_α	volume fraction of monomer species α
$\varphi_\alpha^{(i)}$	intrachain composition profile of monomer species α
$\phi_\alpha(r)$	volume fraction field of monomer species α
$F_D(\lambda)$	detrended local fluctuations within a window of length λ
$f(\lambda)$	block length distribution function
F_s	sequence free energy
γ	solvation parameter
h	Shannon's entropy
H	hydrophobic monomer (monomeric unit)
H	symmetric matrix of total site-site correlation functions
HA model	hydrophobic-amphiphilic (side-chain) model
HPE	hydrophobic polyelectrolyte
HP model	hydrophobic-hydrophilic(polar) model
$h(r)$	total site-site pair correlation function
JS	Jensen–Shannon divergence measure
K	association equilibrium constant
k_B	Boltzmann constant
ℓ	block length
L	average block length
λ	length of sliding window along copolymer sequence
L_H	average length of hydrophobic blocks
L_P	average length of hydrophilic (polar) blocks
LRC	long-range correlation
m	average number of copolymer chains per aggregate
MAST	macrophase separation transition
MFPT	mean first passage time
MIST	microphase separation transition
N	total chain length, number of repeat units
N_α	number of repeat units of type α (α = A, B) in the chain
NIPA	poly(N-isopropylacrylamide)
n_J	number of crosslinks
N_τ	current chain length

Ω_α	volume occupied by monomer species α
$\omega_\alpha(r)$	average chemical potential field of monomer species α
ODT	order-disorder transition
P	hydrophilic (polar) monomer (monomeric unit)
PA	hydrophilic-amphiphilic copolymer
$p_\alpha^{(i)}$	probability that monomer α is located at the ith position in the chain
PCF	site-site pair correlation function
PEO	poly(ethylene oxide)
PM	probabilistic model
PMF	potential of mean force
PMMA	poly(methylmethacrylate)
PRISM	polymer-reference-interaction-site model
PrP	prion protein
PS	polystyrene
$P(\sigma,T)$	probability of copolymer/particle complex
q	wave number
$\boldsymbol{q}$	wave vector
q^*	wave number of maximum instability (peak in the structure factor)
$q(r,s)$	propagator
$q^\dagger(r,s)$	conjugate propagator
R	size of micelle
ρ	monomer number density
r^*	spatial scale of microdomain structure (domain size)
R_g	radius of gyration
R_g^2	mean-square gyration radius
R_g^{app}	apparent radius of gyration
$R_g(t)$	time-dependent radius of gyration
R_{gH}^2	partial mean-square gyration radius of hydrophobic monomers
R_{gP}^2	partial mean-square gyration radius of hydrophilic monomers
$R_{g\Theta}^2$	mean-square gyration radius of unperturbed chain
RISM	integral equation reference-interaction-site model
RPA	random phase approximation
r_s	distance between nearest adsorption sites
s	contour length of chain
σ	monomer size
σ_α	effective size of monomer species α
SASA	solvent-accessible surface area
$S_\alpha(q)$	partial scattering function for monomer species α
SCF	self-consistent-field
SCMF	self-consistent mean-field
SFRP	stable free-radical polymerization
σ_p	size of "parental" particle
T	absolute temperature
τ	reduced temperature
Θ	Flory theta temperature
T^*	temperature of spinodal instability
$\Theta_{\alpha\beta}^{(i)}$	chemical correlator

T_c	critical temperature of counterion condensation
τ_D	characteristic diffusion time
τ_R	reaction time characterizing polymerization rate
τ_{rel}	chain relaxation time
T_s	sequence design temperature
T_s^*	critical sequence design temperature
$u(r)$	site-site potential describing interaction between nonbonded monomer units
v	probability of location of terminal reactive site in a given volume
W	matrix of intramolecular correlation functions
$w(q)$	single-chain form-factor
$w(r)$	intramolecular site-site correlation function
$W(r)$	radial distribution function of monomeric units
$W(r^*)$	distribution of domain sizes
z	counterion charge (valence)

1
Introduction: Two Paradigms in Sequence Design

Copolymers have been studied extensively for several decades, partly because of their industrial and biological importance, and partly because of their interesting and sometimes perplexing properties. Many physical and mechanical properties of copolymers, which comprise two or more covalently bonded sequences of chemically distinct monomer species, depend on both the comonomer composition and the arrangement of these comonomers in the polymer chain. There may be significant differences, for example, between two polymer systems with the same chemical composition, but one of which has the comonomers randomly distributed in the chain while the other has long blocks of each monomer type.

One may say that in many cases, sequence dictates structure and properties. To illustrate this, we will mention only two familiar examples.

Synthetic block copolymers can spontaneously self-assemble into highly ordered patterns of supramolecular structures (condensed modulated phases), showing a surprisingly rich morphological behavior. These modulated phases with length scales on the order of 1 to 10^3 nm can potentially form the basis for various nanotechnology applications, including the design of synthetic hierarchical materials, and may be effectively controlled by changing the lengths of blocks or their distribution along the chain [1].

Typically, proteins fold to organize a very specific globular conformation, known as the protein's native state, which is in general reasonably stable and unique. It is this well-defined three-dimensional conformation of a polypeptide chain that determines the macroscopic properties and function of a protein. The folding mechanism and biological functionality are directly related to the polypeptide sequence; a completely random amino acid sequence is unlikely to form a functional structure. In this view, polypeptide sequence forces a protein to be more than a collapsed heteropolymer, but rather to assume a highly specific three-dimensional structure. Hence, a fundamental issue is how functional protein sequences, which determine biologically active structures, differ from random sequences. Understanding the relationship between a protein's sequence and its native structure is one of the key problems in modern science [2–4].

In recent years we have seen intense interest in developing new types of functional polymer macromolecules via clever design of sequences of monomeric units in a copolymer chain. Broadly speaking, sequence design may be defined as an approach aimed at finding the optimum sequence that provides desired properties of the resultant polymer. This requires a scoring function which may typically be based on physical principles, knowledge-based approaches, or a specifically designed function. Alas, insofar as the terms "sequence design" and "sequence engineering" imply a rational, planned approach to the creation or modification of copolymer structure and function, both still remain beyond our capabilities in a general way.

There are two main paradigms in the sequence design problem.

In protein science, the *de novo* sequence design problem consists of finding a sequence of amino acids that fold into a *target* globular structure. This problem is sometimes called the inverse protein folding problem. Many current methods for *de novo* protein sequence design consist of numerically mutating a sequence until a maximum stability is achieved for the target structure that is usually considered as a ground state. There are a number of reviews that cover this subject [5–9]. In polymer chemistry and physics, emphasis is on the development of new methods of synthesis, on the control of (co)polymer stereochemistry and architecture, and on the design of high performance polymeric materials tailored for specific uses and properties.

The difference between the two sequence design concepts is related to several essential differences between natural and non-natural copolymers. We mention here only a few of them.

The order of amino acids in a polypeptide chain produced by the synthetic apparatus of the living cell is always the same for a particular protein so that all the protein sequences of a given type are structurally identical copies in every cell in a living organism. We cannot distinguish one individual protein sequence from another. For most synthetic copolymers, produced industrially or synthesized in research labs, the occurrence of a certain degree of sequence disorder is almost inevitable. Therefore, if we are speaking about a synthetic copolymer sequence, we mean, explicitly or implicitly, that averaging over many different sequences has been carried out. For a protein to function, it must be in its highly specific native conformation that is stable only in a narrow temperature region. On the other hand, the properties and functions of synthetic copolymers are not so tightly related to their conformation. Moreover, we are mainly interested in the nonunique copolymer conformations. The interior of proteins has the packing density of a molecular crystal while synthetic globules are typically liquid-like. This list is of course far from exhaustive, but should rather be taken as simply a set of examples.

In the present article we will deal mainly with synthetic macromolecules and practically will not touch on biopolymers.

Diblock and repeated-block AB copolymers are the simplest examples of two-letter copolymers made up of two different monomer species, denoted by letters A and B. More sophisticated distributions of chemically different groups along the chain are characteristic of random and random-block copolymers, including uncorrelated or ideal random copolymers and so-called correlated random copolymers. In the former class, the corresponding chemical sequences are uncorrelated and this corresponds to Bernoulli or zeroth-order Markov processes [10]. In the latter class, the correlation in the sequences of both types of A and B segments is defined by means of a first-order Markov process. It is important to emphasize that in both cases the correlations characterizing the distribution of monomeric units along copolymer chains decay exponentially. There are, however, copolymers for which this is not the case. In this review, we will consider just these copolymers, focusing on the computer-aided design of their chemical sequences as well as on the properties of designed polymers.

Although recent years have witnessed an impressive confluence of experiments and statistical theories, presently there is no comprehensive understanding of the interrelation between chemical sequences in synthetic copolymers and the conditions of synthesis. One has merely to glance at recent literature in polymer science and biophysics to realize that the problem of sequence-property relationship is by no means entirely solved. As always, in these circumstances, an alternative to analytical theories is computer simulations, which are designed to obtain a numerical answer without knowledge of an analytical solution.

The computer simulations are likely to be useful in two distinct situations—the first in which numerical data of a specified accuracy are required, possibly for some utilitarian purpose; the second, perhaps more fundamental, in providing guidance to the theoretician's intuition, e.g., by comparing numerical results with those from approximate analytical approaches. As a consequence, the physical content of the model will depend upon the purpose of the calculation. Our attention here will be focused largely on the coarse-grained (lattice and off-lattice) models of polymers. Naturally, these models should reflect those generic properties of polymers that are the result of the chain-like structure of macromolecules.

Apart from the introductory section, the article is subdivided into two major sections: Synthesis and Properties of designed copolymers.

2
New Synthetic Strategies in Sequence Design

2.1
Preliminary Remarks

Today, the majority of all polymeric materials is produced using the free-radical polymerization technique [11–17]. Unfortunately, however, in conventional free-radical copolymerization, control of the incorporation of monomer species into a copolymer chain is practically impossible. Furthermore, in this process, the propagating macroradicals usually attach monomeric units in a random way, governed by the relative reactivities of polymerizing comonomers. This lack of control confines the versatility of the free-radical process, because the microscopic polymer properties, such as chemical composition distribution and tacticity are key parameters that determine the macroscopic behavior of the resultant product.

The absence of control of the incorporation of monomers into the polymeric chain implies that many macroscopic properties cannot be influenced to a large extent. Therefore, in recent years much effort has been directed towards the development of controlled radical polymerization (CRP) methods for the preparation of various copolymers (for a recent review, see [17]).

These methods are based on the idea of establishing equilibrium between the active and dormant species in solution phase. In particular, the methods include three major techniques called stable free-radical polymerization (SFRP), atom transfer radical polymerization (ATRP), and the degenerative chain transfer technique (DCTT) [17]. Although such syntheses pose significant technical problems, these difficulties have all been successively overcome in the last few years. Nevertheless, the procedure of preparation of the resulting copolymers remains somewhat complicated.

On the other hand, it should be realized that radical copolymerization at heterogeneous conditions offers additional unique opportunities not available in homogeneous (solution) copolymerization. These include the intrinsic possibilities of exploiting the heterogeneities of the reaction system to control the chemical microstructure of the synthesized copolymers, making possible new paradigms for synthesis and production of polymeric materials. In this contribution, we discuss some new synthetic strategies, which have been developed in recent years to provide effective control of the chemical sequences.

In a series of publications [18–20], the concept of conformation-dependent sequence design (CDSD) of functional copolymers has been introduced (for recent reviews, see [21–25]). The essence of the proposed approach is based on the assumption that a copolymer obtained under certain preparation conditions is able to remember features of the original ("parent") conformation in which it was built and to store the corresponding information in the resulting chemical sequence. In other words, this concept takes into account a strong coupling between the conformation and primary structure of copolymers during their synthesis. Ideologically, the approach [18–20] bears some similarities with that proposed earlier in the context of the problems of protein physics [26–30], however, it is aimed at synthetic copolymers rather than biopolymers. The original idea for protein design [26–30] consisted of running through sequences of amino acids to determine which sequence (or sequences) had the lowest energy in a unique (target) conformation. In the Introduction, we stressed the differences between this approach and CDSD.

In polymer chemistry, there are two known CDSD techniques: (i) the chemical modification (polymer-analogous transformation) of homopolymers and (ii) the step-growth copolymerization of monomers with different properties under special conditions. We will address both these techniques.

Polymer simulations are being done on different levels. An atomistic model of a polymer contains all the atoms that are present in the real polymer. Coarse-grained models simplify the problem by combining atoms into effective united atoms. In this way, only significant microscopic information— "essential features" of the real system —is retained. The unit can represent a chemical group of a few atoms, a monomeric unit in a polymer, groups of monomeric units, or chain segments of various lengths. Certainly, whether a coarse-grained description is adequate for understanding a particular polymer system depends very much on the problem being studied. The challenge lies in selecting just the right amount of atomistic detail to build coarse-grained models with maximum generality. To a certain degree, of course, that is precisely what physics is about. Therefore, the question is how can we construct copolymer models that are sufficiently realistic to capture the essential features of real macromolecules yet simple enough to allow large-scale computations of polymer conformation and dynamics?

There are many kinds of polymerizing monomers used to make up copolymers. These differ in physical and chemical properties. One of the most important differences (essential features) is their solubility, that is, how much they like or dislike a solvent, e.g., water. Hence the chemical and atomistic details of different monomeric units may not be necessary to understand the properties of many "two-letter" copolymers. In what follows, we will mainly use the so-called HP model [31]. This two-letter model of a linear hydrophobic/hydrophilic macromolecule reflects the spirit of minimalist models, in that it is simple yet based on a physical principle.

The HP model is a coarse-grained (lattice or off-lattice) polymer model that abstracts from real polymers in two important ways: (i) Instead of modeling the positions of all atoms of the polymer, it models only the backbone structure of the polymer, i.e., one position for each monomeric unit. (ii) Usually, only the hydrophobic interaction between the monomeric units is modeled, therefore the model distinguishes only two kinds of monomeric units, namely hydrophobic (H) and hydrophilic (or polar, P).

2.2
CDSD Via Polymer-Analogous Modification

2.2.1
Protein-like Copolymers: Structure Dictates Sequence

The studies of structures formed by copolymers consisting of two kinds of monomeric units constitute a rather large field of polymer chemistry and physics [32]. The systems that are most extensively studied are block-copolymers (with a block primary structure) and random copolymers (with statistical primary structure). Sometimes, copolymers with some short-range correlations along the chain are also investigated. Such correlations will always show up after the copolymerization process, if the probability of addition of unit A or B to the growing chain depends on the type of unit that was added on the previous polymerization step [17]. The type of primary structure that emerges in this case can be characterized as "random with short-range correlations". On the other hand, globular proteins can also be regarded from a very rough viewpoint as a kind of binary copolymer. Indeed, the most important difference between the monomeric units of globular proteins is that some amino acid residues are hydrophobic, while others are hydrophilic or charged. We can very roughly attribute the index H to the former type of units and the index P to the latter ones [33]. If we then analyze the primary structures of the globular proteins obtained in this way and compare them with the simple primary structure of conventional synthetic copolymers, we should draw the conclusion that protein-originated AB texts are much more informative and specific.

It is generally believed that in globular proteins the hydrophilic P units mainly cover the surface of the globule, giving rise to their stability against intermolecular aggregation, while hydrophobic H units main form the core of the globule [33]. It can be assumed that such a requirement (in the dense globular state the P amino acids should be on the surface and the H amino acids in the core) is rather restrictive, i.e., it is satisfied only for a very small fraction of all possible primary structures. Moreover, since the HP correlations defined in such a way depend on the conformation of the globule as a whole (i.e., on the ternary structure), they should be characterized as long-range ones.

The question is, whether such primary structures can be obtained for binary copolymers, not obligatorily of biological origin. It is easy to do this by computer simulation [18], and much more difficult in real experiments. However, in both cases the corresponding procedure should involve the following stages that are schematically depicted in Fig. 1:

Stage 1. We take a homopolymer coil with excluded volume interactions in a good solvent.

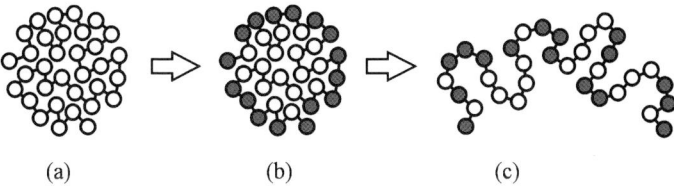

Fig. 1 Schematic representation of the CDSD chemical modification of a homopolymer chain: **a** initial conformation formed by a homopolymer globule; **b** chemical modification (surface "coloring"); **c** resulting copolymer. Modified chain segments are shown in *gray*

Stage 2. Strong attraction between all monomeric units is switched on and a homopolymer globule (parent conformation) is formed (Fig. 1a). Of course, when we are speaking about real experiments, by switching on the attraction we should understand the jump of temperature, addition of poor solvent, etc.

Stage 3. This step is more easily realizable in computer experiments. We should simply consider the "instant photo" of the globule and "color" the units on the surface and in the core in different "colors", i.e., we assign the index P to those units that are on the surface of the globule and call these units hydrophilic and assign the index H to the units in the core of the globule and call these units hydrophobic. Then we fix this primary structure (Fig. 1b). In real experiments the "coloring" of the surface can be done with a chemical reagent Z entering the reaction with monomeric units and converting them from hydrophobic to hydrophilic, $H + Z \rightarrow P$. In chemical language, this is called polymer-analogous transformation. If the amount of reagent is small enough, only surface monomeric units will be contaminated, the core remaining hydrophobic. Another important feature is to have a fast-enough coloring reaction and slow-enough intermolecular aggregation (which will always take place under the conditions when globules are formed). To slow down the aggregation, the neutral thickeners of the aqueous solutions may be used.

Stage 4. This last step is necessary for computer realization. The uniform strong attraction of units should be switched off, and different interaction potentials should be introduced for H and P units.

Initially, the protein-like HP sequences were generated in [18] for the lattice chains of $N = 512$ monomeric units (statistical segments), using for simulations a Monte Carlo method and the lattice bond-fluctuation model [34]. When the chain is a random (quasirandom) heteropolymer, an average over many different sequence distributions must be carried out explicitly to produce the final properties. Therefore, the sequence design scheme was repeated many times, and the results were averaged over different initial configurations.

It turned out that many statistical properties of protein-like and random copolymers with the same HP composition are very different. In order to be able to distinguish whether this difference is due to the special sequence design described above, or just due to the different degree of blockiness, one can introduce for comparison also the random-block primary sequence. The random-block HP copolymers have the same chemical composition and the same average length L of uninterrupted H or P sequence as protein-like copolymers, but in other respects the HP sequence is random. In [18], the distribution of block length λ was taken in the Poisson form: $f(\lambda) = e^{-L} L^{\ell} / \ell!$.

In Fig. 2 we present the typical distributions of H and P monomers along the chain for regular (multiblock) copolymers as well as for purely random, random-block, and protein-like copolymers. From the comparison of the primary structures of the random and protein-like copolymers we can see that the average lengths of H and P blocks in the protein-like copolymer are notably longer. On the other hand, the copolymer with the random-block architecture, having the same average length of H and P sections, L_H and L_P, as for the protein-like copolymer, exhibits a different distribution of these blocks

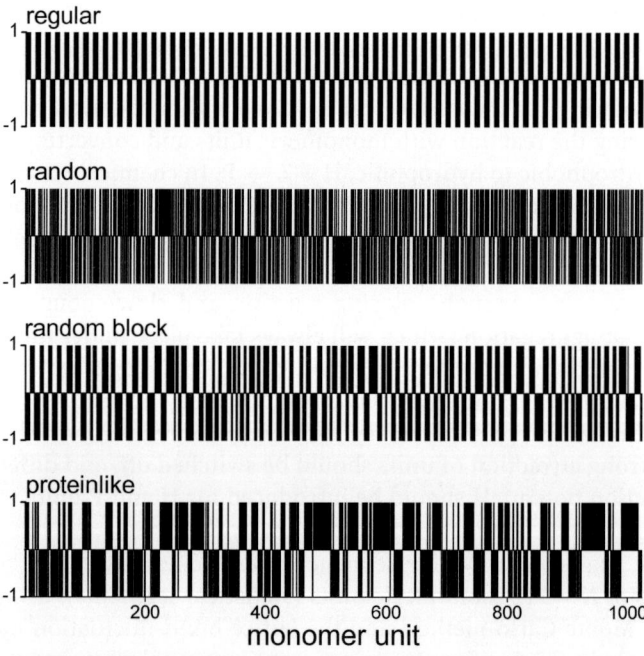

Fig. 2 Typical distributions of H and P monomeric units along the chain for a regular (multiblock) sequence with a fixed block length of 8 units, as well as for a purely random sequence with an average block length $L = 2$, a random-block sequence with $L = 6.4$, and a protein-like sequence with $L = 6.4$. The H units are denoted as +1 and the P units as −1. The sequence length is $N = 1024$

along the chain. A main feature of the protein-like sequences is the presence of rather long uniform H and P sections.

Also, it was found that the coil-to-globule transition for protein-like copolymers, induced by the strong attraction of H segments, occurs at higher temperatures, leads to the formation of denser globules and has faster kinetics than for random and random-block counterparts [18–20]. The reason for this is illustrated in Fig. 3 where the typical snapshots of globules formed by protein-like and random copolymers are shown. One can see that the HP heteropolymer obtained as a result of the simple one-step coloring procedure can self-assemble into a segregated core-shell microstructure thus resembling some of the basic properties of globular proteins. The core of the protein-like globule is much more compact and better formed compared to that observed for random copolymers; it is surrounded by the loops of hydrophilic segments, which stabilize the core.

Apparently, this is due to some *memory effect*: the core, which existed in the parent globular conformation (Fig. 1b), was simply reproduced upon refolding caused by the attraction between H units. One may say that the features of the parent conformation are "inherited" by the protein-like copolymer. Looking at the conformations of Fig. 3, it is natural to argue that protein-like copolymer globules could be soluble in water and thus they are open to

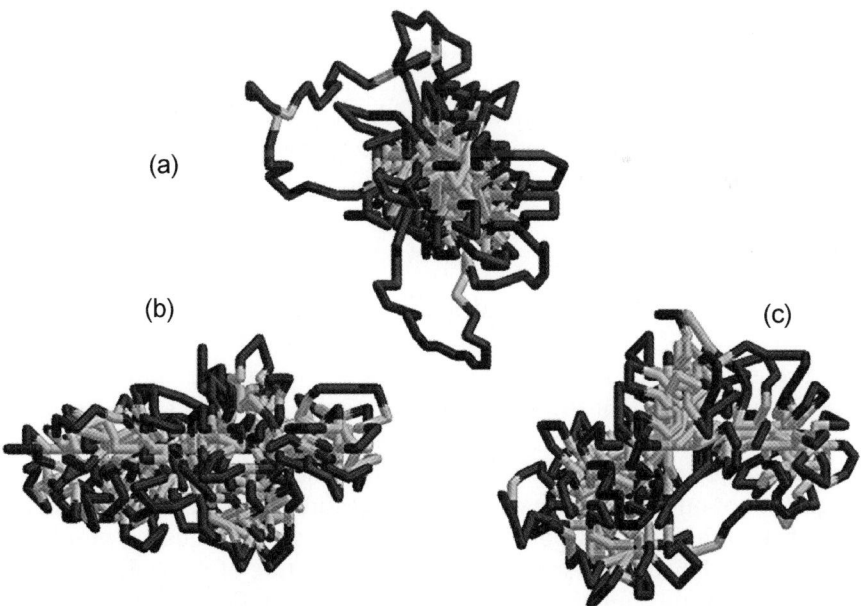

Fig. 3 Typical snapshots of the globular conformations of **a** protein-like, **b** random, and **c** random-block copolymers of the same length ($N = 512$). Hydrophobic segments are shown in *light gray* and hydrophilic segments in *dark gray*. Adapted from [18]

further chemical modification, while random copolymer globules will most probably precipitate. We will address these statements in more detail later.

It is clear that the main reason for the deviation of the properties of protein-like copolymers from those of random copolymers is the special sequence design scheme [18], not just differences in the degree of blockiness.

Govorun et al. [35] have shown, both by exact analytical theory and by computer simulations, that the corresponding chemical sequence is nonalternating and demonstrates the specific long-range correlations (LRCs), which can be described by the statistics of the Lévy-flight type [36]. For such probabilistic processes an observable stochastic variable x exhibits large jumps ("flights"), called "Lévy flights", characterized by a power-law (rather than exponential) probability distribution function, $f(x) \propto x^{-\mu}$ ($1 < \mu < 3$). The possibility for exact analytical description of sequences resulting from surface coloring (Fig. 1b) comes from the fact that the statistics of polymer chains inside dense globules is Gaussian, i.e., it is described by the ordinary diffusion equation. One has only to worry about correct boundary conditions, and this problem was resolved in [35]. An analogous result was obtained in a later work [37] for the model that takes into account more chemical details.

Experimentally, the conformation-dependent design described above was first realized in a series of papers, where the lyophilization (hydrophilization) of a homopolymer was achieved by the grafting of short poly(ethylene oxide) chains to the surface of globules formed by long poly(N-isopropylacrylamide) (NIPA) and glicydil methacrylate chains [38–40]. This synthesis resulted in the formation of a nonrandom copolymer, apparently having a core-shell morphology, in qualitative agreement with simulation data [18–20]. It was shown that grafting to the more compact conformation results in globules that are more stable to precipitation than random grafting to a coil conformation.

2.2.2
Long-Range Correlations and Their Measure

The presence of LRCs in designed sequences is a very important feature. It is easy to understand that these correlations are due to the fact that assigning of the type of chain segments (H or P) under the preparation conditions described above depends on the conformation of the parent globule as a whole (Fig. 1b), not on the conformation of small sections of the initial homopolymer chain. From this viewpoint one may say that such a sequence *encodes* in a two-letter alphabet the spatial (core-shell) structure of a copolymer globule. Obviously, this functional feature can be realized if and only if a general statistical pattern is attributed to the sequence as a whole, and cannot be obtained by the joining of independent statistical patterns of two or more subsequences of smaller lengths.

It may well be that Nature also chose such a path in the evolution of the main biological macromolecules: proteins, DNA, and RNA. These polymers in living systems are responsible for functions, which are incomparably more complex and diverse than the functions that we normally discuss for synthetic polymers. The molecular basis for this ability to perform sophisticated functions is associated with the unique chemical sequences of these biopolymers. In particular, a protein sequence as a whole determines the globular conformation and hence biological function, whereas if this sequence is cut into two pieces, those pieces normally neither correspond to a soluble globule nor have any biological function (Fig. 4). The same is true for statistically complex DNA sequences, which encode in a four-letter alphabet all genetic information and exhibit significant correlation on different scales [41]. All these peculiarities are connected to LRC effects, and the corresponding sequence, which cannot be divided into shorter subsequences with similar statistical patterns and functional features, may be termed an *inseparable sequence*. Such sequence integrity and LRCs are not characteristic of the majority of synthetic linear copolymers, the primary structure of which is chemically

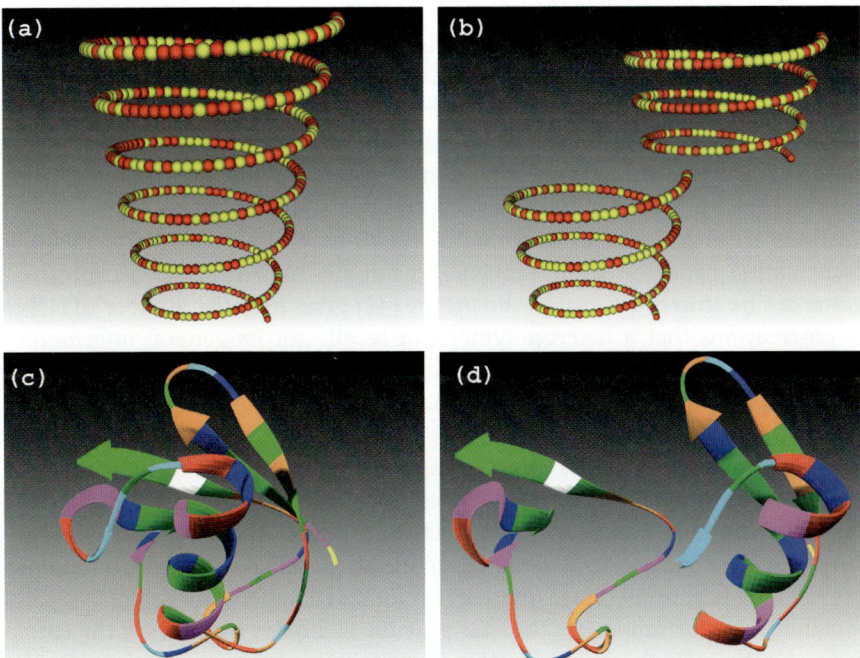

Fig. 4 a,b The sequence distribution for rather large sections of a synthetic random copolymer is practically identical to that of the whole polymer. **c,d** If a protein sequence is cut into two pieces, those pieces neither correspond to a native globule nor have any biological function

homogeneous on a large scale. Indeed, the sequence distribution for rather large sections of synthetic copolymers is normally practically identical to that of the whole polymer (Fig. 4). Certainly, because of the finite size of the bare globule subjected to CDSD, the longest correlations that can be found in this case are also finite; however, the range of these correlations is much larger than for usual synthetic copolymers obtained, e.g., via radical copolymerization under homogeneous conditions [17]. On the other hand, one may anticipate that if scale-invariant correlations extend along the entire copolymer sequence, that is, the sequence is a "true" fractal object, its sufficiently large parts would have the same statistical pattern and large-scale compositional inhomogeneities. In the present article, we will focus just on such sequences, which show both strong chemical inhomogeneity and LRCs.

How many different, thermodynamically stable structures can be encoded by a long heteropolymer sequence? To answer this question, Fink and Ball [42] have used arguments based on energy fluctuations and information theory. They have found that the maximum number of compact conformations, p_{max}, which are simultaneously thermodynamically stable, depends only on the number of chemically different comonomers in the chain, s, but not on its length N. If a chain is placed on a lattice with an effective coordination number κ (for a cubic lattice, $\kappa \approx 1.85$), the theory predicts: $p_{max} = \ln(s)/\ln(\kappa)$, that is, for the two-letter alphabet, $p_{max} \approx 1$. This means that for a binary (two-letter) copolymer, only one nondegenerate ground-state conformation can exist. It is probably not too surprising that binary models are not accurate representations of real proteins. On the other hand they are well suited in polymer chemistry. For a homopolymer, zero conformations are encodable, while for a protein 20-letter amino acid alphabet, the information capacity of the sequence is notably higher: about 5 conformations can be stored. Of course, the theory [42] gives only the number of stable conformations that a heteropolymer can recall, not its general *information complexity* that can be infinitely higher at the $N \to \infty$ limit. Indeed, using the Morse alphabet, we can write all the human history in a message of reasonable length.

For a statistical analysis of copolymer sequences, different mathematical techniques are used. For mathematically oriented researchers, a copolymer sequence might be considered as a string of symbols whose correlation structure can be characterized completely by all possible monomer-monomer correlation functions. Since the correlations at long distances are typically small, it is important to use the best possible estimates to measure the correlations, otherwise the error due to a finite sample size can be as large as the correlation value itself.

To monitor the long-range statistical properties of computer-generated sequences, the method developed by Stanley and co-workers [43, 44] in their search for LRCs in DNA sequences is usually employed. In this approach, each

HP copolymer sequence is transformed into a sequence of + 1 and − 1 symbols (as used in Fig. 2), which are considered as steps of a one-dimensional random walk. Shifting the sliding window of length λ along this sequence step by step, the number of + 1 and − 1 symbols inside the window is counted at each step. This number $\gamma_k(\lambda)$ is a new random variable that depends on the position k of the window along the sequence. The variable $\gamma_k(\lambda)$ has a certain distribution. Its average is determined by the overall sequence composition, and its dispersion is given by $D_\lambda^2 = \langle \gamma^2(\lambda) \rangle - \langle \gamma(\lambda) \rangle^2$, where $\langle ... \rangle$ represents the average over all windows of size λ and generated sequences. If the sequence is uncorrelated (normal random walk) or there are only local correlations extending up to a characteristic range (Markov chain), then the value of D_λ scales as $\lambda^{1/2}$ with a window of a sufficiently large λ. A power law $D_\lambda \propto \lambda^\alpha$ with $\alpha > 1/2$ will then manifest the existence of LRCs.

In some cases, because of large fluctuations, conventional scaling analysis cannot be applied reliably to the relatively short sequences generated in typical simulations. To avoid this problem, one can use the so-called detrended fluctuation analysis (DFA), the method specifically adapted to handle problems associated with short nonstationary sequences [43, 44]. This approach leads to a special function $F_D(\lambda)$ that characterizes the detrended local fluctuations within a window of length λ. Generally, $F_D(\lambda)$ shows the same behavior as D_λ. There are many other methods for monitoring LRCs in copolymers [44–47].

The result of calculations [35] averaged over 2000 independent protein-like HP-sequences of $N = 1024$ monomeric units with a 1 : 1 HP composition is presented in Fig. 5. For comparison, the data for two other types of sequences are also shown. One of them is a purely random 1 : 1 sequence; it demonstrates $D_\lambda \propto \lambda^{1/2}$ scaling, as expected. Comparing this curve with

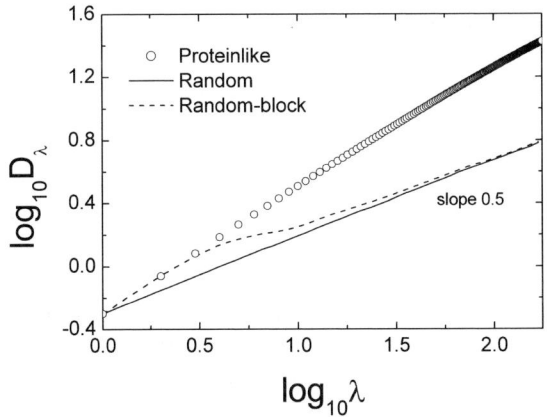

Fig. 5 Dispersion of the number of different monomeric units in a fragment of sequence size λ for protein-like, random, and random-block copolymers. Adapted from [35]

Monte Carlo results we see immediately that the protein-like sequence is not random and well-pronounced correlations do exist in it. Thus, it is interesting to compare the simulation data with the Poisson distribution adjusted to achieve the same 1:1 composition and the same degree of blockiness as for a protein-like copolymer. This model sequence exhibits a somewhat more rapid variation of D_λ at small λ, but ultimately the law $D_\lambda \propto \lambda^{1/2}$ is obeyed for large values of λ. Nevertheless, this random-block model is also seen to be unsatisfactory for the statistical behavior of a protein-like sequence throughout the interval of λ examined, $2 < \lambda < 512$. Although the simulation data do not fit accurately to any power law $D_\lambda \propto \lambda^\alpha$, the slope of the observed D_λ dependence corresponds to a significantly larger value than $1/2$, up to about $\alpha = 0.85$, thus indicating pronounced long-range correlations in a protein-like sequence. Thus, the primary structure emerging in the case of protein-like copolymers can be characterized as "quasirandom with long-range correlations". Analytical theory [35] suggests that the Lévy-flight statistics, albeit with a broader crossover region, is expected even if parental (globular) conformation used to generate protein-like macromolecules is not maximally compact, but rather a globule somewhat closer to the Θ-point.

These findings are surprisingly similar to those known for DNA sequences, which appeared as a mosaic of coding and noncoding patches [41, 43, 44]. Indeed, like DNA chains containing coding and noncoding regions, the copolymer under consideration also contains two types of alternating sections forming a certain pattern. It is known that the noncoding regions in DNA do not interrupt the correlation between the coding regions (and vice versa), and the DNA chain is fully correlated throughout its whole length. As a result, the D_λ^2 (or F_D^2) curve does not contain the linear portion $D_\lambda^2 \propto \lambda$. In principle, the same behavior is observed for protein-like sequences [18–20].

The question of whether proteins originate from random sequences of amino acids was addressed in many works. It was demonstrated that protein sequences are not completely random sequences [48]. In particular, the statistical distribution of hydrophobic residues along chains of functional proteins is nonrandom [49]. Furthermore, protein sequences derived from corresponding complete genomes display a distinct multifractal behavior characterized by the so-called generalized Rényi dimensions (instead of a single fractal dimension as in the case of self-similar processes) [50]. It should be kept in mind that sequence correlations in real proteins is a delicate issue which requires a careful analysis.

To end this section, it is worthwhile to note that long-range dependence processes (also called long-memory processes) and their statistics have many areas of application: statistical physics, neuroscience, communication networks, turbulence, hydrology, meteorology, geophysics, finance, econometrics. The literature on the subject is vast (see, e.g., [51]).

2.2.3
Hydrophobic Modification of Hydrophilic Polymers

The segregated core-shell microstructures, consisting of a hydrophobic core surrounded by a hydrophilic shell, are of great practical interest as their mechanical properties are mainly influenced by the core polymer and the chemical properties mainly by the shell monomer units. Recently, a further development of the approach to synthesis of copolymers with a specific primary sequence capable of forming core-shell microstructures has been suggested [25, 52]. The idea is rather similar to the previous approach but has an important difference: instead of the hydrophilization of the globular surface, one can perform a sequential hydrophobization of a hydrophilic polymer chain (poly-P) dissolved in a good solvent, using a low-molecular-weight hydrophobic modifier (H) poorly soluble in this solvent.

In a dilute solution, when the polymer is in a coil state (Fig. 6a), the diffusion of hydrophobic particles into the coil is normally faster than the chemical reaction [53]. In this case, the local concentration of particles H inside the coil is practically the same as in the bulk. Therefore, we expect that at the initial stage, the reaction will lead to a random copolymer: some of the P monomeric units will attach to H reagent and thereby they will acquire amphiphilic (A) properties: P + H → A (Fig. 6b). As long as the number of modified A units is not too large, the chain remains in a swollen coil-like conformation (Fig. 6b). However, when this number becomes sufficiently large, the hydrophobically modified polymer segments would tend to form

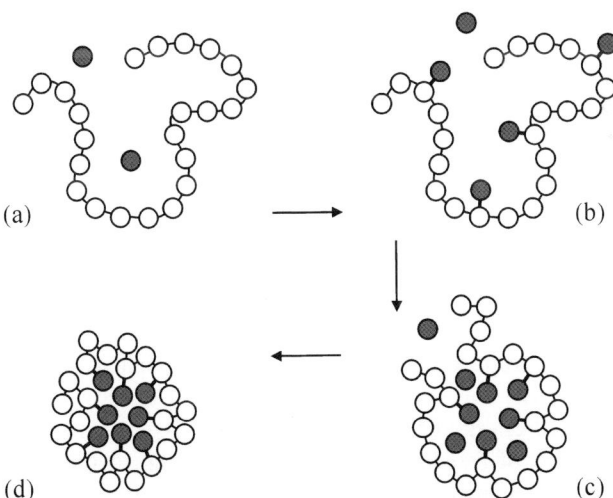

Fig. 6 Schematic representation of the hydrophobic modification of a hydrophilic polymer in a solvent, which is selectively poor for hydrophobic modifier and modified chain segments. The modifying agent and hydrophobic monomers are shown in *gray*

intrachain micelle-like aggregates (Fig. 6c). This is due to the loss in translational entropy of covalently bonded H species after their grafting to the hydrophilic backbone. Structurally, the intramolecular micelles are similar to reverse micelles formed by free low-molecular-weight surfactants, and like ordinary micelles, they should solubilize hydrophobic species.

The presence of the intrachain aggregates can dramatically change the reaction conditions. Because of preferential adsorption, the poorly soluble modifier will diffuse inside the A-rich regions thus leading to spatially inhomogeneous distribution of the concentrations (Fig. 6c). One can expect that further modification of the hydrophilic sections of the polymer chain will occur predominantly within these microglobular regions. In the course of chemical reaction, they would progressively increase in size and then coalesce. As a result, a nonrandom microblock distribution of the chemically modified chain segments would emerge. Finally, we should have hydrophobically modified segments inside rather compact conformation formed by the resulting hydrophilic-amphiphilic (PA) copolymer. In that way, one can expect the formation of core-shell morphologies with inner (poorly soluble) core and outer (well-soluble) hydrophilic cover (Fig. 6d). In principle, the polymer-analogous reaction can be terminated at any desirable time by quenching (e.g., by temperature lowering) or by stopping the supply of reactive compounds.

Given the shortcomings of an approximate analytical treatment and the difficulties with the laboratory measurements, it is conceivable that computer simulations might help greatly in verification of the qualitative arguments presented above.

In the simulation [52], a continuous (off-lattice) model of a polymer and the method of stochastic molecular dynamics were employed. The nonmodified polymer was treated as a flexible chain consisting of N hydrophilic (P) segments (monomeric beads), with adjacent beads connected by a rigid rod of a fixed length b. Each hydrophobic modifier (H) was considered as a single-site particle (monomer). Each amphiphilic group (A) arising after the attachment of H monomer to the hydrophilic backbone was modeled by a two-site "dumbbell" consisting of H and P sites linked by a rigid bond of length b. The hydrophilic sites were connected with each other in a linear fashion and formed the backbone of the hydrophobically modified PA copolymer.

The total potential energy of the system consisting of the solute and solvent molecules was decomposed in the following three parts: (i) the solute-solute potential energy, (ii) the solvent-solvent potential energy, and (iii) the interaction potential energy of the solute and the solvent. An integration over all solvent degrees of freedom leads to the potential of mean force (PMF). This effective solvent-mediated potential can be written as the sum of the solute-solute potential energy and the mean solvation term that describes the solvent-induced effects. The hydrophobic effect is the most important force in stabilizing globular structures. It is believed that the corresponding attractive interaction is primarily entropy driven and it should be short-range and should depend on a shell of surrounding solvent molecules. Experimental data show that around room temperature for a wide range of different hydrophobic molecules, the hydrophobic interaction energy depends linearly on the burial of solvent-accessible surface area (SASA). Taking into account this fact, to treat the hydrophobic interactions in the simulation [52], an approach based on the SASA model [54] was used.

In this model, the solvation free energy, $\Delta\mu_s$, required to transfer an N-site molecule from vacuum to polar solvent (e.g., in water) is approximated by the sum of linear terms $\Delta\mu_s = \sum_{i=1}^{N} \gamma_i S_i$, where S_i is the solvent-accessible surface area for site i and γ_i is the solvation parameter for the site. In this approximation, $\Delta\mu_s$ can also be considered as the free energy cost for the transfer of a particle from the interior of a hydrophobic cluster to the solvent. The solvation parameter γ is an estimate of the free energy of transfer divided by exposed surface area. For hydrophilic particles, $\gamma \leq 0$. For hydrophobic particles, solvent quality becomes poorer with increasing γ. In reality, when the temperature is fixed, the change in γ can be due to variation of the solvent composition.

The algorithm used in [52] simulated a reaction in a 3D cube utilizing periodic boundary conditions. Initially, an N-unit hydrophilic homopolymer (poly-P) in a coil state and N free hydrophobic H monomers were placed in the cube. For the conditions considered, despite the attraction between H monomers, they were soluble due to their high translational entropy. If a freely diffusing monofunctional H monomer approached within the prescribed distance (reaction radius) to a P bead on the chain, a bond could be formed between the two. This led to the formation of an amphiphilic monomer unit: P + H → A. The reaction was considered as a sequence of the alternating steps: grafting of a new H monomer to the chain and the subsequent long relaxation of the chain. The process was terminated when the required number of the hydrophilic monomer units was transformed into the A type. In the study [25, 52], the composition of the resulting copolymer was constrained so that there were 75% hydrophilic and 25% amphiphilic monomer units. As a result, a hydrophilic-amphiphilic PA copolymer with a certain distribution of P and A units along the hydrophilic polymer backbone was obtained. To gain better statistics, ca. 10^3 hydrophobically modified copolymer chains for each set of the parameters were independently generated and then required average characteristics were found.

To give a visual impression of the simulated system, Fig. 7 presents a typical snapshot of an amphiphilic copolymer having a 256-unit hydrophilic backbone with 64 attached hydrophobic side groups. Already from this picture, it is seen that, using the synthetic strategy described above, one can indeed end up with a copolymer having a dense hydrophobic core surrounded by a hydrophilic shell.

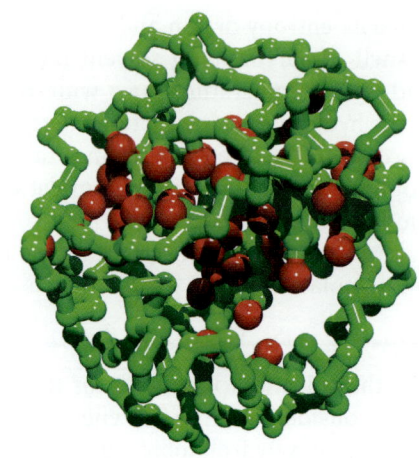

Fig. 7 Typical snapshot of a hydrophobically modified copolymer. Hydrophilic chain segments are shown in *green* and hydrophobic side groups in *red*

From the picture presented in Fig. 7, one can expect that the sequential hydrophobization of a polymer coil should lead to a copolymer with a nonrandom sequence distribution. This is indeed the case. As an example, let us consider the average number fractions of blocks consisting of ℓ neighboring amphiphilic monomers, $f_A(\ell)$, occurring in a copolymer chain. Some results are shown in Fig. 8 on a semilogarithmic scale.

We expect that for a random distribution of monomers A incorporated into the polymer chain, the function $f_A(\ell)$ should decay exponentially with increasing ℓ. In fact, such a behavior is observed for the copolymer modified in a solvent, which is good for the low-molecular-weight modifier, that is, at $\gamma = 0$. When the solvation parameter γ is increased and the solvent becomes selectively poorer for the modifying agent, the values of $f_A(\ell)$ are skewed toward A sections of greater length, which implies a copolymer with blocky tendencies [10]. A further worsening of the solvent quality leads to a strong deviation from the exponential decay of $f_A(\ell)$ thus indicating a nonrandom copolymer sequence distribution.

Thus, copolymers of the same composition can have qualitatively different sequence distributions depending on the solvent in which the chemical transformation is performed. In a solvent selectively poor for modifying agent, hydrophobically-modified copolymers were found to have the sequence distribution with LRCs, whereas in a nonselective (good) solvent, the reaction always leads to the formation of random (Bernoullian) copolymers [52]. In the former case, the chemical microstructure cannot be described by any Markov process, contrary to the majority of conventional synthetic copolymers [10].

In general, the microsegregated structures observed for hydrophobically modified polymers (Fig. 7) are similar to core-shell globules obtained via

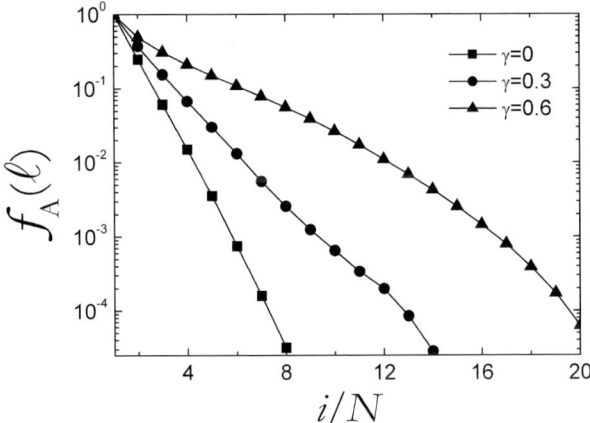

Fig. 8 Average number fraction of blocks consisting of ℓ neighboring amphiphilic monomers occurring in a 256-unit copolymer chain hydrophobically modified in a solvent having different selectivity for modifying agent. Adapted from [25]

the coloring procedure (Fig. 3). However, it should be kept in mind that the experimental realization of the hydrophilic modification of the surface of a hydrophobic globule was shown to be rather unreliable, because of the difficulty of stabilizing dense globules in the solution for the time sufficient to implement a polymer-analogous transformation [23]. On the other hand, the simulations [25, 52] show that the method based on the hydrophobic modification of soluble polymers should be quite universal and robust.

2.2.4
Adsorption-Tuned Copolymers

The idea of conformation-dependent sequence design via polymer-analogous transformation can be generalized in many respects [23]. Indeed, a special chemical sequence can be obtained not only from a globular conformation; any specific polymer chain conformation can play the role of parent.

The simplest example of this kind is connected to the conformation of a homopolymer partly adsorbed onto a flat substrate (Fig. 9). Let us assume that the chain segments being in direct contact with the surface in some typical instant conformation (Fig. 9a) are chemically modified (Fig. 9b). This can take place when the surface catalyzes some chemical transformation of the adsorbed segments. One can expect that after desorption (Fig. 9c), such a copolymer will have special functional properties: it will be "tuned to adsorption".

Following this line, Zheligovskaya et al. [55] compared the adsorption properties of copolymers with special "adsorption-tuned" primary structures (adsorption-tuned copolymers, ATCs) with those of truly random copoly-

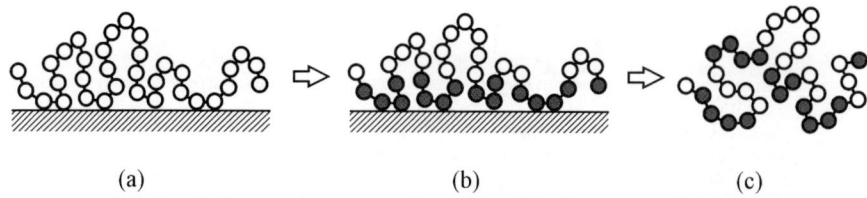

Fig. 9 Schematic representation of the sequence design procedure leading to an adsorption-tuned copolymer: **a** initial partly adsorbed homopolymer, **b** chemical modification of adsorbed chain segments, **c** resulting copolymer. Modified segments are shown in *gray*

mers and random-block copolymers. Monte Carlo simulations revealed that specific features of the ATC primary structure promoted the adsorption of ATC chains, in comparison with their random and random-block counterparts under the same conditions. In other words, the resulting copolymer sequence "memorizes" the original state of the adsorbed homopolymer chain. Its statistical properties exhibit LRCs of the Lévy-flight type similar to those found for copolymers obtained via coloring of a homopolymer globule [56].

Recently, Velichko et al. [57] suggested the model of a so-called *molecular dispenser*. This idea is a further development in the direction of conformation-dependent sequence design. Namely, they considered the conformation of a homopolymer chain adsorbed on a spherical colloidal particle (Fig. 10) and performed design of sequence for this state of macromolecule. The motivation behind this design procedure is that if we eliminate the parent colloidal particle after the design is completed (e.g., by etching), the resulting copolymer will be hopefully tuned to selectively adsorb another colloidal particle of a parental size σ_p. For instance, if such a copolymer is exposed to a polydisperse colloidal solution of particles of different size, it will selectively choose to form a complex with the particle having the same radius as that in parental conditions. That is why such a macromolecular object can be called a molecular dispenser.

It should be noted that the development of such polymer systems is stimulated by existing experimental works. In particular, the experimental methods of preparation of nanometer-sized hollow-sphere structures have been suggested [58–63] because of their possible usage for encapsulation of molecules or colloidal particles. The preparation of hollow-sphere structures, generally, is based on self-assembling properties of block copolymers in a selective solvent, i.e., on the formation of polymer micelles with a nanometer-sized diameter. Further cross-linking of the shell of the micelle and photodegradation [64] of the core part produce nanometer-sized hollow cross-linked micelles.

The sequence design procedure proposed in [57] can be described in more detail as follows.

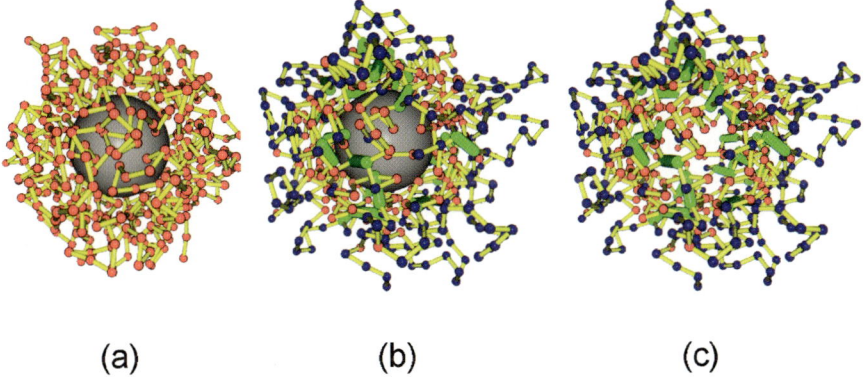

Fig. 10 Stages of preparation of a copolymer envelope: **a** adsorption of homopolymer chain on a colloidal particle; **b** "coloring" of the polymer chain (*blue* corresponds to chemically modified monomer units and *red* to adsorbed units) and introduction of crosslinks (shown as *green sticks*) to stabilize the hollow-spherical structure; **c** elimination of the core particle. Adapted from [57]

First, we consider a homopolymer chain attracting to a colloidal nanoparticle. Such a chain forms an adsorbed complex with the particle (its typical conformation is shown in Fig. 10a). Only part of the chain segments are in direct contact with the colloidal particle, while other segments form flower-like loops. Then we color the segments in the loops in "blue", while the segments near the colloidal particle remain "red", i.e., they are attracting to this particle (Fig. 10b). If the sequence design is stopped at this stage, the pronounced selectivity of the complex formation with another particle of parental size σ_p is not reached [57]. However, if additional crosslinks are introduced between adsorbed (red) units, thus fixing the cage structure of the central cavity (Fig. 10b), the macromolecule emerging after elimination of the colloidal particle (Fig. 10c) does indeed show the features of a molecular dispenser. This sequence design scheme was realized in the computer simulations [57]. We will address the results of this work in Sect. 3.5.2.

2.2.5
Design as a Simulation of Evolutionary Process

The concept of evolution of primary sequences of biopolymers has attracted great interest from biologists, chemists and physicists for a long time [65–68]. As has been discussed, it is natural to expect that the content of information in the sequences of biopolymers (proteins, DNA, RNA) is relatively high in comparison with random sequences where it should be almost zero [69]. Presumably, the information complexity of early ancestors of present-day biopolymers has been increased in the course of molecular evolution when the copolymer sequences became more and more complicated [66]. The study of various possibilities of this evolution of copolymer sequences is just the area where the evolution concept can be used in the context of polymer science.

It is worthwhile to note that since the information content of a sequence can be represented as a mathematically defined quantity, the whole process of evolution of biopolymer sequences can be specified in exact mathematical terms. The formulated fundamental problem is extremely difficult because of the absence of direct information on the early prebiological evolution. Therefore, of particular interest are "toy models" of evolution of sequences that show different possibilities for appearance of statistical complexity and of long-range correlations in the sequences.

It is clear that information complexity cannot emerge just as a result of random mutations. Some coupling of mutations to other factors is necessary. It is most straightforward and natural to introduce the interrelation of mutations and conformations, i.e., to consider conformation-dependent sequence design in the context of evolution.

Evolutionary computation approaches are optimization methods. They are conveniently presented using the metaphor of natural evolution: a randomly initialized population of individuals evolves following a crude parody of the Darwinian principle of the survival of the fittest. New individuals are generated using simulated evolutionary operations such as mutations. The probability of survival of the newly generated solutions depends on their fitness (how well they perform with respect to the optimization problem at hand): the "best" are kept with a high probability, the "worst" are rapidly discarded.

In the literature, some computer models describing the evolution of copolymer sequences have been proposed [26, 28]. Most of them are based on a stochastic Monte Carlo optimization principle (Metropolis scheme) and aimed at the problems of protein physics. Such optimization algorithms start with arbitrary sequences and proceed by making random substitutions biased to minimize relative potential energy of the initial sequence and/or to maximize the folding rate of the target structure.

It should be emphasized that the problem, which we address here, is somewhat different from that usually discussed in the context of protein physics [5–9]. We are not aiming at the search for a unique three-dimensional (native) conformation with a fast folding rate. On the contrary, we are interested in a state with a large entropy. In general, our aim is to learn whether it is possible to make with synthetic copolymers a step along the same line as molecular evolution.

Ascending and descending branches of sequence evolution. The aim of the study [70] was to introduce explicitly the concept of sequence evolution into the CDSD scheme.

Using a molecular-dynamics-based algorithm, the conformation-dependent evolution of model HP copolymer sequences was simulated [70]. The sequence evolution mechanism involved the generation of an initial protein-like sequence by inspecting of a homopolymer globule and by attributing the H-type to the monomeric units in the core of this globule and the P-type to the units on the surface of the globule. The resulting copolymer was then transferred to a coil conformation and then refolded due to the strong attraction between the H units. The HP sequence was then further modified, depending on the position of a monomer in the core or on the surface of a newly formed globule. Such modifications, leading to a change in the primary HP sequence, are repeated many times (ca. 10^3). With this evolutionary process, which can be called "repeated coloring", structures and sequences are formed self-consistently.

A 128-unit flexible-chain heteropolymer with a HP composition fixed at 1 : 1 was simulated for the condition when hydrophobic H monomers strongly attract each other, thus stabilizing a dense globular core, whereas the attraction energy ε_{PP} between hydrophilic P monomers was considered as a parameter (the interaction between H and P monomers is given by $\varepsilon_{HP} = \sqrt{\varepsilon_{HH}\varepsilon_{PP}}$). For this model system, various conformation-dependent and sequence-dependent properties, including information-theoretic-based quantities, can be calculated.

Depending on the attraction energy between polar segments ε_{PP}, it is possible to find two regimes (branches) of evolution (regimes I and II) [70]. If ε_{PP} is smaller than some crossover energy ε_{PP}^* (regime I), the evolution can lead to a second-order-like transition in sequence space from the sequences with a protein-like primary structure capable of forming a core-shell globule to the degenerated (nonprotein-like) sequences having long uniform H and P blocks. This transition is also accompanied by strong changes in the conformational properties of the copolymer.

Figure 11a shows the mean square gyration radius, R_g^2, plotted vs. ε_{PP}. As seen, R_g^2 is a weakly decreasing function of ε_{PP} in the range $\varepsilon_{PP} > \varepsilon_{PP}^*$ and demonstrates a rapid growth when ε_{PP} decreases and becomes less than ε_{PP}^*. The critical value ε_{PP}^* is found to be smaller than the critical energy at which the coil-to-globule transition takes place in a homopolymer chain of the same length.

The degenerated primary structure looks like a di- or triblock sequence ("core-tail" or "tadpole-like" conformation).

Thus, when the attraction between hydrophilic segments is not sufficiently strong, we deal with the *descending branch* of the evolution, which leads to nonprotein-like sequences having low information content and low complexity. On the other hand, in the second regime (at $\varepsilon_{PP} \geq \varepsilon_{PP}^*$) the complexity of protein-like structures is found to increase and therefore we have the *ascending branch* of the evolution.

Information complexity of copolymer sequences. A common approach to the analysis of the complexity of a system is to use concepts from information theory and information-theoretic-based techniques.

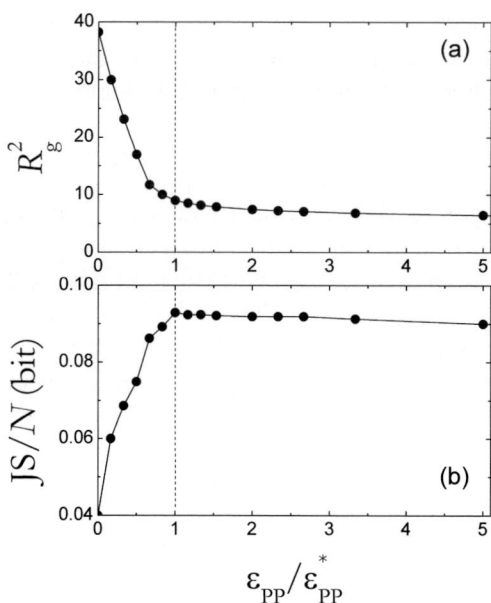

Fig. 11 a Mean square gyration radius and **b** Jensen–Shannon divergence measure as a function of the attraction energy ε_{PP} between hydrophilic segments, after the sequence evolution procedure. The characteristic energy of H – H interactions is fixed at $\varepsilon_{HH} = 2k_B T$, thus stabilizing a dense globular core. Adapted from [70]

What is complexity? There is no good general definition of complexity, though there are many. Intuitively, complexity lies somewhere between order and disorder, between regularity and randomness, between perfect crystal and gas. Complexity has been measured by logical depth, metric entropy, information content (Shannon's entropy), fluctuation complexity, and many other techniques; some of them are discussed below. These measures are well suited to specific physical or chemical applications, but none describe the general features of complexity. Obviously, the lack of a definition of complexity does not prevent researchers from using the term.

In general, the aim here is to find a measure capable of indicating how far copolymer sequences generated during the evolutionary process differ from each other and from random or trivial (degenerate) sequences. It turned out that the usual measures of the degree of complexity (based, e.g., on Shannon's entropy and related characteristics) are nonadequate [70]. To overcome this problem, it was proposed to use the so-called Jensen–Shannon (JS) divergence measure [70]. Let us explain how it can be defined.

Let $S = \{s_1, ..., s_N\}$ be a sequence of N symbols. For two subsequences $S_1 = \{s_1, ..., s_n\}$ and $S_2 = \{s_{n+1}, ..., s_N\}$ of lengths n and $N - n$, the difference between the corresponding discrete probability distributions $f_1(s_1, ..., s_n)$ and

$f_2(s_{n+1}, ..., s_N)$ is quantified by the Jensen–Shannon divergence

$$JS(\mathbf{S}_1, \mathbf{S}_2)/N = h(\mathbf{S}) - \left[\frac{n}{N}h(\mathbf{S}_1) + \frac{N-n}{N}h(\mathbf{S}_2)\right], \quad (1)$$

where $\mathbf{S} = \mathbf{S}_1 \oplus \mathbf{S}_2$ (concatenation) and $h(\mathbf{S})$ is Shannon's entropy of the empirical probability distribution obtained from block frequencies in the corresponding subsequence. Of course, Shannon's entropy depends on the definition of a set of words in the sequence. For two-letter HP copolymers, one can adopt the following set of words (uniform blocks): H, HH, HHH, ..., P, PP, PPP, ...; that is, the word (block) is defined by its length ℓ and type. In this case, Shannon's entropy per monomer can be written as

$$h = -\frac{N_w}{2N}\sum_{\ell}\left[f_H(\ell)\log_2 f_H(\ell) + f_P(\ell)\log_2 f_P(\ell)\right], \quad (2)$$

where $f_H(\ell)$ and $f_P(\ell)$ are the frequencies of words of length ℓ composed of letters H and P, respectively, and N_w is the total number of words.

The Jensen–Shannon divergence JS is zero for subsequences with the same statistical characteristics; it takes higher values for increasing differences between the statistical patterns in the subsequences, and reaches its maximum value for a certain set of distributions. In particular, both random and any regular (multiblock) copolymers of infinite length show $JS = 0$. We normally expect that a completely random sequence or a sequence with long uniform blocks contain less information than a sequence containing many different blocks (words) of medium length. Of course, only using sequence analysis, we cannot unambiguously distinguish between what might be called quantity and quality of information.

Using the Jensen–Shannon divergence, JS, as a measure of complexity for the generated sequences, one can obtain an interesting result (see Fig. 11b). The most important feature is that the JS value is a *nonmonotonous* function of ε_{PP}, whereas Shannon's entropy, Shannon's index, and many other sequence-dependent parameters change always *gradually* [70].

For the sequences generated in the evolutionary process described above, it was shown that at $\varepsilon_{PP} \geq \varepsilon_{PP}^*$ (regime II) the degree of complexity, as measured by JS divergence, can be considerably higher as compared to that observed for regime I, at $\varepsilon_{PP} < \varepsilon_{PP}^*$. The complexity slightly increases with ε_{PP} decreasing, reaches its maximum just on the boundary of regimes I and II, and then sharply drops (Fig. 11b). Therefore, in regime II, the evolution preserved the copolymer sequence of high complexity, whereas in regime I, the information content of the sequence has degenerated in the course of evolution.

Simultaneous evolution of sequences and conformations. In the work [71], the simultaneous evolution of sequences and conformations was studied. This design procedure leads to the final state that depends on the set of interaction parameters and on the rearrangements both in conformational space and in sequence space. These rearrangements are characterized by the usual thermodynamic temperature, T, for conformational space as well as effective sequence rearrangement temperature, T_s. Namely, after a certain number of Monte Carlo steps in conformational space (in the course of this process the system equilibrates at temperature T) the possibility of mutation of two randomly chosen monomeric units is tried: monomeric unit H converts into P and vice versa. If this leads to a decrease in the energy of a protein-like globule, this move is accepted. If the globular energy is increasing by the amount ΔE, this move is accepted with the probability defined by $\exp(-\Delta E/k_B T_s)$.

In general, we can define evolution of sequences for any value of T_s. However, the simulation for three characteristic cases can be most easily understood.

(i) The inequality $T \ll T_s$ means that all the moves in sequence space are accepted independent of conformation. This corresponds to random mutations, and the final sequence (after long evolution) will be that of a random HP copolymer (no information complexity).

(ii) The inequality $T \gg T_s$ means that only the moves leading to a decrease in the globular energy E are accepted. Such evolution should lead to a sequence corresponding to a minimum of globular free energy in conformational space. For the case of the absence of any attractive interactions between P units, it was shown [70] that the final sequence after evolution should have a hydrophobic core with very few hydrophilic (or polar) loops and a long hydrophilic tail. This sequence is close to that of a HP diblock copolymer, and should not exhibit any information complexity, as has been stated above.

(iii) The case $T = T_s$ corresponds to an annealed HP sequence. This case is equivalent to the situation when monomeric unit H can be converted into P by attaching some ligand L: P $\rightleftarrows$ H + L. We assume that the number of ligands is fixed to maintain 1 : 1 HP composition, however, they can choose which monomeric unit to bind. This defines, in particular, the chain sequence. Annealed HP sequences in the context of polymer globules were first considered by Grosberg [72].

When T_s does not correspond to any of the characteristic cases described above, the evolution of sequences coupled with conformations can be still defined in the same way. One has only to remember that if $T \neq T_s$ and both T and T_s are finite, there is a flow of heat between conformational space and sequence space, so that full thermodynamic equilibrium is impossible. Still, we can be in a stationary regime corresponding to the sequences tending to a certain fixed point, and possessing (or not possessing) information complexity.

The simulations and theoretical arguments [71] predict that at high T_s, the sequence free energy F_s dominates and the sequence tends to be completely random, corresponding to the minimum of F_s. At low T_s, the evolution selects those sequences, which correspond to the low value of the conformation free energy (structure of the core-tail type). In the intermediate regime, the conformational thermodynamic force and the sequence contribution (evolution pressure) interplay. Therefore, the formation of nontrivial structures and sequences is possible. Even in the absence of attraction between P units the final sequences remain protein-like (i.e., the final copolymer has the core-shell structure of a globule) and therefore maintains certain information complexity.

2.3
CDSD Via Copolymerization

2.3.1
Conditions for CDSD

Polymer growth is an example of a hybrid stochastic process, which involves both discrete and continuous random variables; the position of the polymer being continuous, while the number of monomers in the growing polymer is discrete. In this section, we will discuss just those processes.

The CDSD copolymerization essentially differs in principles and in results from the conformation-dependent polymer-analogous transformation [25]. Such a process can be considered as a variant of template polymerization (also called molecular imprinting) based on the noncovalent binding of polymerizing monomers to a template.

The essence of this technique consists of the copolymerization of monomers differing in their affinity to the template and therefore differently distributed in the reaction system. In contrast to conventional types of template polymerization [73, 74], in the CDSD copolymerization, all the monomers are bifunctional and form linear polymers, not cross-linked ones. The sequence of the segments in the resulting molecularly imprinted copolymer is determined by the template-controlled conformation of the propagating macroradical [24, 25].

Thus, the CDSD copolymerization regime is possible only in reaction systems with a strongly inhomogeneous spatial distribution of monomer concentrations [24, 25]. As an origin of inhomogeneities, one can mention interfaces, nanoparticles, molecular and supramolecular aggregates (e.g., for macroradicals capable of forming globular conformations), and so forth. They should selectively adsorb one of the comonomers (Fig. 12). Moreover, the spatial scale of the concentration inhomogeneities should be comparable to the size of the growing macroradical. Also, the polymerizing monomers should retain their properties after they are incorporated into a polymer chain. In the case of a copolymerization near an interface, one of the types of monomers and chain segments should be preferentially distributed near the adsorbing surface, whereas other monomers and chain segments should be preferentially located in the solution. As a result, the chemical sequence and

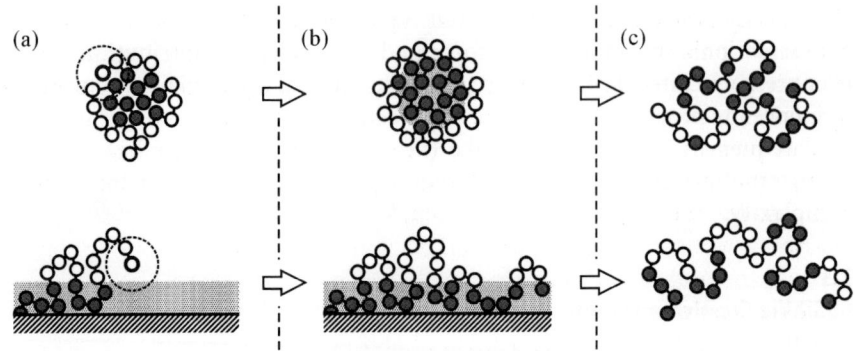

Fig. 12 Schematic representation of the CDSD copolymerization process in the cases when one of the comonomers is selectively absorbed by a polymer globule (*top*) or adsorbed on a surface (*bottom*). **a** Growing chains during the copolymerization (reaction zone around the growing chain end is marked with a *dashed line*); resulting copolymers in **b** globular (adsorbed) and **c** coil-like states. Regions where absorbed (adsorbed) monomers dominate are shown in *gray*

conformation of the growing macroradical become mutually dependent. Such an interdependence determines the further way of chain growth through the comonomer concentrations in a small reaction volume (see Fig. 12).

Finally, the rate of copolymerization should be slow enough. This guarantees that, during the chemical reaction, the equilibrium concentration fields remain approximately constant, and the growing chain has an equilibrium conformation between successive attachments of the monomers. Therefore, the CDSD regime is realized when

$$\max\{\tau_D, \tau_{rel}\} < \tau_R . \qquad (3)$$

where τ_R is the reaction time characterizing the polymerization rate, τ_D is the characteristic diffusion time for monomers, and τ_{rel} is the chain relaxation time that depends on the current chain length N_τ. For an ideal chain, the Rouse relaxation time scales with N as $\tau_{rel} \sim N^2$ [75]. It is clear that Eq. 3 corresponds to a kinetically controlled regime.

2.3.2
Copolymerization with Simultaneous Globule Formation

Let us assume that we are performing a radical copolymerization of moderately hydrophobic (H) and hydrophilic (P) monomers in an aqueous medium. The conditions for copolymerization (e.g., the temperature and solvent composition) should be chosen in such a way that when the emerging chain is long enough it can form a globule. As long as the current chain length N_τ is not too large, the growing hydrophobic-hydrophilic (HP) macroradical remains in a coil-like conformation. However, when its length becomes sufficiently large, the chain tends to form a two-layer globule with a hydrophobic core and a polar envelope. As has been noted in Sect. 2.2.3, the origin of this effect is connected to the loss in translational entropy of covalently bonded hydrophobic species after their incorporation into the growing chain.

The presence of a hydrophobic-hydrophilic interface can dramatically change the reaction conditions. The hydrophobic core will selectively absorb hydrophobic species from the solution (Fig. 12), and this will result in a redistribution of monomer concentrations between the core and bulk solution. Because the probability of attachment for each comonomer is determined by its concentration in a relatively small reaction volume near an active chain end, the active center inside the hydrophobic core will mainly attach more hydrophobic species; on the other hand, when the active center is located on the globule surface, it will mainly attach polar (soluble) monomers. In this way, the two-layer globule will grow, retaining its core-shell structure with a predominantly hydrophobic core and a hydrophilic outer envelope (see Fig. 12).

Theoretically, chain conformations and chemical sequences obtained via conformation-dependent copolymerization in a selective solvent were studied by Berezkin et al. [76–78]. It was shown that the corresponding polymerization process could not be interpreted in terms of classical kinetic models allowing for only short-range effects [53]. Using Monte Carlo and molecular dynamics simulation techniques, the process of irreversible radical copolymerization of hydrophobic and hydrophilic monomers, which led to globule formation, was studied [76–78]. The polymerization was modeled as a step-by-step chemical reaction of the addition of H and P monomers to the growing copolymer chain, under the assumption that a depolymerization reaction was not allowed. To simplify the analysis, it was assumed that the copolymerization is ideal with equal reactivity ratios. The type of attached monomers and the probabilities of their addition were determined from the average concentrations of the reactive monomers in a reaction volume V_τ around the moving active end of the macroradical of a given length N_τ during a reaction time τ_R. It was also assumed that the reaction bath is sufficiently large and the reactive species are sufficiently dilute so that there is significant time for the growing macroradicals to propagate independently. The preferential sorption of hydrophobic monomers in the core of the arising globule was explicitly taken into account.

The simulations [76–78] show that, using the conformation-dependent polymerization mechanism, one can indeed end up with a copolymer having a dense hydrophobic core surrounded by a hydrophilic shell. Figure 13 presents the typical radial distribution $W(r)$ of hydrophobic and hydrophilic units (with respect to the center of the globule) calculated for a copolymer chain of 512 units with a 1:1 HP composition. In Fig. 14, we show a typical snapshot. These findings explicitly prove the core-shell microstructure of the globule obtained via irreversible copolymerization in the solvent, which is moderately poor for hydrophobic species. In general, the same microsegre-

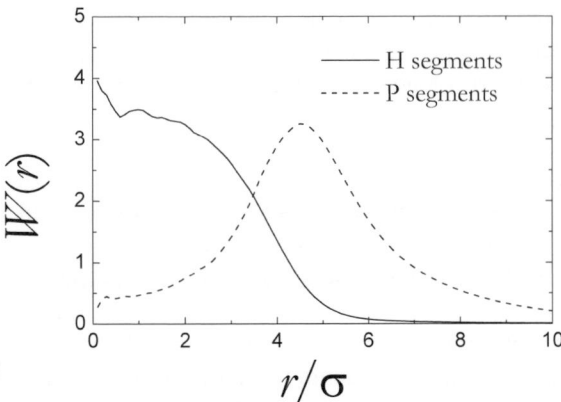

Fig. 13 Radial distribution of hydrophobic and hydrophilic segments (with respect to the center of the globule) for a copolymer chain of 512 segments with a 1 : 1 HP composition. σ is the size of the chain segment. Adapted from [76]

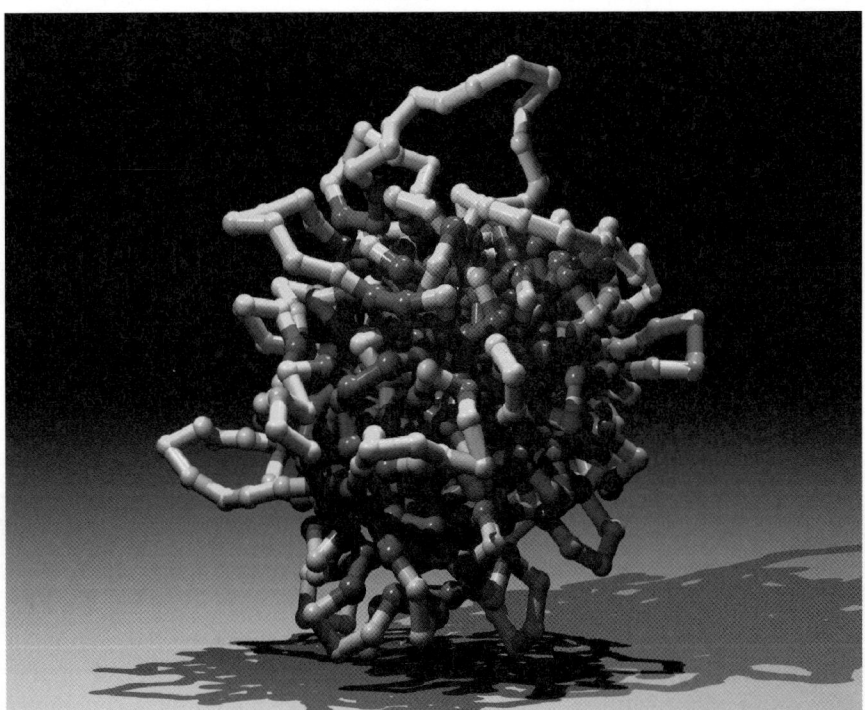

Fig. 14 Typical core-shell microsegregated structure that is obtained via conformation-dependent copolymerization with simultaneous globule formation in a selective solvent [76]. Hydrophobic chain segments are shown in *dark gray* and hydrophilic segments are shown in *light gray*

gated structure is observed for core-shell globules obtained via the coloring procedure [18–20].

The average chemical composition of the resulting HP polymer is determined by the bulk concentrations of the monomers, C_H^0 and C_P^0, and by the solubility of hydrophobic monomers in a globular core. For approximately equal contents of H and P segments in the growing copolymer, the solution should contain an excess of P monomers. In general, the chemical composition of the synthesized copolymer strongly deviates from the statistical one. Thus, the change in the solution concentration of polymerizing monomers allows the copolymer composition to vary over a wide range. Although the reactivities of both monomers are equal [76–78], their copolymerization leads to a copolymer that is preferentially enriched with hydrophobic segments. Therefore, a strong deviation from the principle of equal reactivity of Flory [79] is observed for the CDSD polymerization regime.

The type of monomer attached to the growth center during the simulation under kinetic control (at large τ_R values, see Eq. 3) is determined by the conformation and primary structure of the growing chain as a whole, not only by the local concentration of reactive monomers near the active end of the macroradical. As a result of such cooperativity, the formation of sequences with specific LRCs of the Lévy-flight type was observed [76–78].

Because of the specific mechanism of the chain propagation, the synthesized copolymers can have a gradient primary structure; that is, the HP composition can change along the growing chain during copolymerization [77]. This interesting result follows from the analysis of the distribution of monomers along the growing chain. Very short chains do not contain a sufficiently large number of hydrophobic units to form a core-shell conformation, and the growth of these chains is similar to that observed for random free radical copolymerization [17]. The composition of these chains is also close to the monomer composition in solution. Then, as the macroradical becomes longer and the number of connected H units increases, a dense hydrophobic core can be formed. This core absorbs H monomers, and their fraction in the resulting copolymer increases. In this case, the probabilities of monomer addition are also not constant because of the changes in the ratio between the volumes of the hydrophobic core and the polar shell during the chain growth. Therefore, the gradient primary structure is formed because of a change in the chain conformation and a continuous redistribution of comonomers between the globule and the solution in the course of the polymerization; in this way, a compositional drift is produced along the chain.

Experimentally, the described synthetic strategy was first realized by Lozinsky et al. [80, 81], who studied the redox-initiated free-radical copolymerization of thermosensitive N-vinylcaprolactam with hydrophilic N-vinylimidazole at different temperatures. These and other experimental studies [82–84] showed the universality of this approach of obtaining copolymers capable of forming nanostructures with a core-shell morphology.

2.3.3
Emulsion Copolymerization

The emulsion polymerization methodology is one of the most important commercial processes. The simplest system for an emulsion (co)polymerization consists of water-insoluble monomers, surfactants in a concentration above the CMC, and a water-soluble initiator, when all these species are placed in water. Initially, the system is emulsified. This results in the formation of thermodynamically stable micelles or microemulsions built up from monomer (nano)droplets stabilized by surfactants. The system is then agitated, e.g., by heating it. This leads to thermal decomposition of the initiator and free-radical polymerization starts [85]. Here, we will consider a somewhat unusual scenario, when a surfactant behaves as a polymerizing comonomer [25, 86].

In the numerical Monte Carlo simulations [86], polymerizing monomers were modeled as hydrophobic/hydrophobic (HH) and hydrophobic/hydrophilic (HP) "dumbbells" consisting of H(P) beads linked by rigid bonds of a fixed length. It is assumed that the monomers are placed in a polar solvent and form a mixed spherical micelle, which can be divided into two regions, Ω_{HH} and Ω_{HP}, initially filled with hydrophobic (HH) and amphiphilic (HP) monomers (Fig. 15a). Also, it is assumed that there is a strong incompatibility of H and P sites (the strong segregation regime). This results in a well-defined

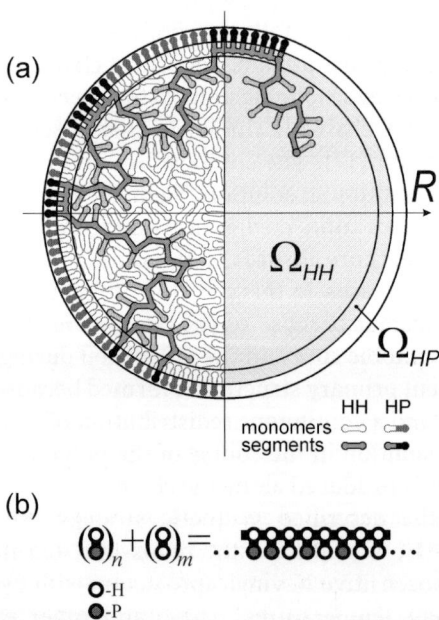

Fig. 15 **a** Schematic representation of the distribution of hydrophobic and amphiphilic monomers polymerizing in micelle and **b** the reaction between the monomers

interface between the Ω_{HH} and Ω_{HP} regions. The polymerization process is modeled as a step-by-step irreversible chemical reaction between H beads, which leads to an amphiphilic copolymer having a hydrophobic backbone with attached hydrophobic and hydrophilic side groups (Fig. 15b). The distribution of the side groups along the hydrophobic backbone defines the primary structure of the resulting amphiphilic copolymer. It is clear that in the model, the copolymer composition should depend on the size of the micelle R and on the ratio of the volumes Ω_{HH}/Ω_{HP}. The maximum length of the chain is limited and defined as $N = 2\pi\rho R^3/3$, where ρ is the total number density of H and P sites in the micelle. Therefore, by changing the micelle size, one can control both chemical composition and chain length simultaneously.

The type of attached monomeric units is determined by the average concentrations of reactive monomers in a reaction volume swept out by the moving active end of the macroradical of a given length N_τ during time τ_R. These concentrations are related to the corresponding instant local concentrations, $C_{HP}(r)$ and $C_{HH}(r)$, in the elementary volume d^3r and to the probability of finding the active end in this volume, $v(r)$. The probability $p_{\alpha\beta}^{(i)}$ that monomeric unit $\alpha\beta$ ($\alpha\beta$ = HP or HH) is located at the ith position from the beginning of a growing macromolecule is given by [86]

$$p_{\alpha\beta}^{(i)} = \frac{\int_{\Omega_{\alpha\beta}} C_{\alpha\beta}(r)v(r)d^3r}{\int_{\Omega_{HP}} C_{HP}(r)v(r)d^3r + \int_{\Omega_{HH}} C_{HH}(r)v(r)d^3r} = \frac{v_{\alpha\beta}C_{\alpha\beta}}{v_{HP}C_{HP} + v_{HH}C_{HH}}, \quad (4)$$

where $C_{\alpha\beta}$ is the time-dependent average concentration of monomer $\alpha\beta$ in the corresponding volume $\Omega_{\alpha\beta}$ and $v_{\alpha\beta}$ is the probability averaged over chain conformations, which define the location of the terminal free-radical reactive site in this volume.

Generally speaking, the values v_{HP} and v_{HH} depend on the conformation of the macroradical as a whole. In the first approximation, however, they can be considered as constant, $\bar{v}_{HP}$ and $\bar{v}_{HH}$. In this case, the polymerization process is described on the basis of the first-order reaction kinetic equations. In particular, one can define the following parameter characterizing the composition change along the chain [86]

$$\eta = \frac{\bar{v}_{HP}}{\bar{v}_{HH}} \frac{C_{HP}^{(0)} n_{HH}^{(0)}}{C_{HH}^{(0)} n_{HP}^{(0)}}, \quad (5)$$

where $C_{HP}^{(0)}$ and $C_{HH}^{(0)}$ are the initial monomer concentrations and $n_{HP}^{(0)}$ and $n_{HH}^{(0)}$ denote the initial numbers of the corresponding monomers in the micelle. It turns out that if $\eta > 1$, then the growing chain end attaches mainly hydrophobic monomers; if $\eta < 1$, it is preferentially enriched with hydrophilic monomers. In both cases, the resulting copolymer has a gradient structure. At $\eta = 1$, different monomers are distributed randomly along the chain. Obviously, the parameter η depends on the micelle size and therefore on N. Also, it can be changed when some amount of chemically inert species (e.g., solvent molecules) is incorporated into the micelle interior.

Let us consider the intramolecular composition profile $\varphi_{HP}^{(i)} = 2p_{HP}^{(i)} - 1$ calculated in [86]. The function $\varphi_{HP}^{(i)}$ characterizes intramolecular chemical inhomogeneity along the chain. The present definition of $\varphi_{HP}^{(i)}$ assumes that the HP-type segments are coded by symbol + 1, whereas symbol – 1 is assigned to the HH-type segments. For an ideal random copolymer in which chemically different segments follow each other in a statistically random fashion, the $p_{HP}^{(i)}$ function should coincide with the average fraction φ_{HP} of HP segments for any i. For a random-block copolymer, the fraction of one component averaged over many generated chains should also be uniform along the chain.

Figure 16 shows $\varphi_{HP}^{(i)}$ as a function of i/N for a few values of N. It is seen that depending on the chain length (and, therefore, on the micelle size), the resulting copolymer can have different gradient primary structures.

Also, in [86], the block length distribution functions $f_{HP}(\ell)$ and $f_{HH}(\ell)$ were calculated. It was found that the $f_{HP}(\ell)$ function decays exponentially with increasing block length ℓ, $f_{HP}(\ell) \propto \exp(-\ell/L_{HP})$, where L_{HP} is the average length of HP blocks. On the other hand, the distribution $f_{HH}(\ell)$ exhibits power-law decay $f_{HH}(\ell) \propto \ell^{-\alpha}$ for not too large ℓ. The exponent α estimated from the simulation data is close to 2. Such a behavior indicates the existence of LRCs. However, for sufficiently long blocks, $f_{HH}(\ell)$ shows an exponential decay that is related to the finite size of the micelle.

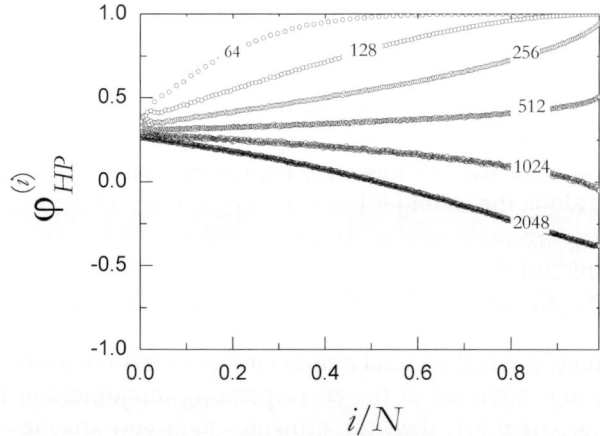

Fig. 16 Intramolecular composition profile presented as a function of monomer number i/N for hydrophobic/amphiphilic copolymer chains obtained via emulsion copolymerization, at a few chain lengths N indicated near the curves. Adapted from [25]

2.3.4
Copolymerization Near a Selectively Adsorbing Surface

Recently, a Monte Carlo simulation technique has been used to study two-letter quasirandom copolymers, which were generated via surface-induced computer-aided sequence design [87, 88]. This approach represents an irreversible radical copolymerization of selectively adsorbed A and B monomers with different affinity to a chemically homogeneous impenetrable surface, allowing for a strong short-range monomer(A)-surface attraction. Thereby, one of the types of monomers and chain segments are preferentially distributed near the adsorbing surface while other monomers and chain segments are preferentially located in the solution. As a result of such concentration inhomogeneities, the chemical sequence and conformation of the growing macroradical become mutually dependent, which corresponds to the CDSD regime [24, 25]. In the simulations, probabilities of A and B monomer additions to the growing polymer were determined by combination of the equilibrium monomer concentration profiles, $C_A(z)$ and $C_B(z)$, normal to the $z = 0$ plane and by the active chain end location. To describe the chain propagation theoretically, a simple analytical model based on stochastic processes and probabilistic statistics was also introduced and investigated in detail [88]. This probabilistic model (PM) provides a close approximation to the simulation data and explains a number of statistical properties of copolymer sequences. The calculations revealed the following conclusions.

In the case of strong adsorption of A species, the average fraction of B units φ_B in the resulting copolymer increases with their average bulk concentration C_B^0, thereby leading to a decrease in the number of strongly adsorbed segments. At some critical value C_B^*, about half of the chain segments are in the adsorbed state, and the average AB composition of synthesized copolymer is close to equimolar ($\varphi_A \approx \varphi_B \approx 1/2$). At $C_B^0 < C_B^*$, the growing chain is enriched with A monomers; when $C_B^0 > C_B^*$ the growing chain end deeply penetrates into the bulk, and the B segments prevail in the generated copolymer. Thus, the change in the solution concentration of unadsorbed (or weakly adsorbed) reactive monomers allows the chemical composition to be varied over a wide range. Figure 17 presents a typical snapshot picture of the synthesized 1024-segment copolymer chain with $\varphi_A \approx 0.5$.

The chain propagation near the adsorbing surface proceeds as a randomly alternating growth, leading to a copolymer with power-law long-range correlations in the distribution of different segments along the chain. Moreover, the statistical properties of the copolymer sequences correspond to those of a one-dimensional *fractal* object with scale-invariant correlations [88].

Figure 18 presents the results of the statistical analysis for the sequences having approximately 1 : 1 AB composition. Also, Fig. 18 demonstrates the fluctuation function $F_D(\lambda)$ predicted by the probabilistic model [88] for $N = 1024$ and $\varphi_A = 1/2$. We see that designed sequences do not correspond

Fig. 17 Snapshot of the synthesized 1024-unit copolymer chain with $\varphi_A \approx 0.5$. The *light gray* and *dark gray spheres* show adsorbed (A) and unadsorbed (B) segments, respectively. The place of connection of the chain with the surface ("initiator") and the chain end are depicted as *larger spheres*

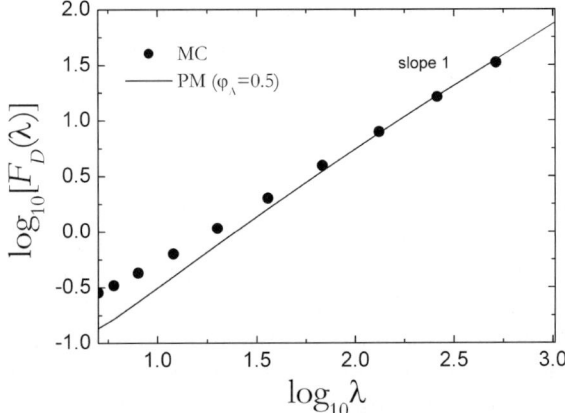

Fig. 18 Detrended fluctuation function for the copolymer sequences of length $N = 1024$, which were generated via surface-induced computer-aided design. *Solid line* shows the analytical (PM) result at $\varphi_A = 0.5$. Adapted from [88]

to random and random-block statistics, and strong correlations exist in these sequences. For sufficiently large λ, very good agreement between the Monte Carlo simulation and the analytical PM result is observed [88]. In both cases, the long-range correlations persist up to the windows with length close to N. Moreover, the correlations turn out to be more pronounced as λ is increased: the dependence of $\log[F_D(\lambda)]$ on $\log \lambda$ becomes a nearly linear function whose slope approaches unity with increasing λ.

From the facts presented above, it is evident that the copolymer sequences discussed here are correlated throughout their whole length. Also, it was found that any sufficiently large part of the averaged sequence has practically the same correlation properties as the entire sequence. This means that the generated sequences show scale invariance, a feature typical of *fractal structures*.

To gain some further insight into the discussed problem, the block length distribution functions $f_A(\ell)$ and $f_B(\ell)$ were calculated, using both Monte Carlo simulations and the analytical probabilistic model [88].

In the $N \to \infty$ limit, the distribution over strongly adsorbed blocks $f_A(\ell)$ is described by the following conditional probability distribution function

$$f_A(\ell) \equiv P_A(\ell|\ell > 0) = (1 - p_{AA})p_{AA}^{\ell-1}, \tag{6}$$

where $\ell = 1, 2, ..., p_{AA} = p_A/[1 - (1 - p_A)(1 - p_B)]$ is the conditional probabilities that the $(i + 1)$th segment in the growing chain is of type A when the ith segment is also of type A, p_A is the probability that the copolymer is lengthened by addition of the next monomeric unit of the type A to a terminal free-radical reactive site, and p_B defines the probability that at each ith step the growing chain goes further from the adsorbing surface.

One may say that the conditional probability p_{AA} defines the strength of persistent correlations in the sequence. If the persistent correlations are extremely strong ($p_{AA} \to 1$), then for the average length of A blocks, we have: $L_A \to \infty$. In the polymerization process, such a situation is realized when the bulk concentration of B monomers approaches zero. If $p_{AA} = 1/2$, one arrives at the known trivial result for the Bernoullian statistics without correlations that corresponds to a random copolymer [10]. In this case, the probability p_A of each segment to be of type A is constant throughout the whole sequence. This means, for example, that the conditional probability p_{AA} is equal to p_A. The average fraction of type A segments φ_A in a purely random sequence is equal to the probability p_A. In particular, for a purely random sequence with $p_A = 1/2$ we have: $L_A = L_B = 2$. Such a copolymer is obtained in the course of solution copolymerization of A and B monomers having identical solubility and reactivity. Finally, at $p_{AA} = 0$ we deal with a regular sequence with alternating distribution of A and B segments in the sequence for which $L_A = L_B = 1$.

For copolymer chains of a finite length N, the probability distribution over the length of blocks is given by [88]

$$f(\ell \leq N) = \begin{cases} f(\ell), & \ell = 1, 2, ..., N-1 \\ 1 - \sum_{n=1}^{\ell-1} f(n), & \ell = N \end{cases} \tag{7}$$

where $f(\ell)$ is defined from Eq. 6 for blocks A, whereas for blocks B, one has:

$$f_B(\ell = 2m) = \frac{p_{AA}}{1 - p_{AA}} \sum_{n_j} \prod_{j=1}^{m} R_{n_j}, \quad \sum_{j=1}^{m} n_j = m \tag{8}$$

$$f_B(\ell = 2m + 1) = 0, \quad m = 1, 2, ... \tag{9}$$

with

$$R_n = (1 - p_{AA})f(k = 2n|k > 0), \quad n = 1, 2, ..., m \tag{10}$$

and $R_0 = 1$. Here the conditional probability $f(k = 2n|k > 0)$ is given by

$$f(k = 2n|k > 0) = \begin{cases} 0, & n = 2m - 1 \\ \phi_{m-1} - \phi_m, & n = 2m \end{cases}, \quad m = 1, 2, \ldots, \quad (11)$$

where

$$\phi_m = \sum_{j=1}^{m} b_{j,m}, \qquad m = 1, 2, \ldots \quad (12)$$

$$b_{j,m} = \frac{j}{j-1} \frac{m-j+1}{m+j} b_{j-1,m}, \qquad j = 2, \ldots, m \quad (13)$$

$$b_{1,m} = \frac{a_m}{m+1} \quad (14)$$

$$a_m = (1 - \frac{1}{2m}) a_{m-1}, \qquad m = 2, 3, \ldots \quad (15)$$

and $a_1 = 1$.

One can show that the block length distribution function $f_B(\ell)$ is characterized for asymptotically large ℓ by the power-law decay of its density $f_B(\ell) \propto \ell^{-\alpha}$ with the exponent $\alpha = 3/2$ [88]. The exponent α estimated from the simulation data ($\alpha \approx 1.6$) is, up to the "experimental" uncertainty, quite close to that predicted by the exact analytical model. As has been noted above, such a power-law decay is a characteristic of the Lévy probabilistic processes [36].

It was found that the resulting copolymer has a specific quasigradient primary structure. Figure 19 demonstrates a typical composition profile calculated as a function of i/N for the sequences, which have approximately 1 : 1 AB composition ($\varphi_A \approx \varphi_B \approx 0.5$). First, we observe rather good (almost quantitative) agreement between Monte Carlo and PM results. Second, both profiles show a monotonous decrease with the segment position i. Therefore, we deal with a copolymer whose primary structure is similar to that known for "tapered" or gradient copolymers exhibiting strong composition inhomogeneity along their chain [17]. It follows from the simulations and the analytical data that the gradient extends along the entire chain for any chain length. Therefore, the copolymer synthesized in this way shows compositional scale invariance. An inhomogeneous spatial distribution of polymerizing monomers and composition constraints in the resulting copolymer are major reasons behind the specific chain growth and the global statistical nature of sequences at large scales.

It is known that the gradient (tapered) nature of copolymers, which can be synthesized in free-radical polymerization processes, is due to a drift in the free monomer composition during solution polymerization [17]. Such copolymers can be considered as a special type of block copolymers in which the composition of one component varies along the chain. With a decreasing difference in the monomer reactivity rations, the formation of gradient sta-

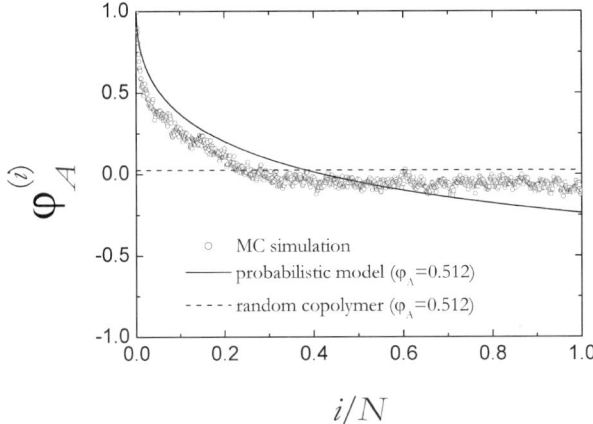

Fig. 19 Intramolecular composition profiles presented as a function of i/N for the sequences having approximately 1 : 1 AB composition ($\varphi_A \approx \varphi_B \approx 0.5$): *circles*, simulation data for the copolymer chains synthesized near a selectively adsorbing surface; *solid line*, probabilistic model, *dashed line*, random sequence. Adapted from [88]

tistical copolymers rather than gradient (tapered) block copolymers occurs. However, in the model polymerization process considered in [87, 88], all the polymerizing species had the same reactivities, and the monomer concentrations remained unchanged during synthesis. Therefore, the changes in the probabilities of the addition of components to the growing macroradical are due to the evolution of its chemical composition; this result is typical for the CDSD regime [24, 25].

2.3.5
Copolymerization Near a Patterned Surface

Because the spatial scale of the monomer concentration gradient in the reaction system discussed in the previous sections was comparable with the macroradical size, the chemical inhomogeneity has been observed for the generated copolymer sequence as a whole. To limit and control the size of the gradient regions, the spatial scale of the concentration inhomogeneities in the reaction system apparently should be limited. The easiest way to achieve this is to perform the copolymerization near a solid patterned surface with discrete adsorption sites. In this case, each of the sites is a small independent source of concentration disturbances lying at a certain distance from other sites. The surfaces with a regular distribution of adsorption sites are of most interest because they can allow a fine control of the primary structure of the copolymers obtained on the basis of the CDSD technique.

Another motivation for this work is the development of copolymers that are tuned to a certain surface, that is, copolymers that have a "memory" of the preparation conditions and are able to reproduce their specific conformation in the vicinity of the surface with predefined chemical heterogeneity. It is thought that copolymers designed in this way have potential for the recognition of patterned substrates through the formation of stable adsorption complexes with planar or spherical substrates composed of two chemically distinct sites, one of which has a preferential affinity for one of the comonomers. Obviously, these copolymers could have enormous potential in molecular technology and biotechnology. In particular, interfacial molecular recognition is ubiquitous and essential in life processes. Examples include enzyme-substrate binding, transmembrane signaling processes, antigen-antibody and protein-receptor interactions [89, 90]. One of the main motivations for studying the pattern recognition mechanism is also related to the design and synthesis of a new generation of copolymers, which carry a target pattern encoded in their sequence distribution and thereby possess the surface recognition ability [91, 92]. Such copolymers can be viewed as macromolecules with pattern-matched sequences.

It should be emphasized that pattern recognition by synthetic flexible polymers is different from biospecific molecular recognition, which is observed, e.g., for folded globular proteins with a specific ternary structure. Indeed, enzyme-substrate or protein-receptor binding can be explained on the basis of the so-called "lock-key" recognition mechanism [90], whereas the recognition by the flexible coil-like polymers is more likely to be described as "pattern-induced conformational fitting" accompanied by strong conformational changes of the chain [93]. In this latter case, the polymer should adjust its conformation such that it touches predominantly the most attractive surface sites [93]. Therefore, the pattern recognition ability of target-imprinted chemical sequences can in principle be optimized via conformation-dependent sequence design [24].

A Monte Carlo simulation of irreversible template copolymerization near a chemically heterogeneous surface with a regular (hexagonal) distribution of discrete adsorption sites was performed in the papers [94, 95]. The sites could selectively adsorb from solution one of the two polymerizing monomers and the corresponding chain segments. The focus of this study was on the influence of the polymerization rate, adsorption energy ε, and distance between adsorption sites r_s on the chain conformation and chemical sequence of the resulting AB copolymers and, specifically, on the coupling between polymerization and selective adsorption.

Under the preparation conditions corresponding to the CDSD regime, the formation of quasiregular copolymers with a blocky primary structure was observed [94]. In such copolymers, there are two types of alternating sections. One of them contains randomly distributed A and B segments. The second one consists mainly of strongly adsorbed A segments. The average length of the random sections is proportional to the distance separating the nearest neighbor adsorption sites r_s. The average length of the A-rich sections is determined by the adsorption capacity of the adsorption sites. Therefore, by varying the interaction parameters and the distribution of adsorption sites on the substrate, one can design and synthesize copolymers with different surface-induced chemical sequences in a controlled fashion.

In particular, the variation of the strength of the effective monomer(A)-substrate interaction ε allows us to determine the asymptotic regimes corresponding to random copolymerization or CDSD. The average intramolecular chemical composition φ_A emerging in the simulation [94] is shown as a function of ε in Fig. 20 for the case in which the solution concentrations of A and B monomers are equal, $C_A^0 = C_B^0$. One can conclude that for the weak adsorption regime, $\varepsilon/k_B T \lesssim 6$, random copolymerization dominates ($\varphi_A \approx 0.5$); in the region $\varepsilon/k_B T \gtrsim 15$, CDSD dominates. The threshold value of the energy parameter ε when random copolymerization and CDSD have about the same probability is $\varepsilon/k_B T \approx 8$ for a given choice of the polymerization parameters and simulation model.

To analyze the correlations in the copolymer sequences, one can use the so-called two-point "chemical correlators" [53]:

$$\Theta_{\alpha\beta}^{(i)} = n_{\alpha\beta}^{(i)} / \sum_{\alpha,\beta} n_{\alpha\beta}^{(i)}, \tag{16}$$

where $n_{\alpha\beta}^{(i)}$ is the number of $\alpha\beta$ (AA, BB, and AB) pairs and i is the chemical distance between these pairs along the chain. Chemical correlator $\Theta_{\alpha\beta}^{(i)}$ is the probability of finding a pair of α and β segments (α,β = A,B) among all the possible pairs separated by i segments along the chain in a given copolymer sequence. Some of the results of the calculation [94] are shown

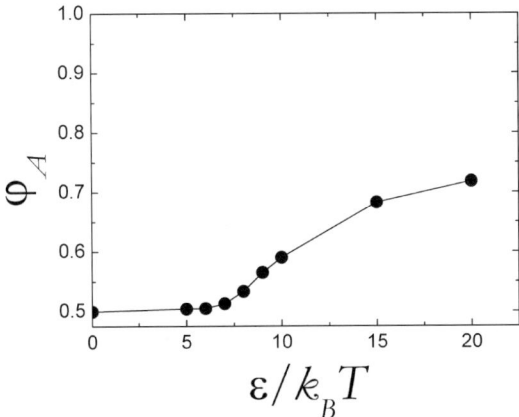

Fig. 20 Average chemical composition as a function of the adsorption energy parameter for the case when the solution concentrations of A and B monomers are equal. Adapted from [94]

in Fig. 21 for a few different values of ε. The dependences of $\Theta_{AA}^{(i)}$ on i indicate that quasiregular copolymers are synthesized in the CDSD regime when the adsorption interaction is sufficiently strong. Indeed, we see that these copolymers are characterized by a periodical variation of the composition along the chain. For the system simulated, the period of this variation is about 20 segments. Therefore, the formation of copolymers with blocky primary structures can be observed.

The following mechanism of the formation of quasiregular copolymers was suggested [94]. Chain growth begins near an adsorption site at which the concentration of strongly adsorbed A monomers is high. These monomers are attached preferably to the end of the growing macroradical and form an initial chain section including mainly A segments. This section gradually covers the nearest adsorption site. Because of the limited adsorption capacity, the screened adsorption site looses the ability to attract the active end group of the macroradical. After that, the reaction volume moves into solution away from the substrate in which random copolymerization occurs. The random chain section grows until it reaches the nearest free adsorption site. Then, the formation of a new adsorbed section begins again. Such cycles are repeated many times. As a result, the formation of a quasiregular copolymer with alternating A-rich and random AB sections is observed.

Thus, the copolymer sequence consists of repeating blocks. Each of these blocks is a short gradient sequence formed by strongly adsorbed sections and random weakly adsorbed sections. The average length of the adsorbed section is almost constant and is related to the adsorption capacity of the adsorption center. On the other hand, the average length of the random section depends on its conformation between the adsorption centers, and in principle, it can vary over a wide range. It should be emphasized that the

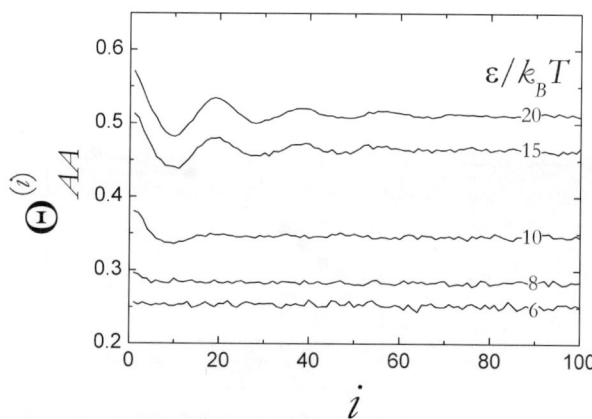

Fig. 21 Chemical correlators for AA diades for different values of the adsorption energy parameter ε, at $N = 512$. Adapted from [94]

quasiregular copolymer is formed only when the random bridges connecting neighboring adsorption sites are strongly stretched and the variation of their length is sufficiently small. The strongly stretched regime leads to the periodical composition variations. Figure 21 shows that this regime is realized for $\varepsilon/k_B T \geq 15$.

The distribution of B blocks, which are included mostly in nonadsorbed chain sections, decays exponentially and thus should obey Bernoullian statistics that correspond to a zeroth-order Markov process [10]. The average length of such blocks is close to 2, that is, the same as that of a random copolymer. In the case of A blocks, the distribution function $f_A(\ell)$ also decays exponentially in the initial region, which corresponds to short blocks included in the random chain sections. For longer A blocks, however, the distribution becomes significantly broader and has a local maximum at $\ell \sim 10$ [95]. Hence, one can conclude that the distribution of A blocks strongly deviates from that known for random sequences.

By varying the distance between nearest adsorption sites, r_s, one can control the composition variation period of the synthesized copolymer. From the chemical correlators defined by Eq. 16, it is easy to find the average number of segments in the repeating chain sections, N, for different r_s values. It is instructive to analyze the relation between N and r_s. As expected, a power law $N \propto r_s^\mu$ is observed. It is clear that exponent μ in this dependence should be between $\mu = 1$ (for a completely stretched chain) and $\mu = \nu^{-1}$ with $\nu \approx 0.6$ (for a random coil with excluded volume [75]). The calculation [95] yields $\mu \approx 1.33$ for $N \geq 15$. This supports the aforementioned assumption that the repeating chain sections are strongly stretched between the adsorption sites. The same conclusion can be drawn from the visual analysis of typical snapshots similar to that presented in Fig. 22.

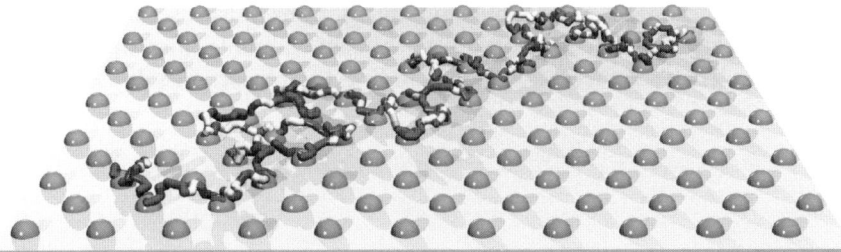

Fig. 22 Snapshot illustrating a typical conformation of 512-unit copolymer chain synthesized near a patterned surface in the strong adsorption regime. Chain segments of A and B type are shown as *dark gray* and *light gray sticks*, respectively. Spheres depict adsorption sites

2.4
Design of Monomeric Units

2.4.1
Amphiphilic Polymers

It is commonplace to say that the properties of a copolymer depend not only on its chemical sequence but also on the chemical structure of its monomeric units. Therefore, the second important route in molecular design can be connected with designing monomeric units of the copolymer having a given sequence distribution. One of the promising ways in this direction is to adjust the *amphiphilic* properties of the copolymer chain.

A large number of macromolecules possess a pronounced amphiphilicity in every repeat unit. Typical examples are synthetic polymers like poly(1-vinylimidazole), poly(N-isopropylacrylamide), poly(2-ethyl acrylic acid), poly(styrene sulfonate), poly(4-vinylpyridine), methylcellulose, etc. Some of them are shown in Fig. 23. In each repeat unit of such polymers there are hydrophilic (polar) and hydrophobic (nonpolar) atomic groups, which have different affinity to water or other polar solvents. Also, many of the important biopolymers (proteins, polysaccharides, phospholipids) are typical amphiphiles. Moreover, among the synthetic polymers, polyamphiphiles are very close to biological macromolecules in nature and behavior. In principle, they may provide useful analogs of proteins and are important for modeling some fundamental properties and sophisticated functions of biopolymers such as protein folding and enzymatic activity.

Since amphiphilic polymers contain monomeric units having hydrophobic/hydrophilic character, they can exhibit conformational transitions induced by temperature, solvent composition, or pH variation [96]. Because of the presence of the two opposing interactions towards the solvent in which they are immersed, amphiphiles can self-assemble, forming a variety of supramolecular structures. Understanding the physics of self-association

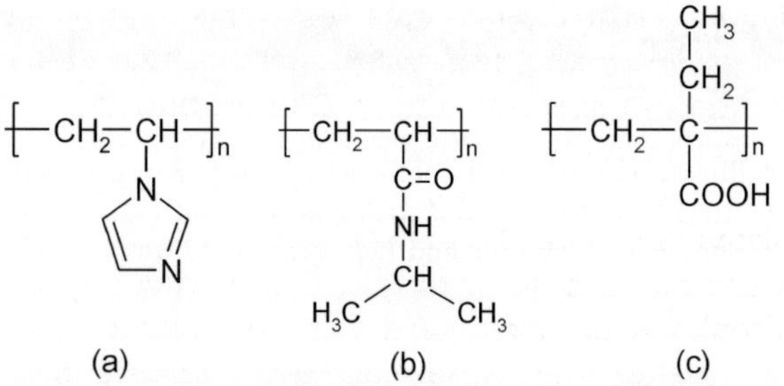

Fig. 23 Examples of amphiphilic polymers: **a** poly(1-vinylimidazole), **b** poly(N-isopropyl acrylamide), and **c** poly(2-ethyl acrylic acid)

of amphiphiles is extremely challenging and also important because the underlying ideas have found connections to other fundamental areas, e.g., phase transitions in membranes, crumpled surfaces, and geometry of random surfaces.

2.4.2
HA Model

The two-letter ("black-and-white") HP model first introduced by Lau and Dill [31] and widely discussed in this article is the simplest model of hydrophobic/hydrophilic polymers. The model is very computationally efficient, but its principal disadvantage is the representation of each monomeric unit of an amphiphilic chain as a point-like interaction site of pure hydrophilic or pure hydrophobic type. At the same time, in the large majority of real amphiphilic polymers, *each* monomeric unit has a *dualistic* (hydrophobic/hydrophilic) character, that is, repeating polymer unit, which is considered as hydrophilic, actually incorporates both hydrophilic and hydrophobic parts concurrently. A typical example is poly(1-vinylimidazole) with the hydrocarbon (hydrophobic) backbone and hydrophilic water-soluble side groups (see Fig. 23). Many of the amino acids also contain both hydrophilic and hydrophobic groups simultaneously and, strictly speaking, the interaction between such amino-acid residues in proteins cannot be literally reduced to pure hydrophilic or pure hydrophobic site-site interactions, as it is presupposed in the HP model by discarding all details of side-group interactions.

One of the possible extensions of the HP model is the HA *side chain model* introduced in [212]. This is a more realistic coarse-grained model of amphiphilic polymers where the dualistic character of each monomeric unit is explicitly represented.

In terms of graph theory, an amphiphilic polymer can be modeled as a caterpillar graph instead of a linear graph corresponding to the standard HP model. More formally, a caterpillar of length N is the **HP** graph in which the set $\{H\}$ represents the nodes in the backbone and the set $\{P\}$ the so-called legs. Each backbone node corresponds to a hydrophobic group (e.g., $CH_2 - CH$ group in Fig. 23) whereas the leg is considered as a polar side group attached to the node. With this representation, each amphiphilic monomer unit (A) is treated as a hydrophobic/hydrophilic HP dumbbell.

Depending on the content of pure hydrophobic and amphiphilic groups, one can simulate amphiphilic homopolymers (poly-A) and copolymers with the same HA composition but with different distribution of H and A units along the hydrophobic backbone, including regular copolymers comprising H and A units in alternating sequence, $(HA)_x$, regular multiblock copolymers $(H_L A_L)_x$ composed of H and A blocks of equal lengths L, and random copolymers having different H and A block lengths (Fig. 24). Gener-

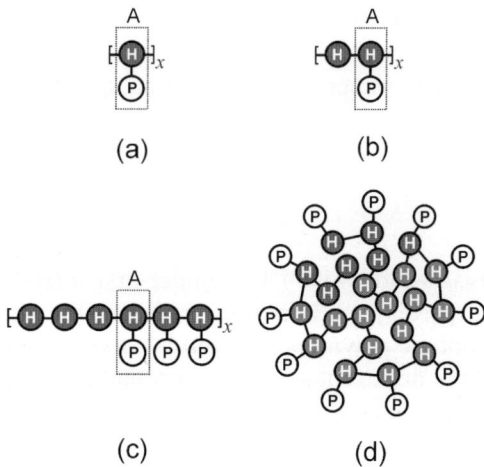

Fig. 24 The "side-chain" models of amphiphilic polymers: **a** amphiphilic homopolymer (poly-A), **b** regular alternating HA copolymer, **c** regular multiblock HA copolymer, and **d** protein-like HA copolymer. Each hydrophobic monomer unit (H) is considered as a single interaction site (bead); each amphiphilic group (A) is modeled by a "dumbbell" consisting of hydrophobic (H) and hydrophilic (P) beads

ally, a random (quasirandom) amphiphilic copolymer is characterized by its composition, by the average lengths of the hydrophobic and amphiphilic blocks, L_H and L_A, and by the specific distribution of H and A units along the chain.

As we will see in Sect. 3.4, such a relatively trivial modification of the standard HP model can lead to some nontrivial consequences when studying the collapse for the single-chain amphiphilic polymers and their aggregation in solution.

There are many further issues that can be addressed by the model of the kind described here. Clearly, the HA model is amenable to a number of generalizations that allow one to study more sophisticated features of amphiphilic copolymers, including, for instance, backbone stiffness, orientational degrees of freedom, or additional structural constraints such as the saturation of monomer-monomer interactions [98], which are crucial, e.g., for the folding of RNA. Also, it is easy to introduce dipole moments for side H – P bonds and specific directional interactions (like hydrogen bonds) for some of the chain units. These additional factors can result in the formation of intramolecular secondary structures and lead to an increase in the stability of globules formed by such polymers.

In the literature, there are several other coarse-grained polymer models in which spherical monomers are replaced by asymmetric objects. Generally, this gets a host of qualitatively new structures, e.g., liquid crystalline phases of helical secondary structures [99]. In those models, including that

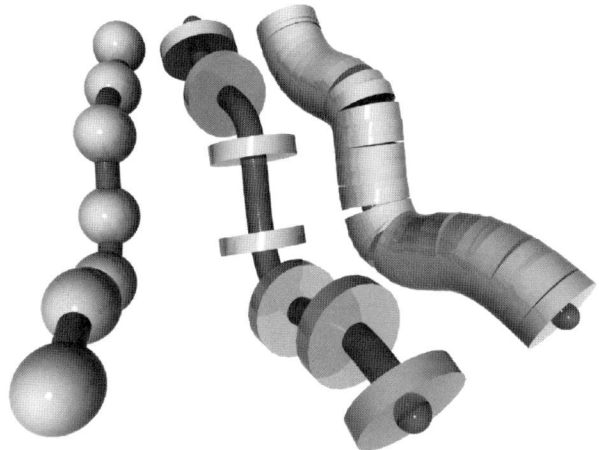

Fig. 25 Chain of beads, chain of coins, and chain of stacked coins

described here, there are two main ingredients: chain units are connected to each other in a linear fashion and each unit of the chain possesses a spatial direction representing the local direction associated with the chain. Familiar examples in protein science include the exotic models that treat the protein backbone not as a chain of spheres but as a chain of anisotropic objects (such as coins) for which one of the three directions differs from the other two (Fig. 25). If such a chain is viewed as being made up of stacked coins instead of tethered spheres, we naturally arrive at the picture of an elastic tube (like a garden hose or spaghetti) whose axis coincides with the chain backbone (Fig. 25). At this coarse-grained level of description, new physics arises from the interplay between two length scales: the range of (anisotropic and many-body) attractive interactions and the thickness of the tube [99].

3
Properties of Designed Copolymers

3.1
Single Chains

3.1.1
Coil-to-Globule Transition

The coil-to-globule transition was studied for designed HP copolymer chains both by means of lattice Monte Carlo simulations using bond fluctuation algorithms and multiple histogram reweighting [100, 101] and by a numerical self-consistent-field (SCF) method [101], assuming that the dominant driving force for polymer collapse is H – H attraction. Copolymer chains of fixed length with H and P monomeric units with regular, random, and specially designed (protein-like) sequences obtained via the coloring procedure were investigated. Qualitatively, the results from both methods are in agreement. The calculations show that it is possible to distinguish four temperature regions.

There exists a low temperature below which the energy of the strongly collapsed chains is constant so that the chains appear to be at the lowest energy. This state may be called the *ground state* or the "frozen glass" phase. Of course, for random copolymers, this state is degenerate, i.e., it is not unique. Its energy is averaged over a few low-energy conformations for each sequence. The type and number of segment-segment contacts in the ground state is well defined for a particular copolymer sequence. With increasing block size L the ground state energy decreases. The reason for this is that small blocks along the chain cannot avoid internal unfavorable H – P contacts. The ground state level of random sequences is significantly higher than that for a protein-like chain. This is because in the protein-like sequence there are longer stretches of hydrophobic units that can efficiently pack in the interior of the dense globule than in the random sequences. Here, we thus note once more that the coloring design procedure [18] results in chains that significantly differ from their random analogues.

At a slightly higher temperature, fluctuations in chain conformations become possible and the compact molecule undergoes internal rearrangements. Under these conditions, the chains remain rather compact but the interaction energy per segment increases more or less linearly with the temperature T. We refer to this regime as the *molten globule state* [102]. There is not a strong L dependence for the onset of the molten globule regime. When more entropic restrictions are locked into the ground state (i.e., for smaller L), the system enters the molten globule state easier. The longer the block length the larger the temperature range over which the molten globules state is found. Both the protein-like and the random copolymers have a well-developed molten globule regime. For protein-like chains this temperature region is wider than the corresponding region for uncorrelated random copolymer chains, because the latter have, on average, shorter stretches of hydrophobic chain segments. The intermediate molten globule state appears most pronounced for the protein-like sequences. Since protein-like sequences behave (quasi)randomly, the calculations [101] thus confirm the previous findings [18–20] that the protein-like copolymers can "inherit" some information from the parent (globular) conformation used to generate its sequence.

With a further increase in temperature, at the end of the molten globule regime, there is a sudden jump in the interaction energy. At this temperature, the coil-to-globule transition is found. The transition temperature increases with increasing block length. For the regular multiblock copolymers the energy jump appears especially large for chains with intermediate block lengths, and the transition looks like a first-order phase transition. Significantly, it was found that the protein-like and the random copolymers have a very small energy jump at the coil-to-globule transition [101]. The transition thus tends to become continuous with increasing levels of heterogeneity along the chain.

Thus, the cooperativity of the coil-to-globule transition for a copolymer with a given block length is highest for regular block copolymer chains. This is explained by renormalization of effective monomeric unit in the case of monodisperse blocks. The cooperativity is lower for the random and protein-like sequences because of the polydispersity of the block length. For regular multiblock copolymers consisting of alternating H and P blocks, the transition temperature is roughly an exponential function of the block length. It should be emphasized that the transition temperature observed for protein-like sequences is much higher than could be expected from their number average block length. Indeed, these transition temperatures are close to those of regular copolymers in which the block length is several folds larger [101]. Because the correlations in the sequence distribution of protein-like copolymers obey the Lévy-flight statistics [35], very long blocks can occur in their chain if the total chain length allows this to happen. Such long blocks increase the sharpness of the transition.

The globular state for protein-like sequences is found to be more stable than that for statistical random copolymers [101]. As has been noted above, the globules of protein-like copolymers exhibit a dense micelle-like core of hydrophobic H units stabilized by the long dangling loops of hydrophilic P units (see Fig. 3a). From the analysis of the collapse transition, one can also conclude that the protein-like sequence is reflected in the shift of the coil-to-globule transition temperature to higher temperatures as compared to the random-block counterparts with the same composition and that same degree of blockiness [18]. Thus, the numerical studies [18, 100, 101] show that specific sequence correlations play an important role for the chain folding and the stability of heteropolymer globules.

Finally, at very high temperature the chains are swollen and behave as a random coil. In such a chain the majority of the interactions are between polymer units and solvent molecules. This state does not depend on the primary structure of the chains and, therefore, the interaction energy levels off to a constant value that depends only on chemical composition.

The effect of copolymer sequence on coil-to-globule transition was also studied using Langevin molecular dynamics [103]. The method for estimation of the quality of reconstruction of core-shell globular structure after chain collapse was proposed. It was found that protein-like sequences exhibit better reconstruction of initial globular structure after the cooling procedure, as compared to purely random sequences.

3.1.2
Kinetics of the Collapse Transition

Experimental observation of the coil-to-globule transition in polymers is very difficult because at concentrations accessible to experiments, the intramolecular collapse of single chains competes with the intermolecular aggregation of macromolecules [104–111]. Consequently, there have been only a few studies on the kinetics of the collapse transition. One possible experimental technique to follow the polymer collapse is to measure the time evolution of the gyration radius by light scattering of a highly dilute solution (in order to get single molecule properties). In computer experiments, we do almost the same observations.

For proteins, the problem is seen as follows: "When an egg is boiled, the proteins it contains unfold. Can this procedure be reversed in theory? Or, in other words, can the encrypted code of protein folding be deciphered from the sequence?" [112, 113].

The "fastest" proteins fold amazingly quickly: some as fast as a millionth of a second. While this time is very fast on a person's timescale, it is remarkably long for computers to simulate. In fact, there is about a 1000-fold gap between the simulation timescales and the times at which the fastest proteins fold. This is why the simulation of collapse kinetics is extremely computationally demanding. Thus, the current challenge lies in understanding how particular chemical sequences in coarse-grained copolymer models lead to particular collapse features. This is a fundamental issue in the problem.

The formation of a compact globule in copolymers requires them to have specific conformations, which are reached through local conformational fluctuations. A characteristic collapse time may be defined as for instance the time for which the gyration radius reaches its equilibrium value. This time measures the approach to equilibrium for the system and is related to the mean first passage time.

The main picture that has emerged from theory and different numerical studies is that polymer collapse proceeds as a series of steps, first the rapid formation of small clusters ("blobs") distributed along the chain that grow and then merge, then the formation of a globular state [114–120]. Of course, the features of the collapse transition depend on the quench depth and the chain rigidity. For heteropolymers, we should also take into account their sequence and chemical composition. Numerical research has attempted to design copolymer sequences with desired properties (e.g., rapid folding to a compact globule) in a manner that mimics biological evolution [26, 121]. An interesting feature of these designed sequences is that they often display a well-defined structure in which similar monomers associate in blocks. Recently, it has been shown that intrachain microphase separation plays an important role in both the thermodynamics and kinetics of collapse for HP heteropolymers with various block sizes [122].

Usually, the simulations of folding kinetics involve an initial equilibration period at sufficiently high temperatures, starting from an initial chain conformation in the strongly swollen state, followed by an instantaneous quench to low temperatures (or by an abrupt decrease in the solvent quality) at which the polymer collapses towards a final state. The collapse is monitored using, e.g., the time-dependent radius of gyration $R_g(t)$ as a function of time measured from the moment of the quench. The degree of compactness of the chain can be monitored by examining how the $R_g(t)$ value changes as folding progresses. It is convenient to use the normalized quantities, $Q^* = Q(t)/Q(0)$, that determine the departure of a given time-dependent property $Q(t)$ from its value in the initial state. To gain better statistics, many independent collapses ($\sim 10^2$) are simulated for each copolymer sequence, each beginning with a unique chain conformation. With a uniform probability density distribution in the initial state, the folding time, t^*, is the mean first passage time (MFPT) to the final compact state, which is defined to be the average over many independent simulations of the time to achieve the final state. Also, one can define the mean intermediate folding times, e.g., $\langle t_{1/4}\rangle$, $\langle t_{1/2}\rangle$, etc., which are required to achieve the corresponding intermediate values $Q^*/4$, $Q^*/2$, etc. For $R_g(t)$, $\langle t_{1/2}\rangle$ is the mean time required to halve the chain size.

Let us compare the kinetics of the selective-solvent-induced collapse of protein-like copolymers with the collapse of random and random-block copolymers [18]. Several kinetic criteria were examined using Langevin molecular dynamics simulations. There are some general results, which seem to be independent of the nature of interactions or the kinetic criteria monitored during the collapse. Here, we restrict our analysis to the evolution of the characteristic ratio $\zeta = \left(R_{gP}^2/R_{gH}^2\right)^{1/2}$ that combines the partial mean-square radii of gyration calculated separately for hydrophobic and hydrophilic beads, R_{gH}^2 and R_{gP}^2. This ratio takes into account both the properties of compactness and solubility for a heteropolymer globule [70] (compactness is directly related to the mean size of the hydrophobic core, whereas solubility should be dependent on the size of the hydrophilic shell).

Using the ζ value, let us define the time-dependent quality $2\left[\zeta(t)/\zeta(0) - 1\right]$ and three intermediate folding times $t_{1/4}$, $t_{1/2}$, and $t_{3/4}$, describing its evolution, as well as the corresponding sequence-averaged probability distribution functions $W_{1/4}$, $W_{1/2}$, and $W_{3/4}$. The distribution of folding times averaged over 1000 different sequences of 128-unit HP copolymers with random, random-block, and protein-like statistics are shown in Fig. 26.

It is seen that among different protein-like sequences, we can find significantly larger fractions of the sequences having small characteristic collapse times than for random and random-block sequences with the same 1 : 1 HP composition. This is true both in the early stages of collapse and in the late stages. Those sequences can be called "fast folders". In a sense, dividing the

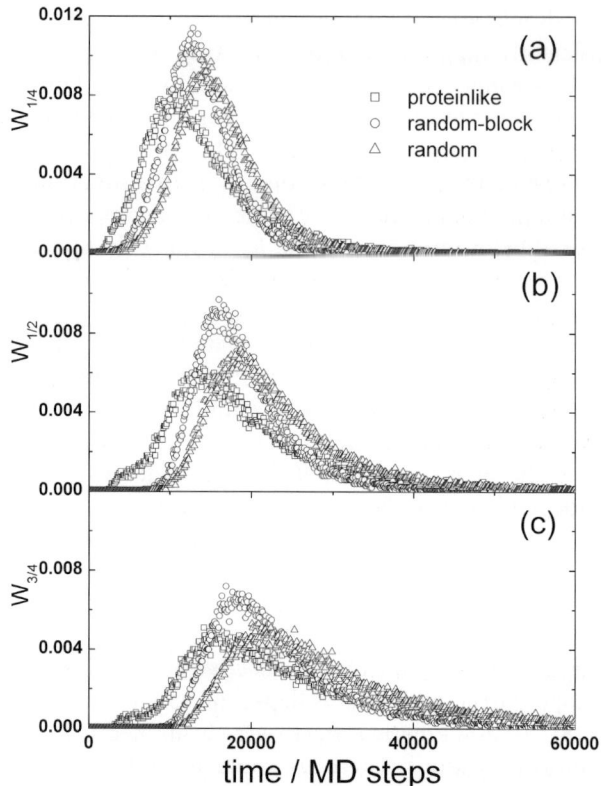

Fig. 26 Distributions of folding times for 128-unit HP copolymers with protein-like, random-block, and random statistics

sequences by fast folders and "slow folders" (that is, those laying before and after the maximum of the W function, respectively) corresponds to a kinetic sequence design or to choosing from the spectrum all possible sequences a set whose kinetic properties possess the attributes we desire. From this viewpoint, our computational procedure is a specific sieve that separates "good" sequences and "bad" ones.

Therefore, we conclude that for the designed copolymers, collapse occurs at a markedly higher kinetic rate, $k \sim 1/\langle t_{1/2} \rangle$. The kinetic properties are clearly related to the sequence distribution. Still it is possible to select "champions" from the protein-like fast folders. Who are those champions? They normally have a larger mean length of hydrophobic and hydrophilic blocks, wider distribution of block length and the average hydrophobicity decreasing towards the chain ends.

Of course, many random processes generating long sequences might in principle produce fast folders, however, there is a negligible chance that we would be able to find their noticeable fraction.

3.2
Phase Behavior: The Sequence-Assembly Problem

Heteropolymers can self-assemble into highly ordered patterns of microstructures, both in solution and in bulk. This subject has been reviewed extensively [1, 123–127]. The driving force for structure formation in such systems is competing interactions, i.e., the attraction between one of the monomer species and the repulsion between the others, on the one hand, and covalent bonding of units within the same macromolecule, on the other hand. The latter factor prevents the separation of the system into homogeneous macroscopic phases, which can, under specific conditions, stabilize some types of microdomain structures. Usually, such a phenomenon is treated as microphase separation transition, MIST, or order-disorder transition, ODT.

For diblock copolymers, periodically arranged spheres (micelles), hexagonally packed cylinders, and a lamellar phase have been observed [1]. A more complex bicontinuous cubic phase with $QI_{a\bar{3}d}$ symmetry (gyroid structure) has also been identified. These supramolecular structures, with length scales on the order of 1 to 10^2 nm, may be controlled by changing the amount of solvent, the length of blocks, or the proportions of A and B monomeric units [128–131].

The conventional theoretical approach to determine the three-dimensional organization of polymers at equilibrium is based on minimizing a suitable Ginzburg–Landau free energy functional with respect to the corresponding lattice parameters for some *a priori* presupposed structure of ideal symmetry (e.g., body centered cubical, hexagonal, lamellar, etc.) [1, 132, 133]. Because of its inherent assumed symmetries, this approach fails to predict the irregular and "defect" microstructures. Also, the computations of the free-energy coefficients within the Leibler-like or random phase approximation (RPA) treatment [132] do not take into account directly the interaction potentials of polymer segments and the system-specific features of polymer architecture on short length scales. Self-consistent mean-field (SCMF) techniques have been proposed in the context of field-theoretic ideas by Matsen and coworkers [134–137] and Fredrickson and coworkers [138, 139]. Microscopic statistical mechanical polymer theories have also been developed, including the polymer RISM theory [140–143], which is an extension to flexible long-chain polymers of the integral equation reference interaction site model (RISM) theory of Chandler and Andersen [144], and density functional (DF) theory [145].

The theoretical approaches mentioned above mostly deal with diblock or regular multiblock AB copolymers [1, 123–125]. It has been shown theoretically [146–149] that for random copolymers the existence of ordered microphases is also not excluded. In particular, it has been shown that in addition to the classical microdomain superstructures, chemical irregularities in multiblock copolymers may result in formation of very exotic positionally ordered structures with many levels (length-scales) of organization (the so-called "secondary superstructures") [150–152]. The phase behavior of random AB copolymers with specific statistics and a strongly correlated distribution of units along the chain is of particular interest. Treating processes taking place in these systems continues to challenge theorists. Even ideal random copolymer mixtures, produced if the reactivities of the two polymerizing comonomers are equal, have unique statistics: each sequence represents a different component. In general case, the possible random sequences generate mixtures of $2^{N-1} + 2$ chemically different N-unit macromolecules.

In this section, we discuss the phase behavior of protein-like copolymers that can be considered as a specific type of correlated copolymers. One can expect that the presence of long-range correlations in protein-like chains will influence the phase behavior of such copolymers.

The main parameters characterizing the thermodynamic states (segregation regimes) of self-assembling AB copolymers are the composition, $\varphi_i = N_i/N$ (i = A, B; $N_A + N_B = N$; N being the total chain length) and the Flory-Huggins parameter χ, which is inversely proportional to temperature, which reflects the interaction between different segments. For random copolymers, we should also introduce an additional parameter, viz., the average length of the segments composed of units A and B, L_A and L_B. We restrict our consideration to the simplest symmetric case: $\varphi_A = \varphi_B = 1/2$ and $L_A = L_B = L$. For comparison, it is instructive to consider a random-block copolymer with the same composition in which the distribution of block length ℓ is described by the Poisson law. In this case, L can vary. For both copolymers, the length of the segment composed by the units of a given type is a random quantity. It is of interest to compare the behavior of such polymers with each other and with regular multiblock copolymers where L is fixed.

3.2.1
The Polymer RISM Theory

To describe the equilibrium structure of the system, one can use the polymer integral equation RISM theory [140, 141], which allows one to find collective correlation functions. For AB copolymers, the polymer RISM equation is represented in the matrix form [142, 143]

$$H(r) = \int_{(r')} \int_{(r'')} W(|r - r'|)C(|r' - r''|)[W(r'') + \rho H(r'')]dr'dr'' . \qquad (17)$$

Here, H and C are symmetric matrices whose elements are the partial total $h_{\alpha\beta}(r)$ and direct $c_{\alpha\beta}(r)$ pair correlation functions (α,β = A,B); W is the matrix of intramolecular correlation functions $w_{\alpha\beta}(r)$ that characterize the conformation of a macromolecule and its sequence distribution; and ρ is the average number density of units in the system. Equation 17 is complemented by the closure relation corresponding to the so-called molecular Percus–

Yevick approximation [141]

$$\int_{(r')}\int_{(r'')} W(|r-r'|)C(|r-r''|)W(r'')dr'dr'' \qquad (18)$$

$$= \int_{(r')}\int_{(r'')} W(|r-r'|)[C^{(0)}(|r'-r''|) + \Delta C(|r'-r''|)]W(r'')dr'dr'',$$

$$r > \sqrt{\sigma_\alpha \sigma_\beta}$$

$$\Delta c_{\alpha\beta}(r) = [1 - e^{-u_{\alpha\beta}(r)/kT}][h^{(0)}_{\alpha\beta}(r) - 1], \quad r > \sqrt{\sigma_\alpha \sigma_\beta} \qquad (19)$$

$$h_{\alpha\beta}(r) = -1, \quad r < \sqrt{\sigma_\alpha \sigma_\beta} \qquad (20)$$

$$c^{(0)}_{\alpha\beta}(r) = 0, \quad r > \sqrt{\sigma_\alpha \sigma_\beta} \qquad (21)$$

Here, $u_{\alpha\beta}(r)$ describes the interaction between nonbonded units and σ_α is the effective unit size ($\sigma_A = \sigma_B = \sigma$). In [153, 154], the Yukawa potential was used

$$u_{\alpha\beta}(r) = \begin{cases} \varepsilon_{\alpha\beta}(\sigma/r)\exp[-2(r/\sigma - 1)], & r \geq \sigma \\ \infty, & r < \sigma \end{cases}, \qquad (22)$$

where $\varepsilon_{\alpha\beta}$ is an energy parameter which is directly related to the χ parameter. We consider the case when A units are attracted to one another ($\varepsilon_{AA} = -1$) while B-B and A-B interactions are the excluded-volume type ($\varepsilon_{AA} = \varepsilon_{AA} = 0$). The functions $h^{(0)}_{\alpha\beta}(r)$ and $c^{(0)}_{\alpha\beta}(r)$ in Eqs. 18–21 correspond to the reference (athermal) system for which all $\varepsilon_{\alpha\beta} = 0$. Within the frame of Gaussian chain statistics, the spatial distribution of the pairs of units i and j separated by $n = |i - j|$ bonds in a given chain is written in the reciprocal q-space as $w_{ij}(q) = [\sin(q\sigma)/q\sigma]^n$. Summation of $w_{ij}(q)$ over i and j ($i, j = 1, 2, ..., N$) gives the matrix elements $w_{\alpha\beta}(q)$ usually termed single-chain form-factors. Below, the σ and ε_{AA} values are used as basic units.

The polymer RISM theory provides the local structural information through the site-site pair correlation functions (PCFs) and the influence of the long wavelength structure for these local properties. Although the phase-separated structures are not available from the polymer RISM theory, one can obtain the structural information for the disordered single-phase at a certain distance from the phase separation point against the athermal reference system. Also, by varying temperature T and monomer density ρ, one can find the conditions under which the spatially homogeneous state of the system becomes unstable. This takes place on a spinodal line, which is determined by the set $\{T^*, \rho^*\}$ or parametrically as $T^* = T^*(\rho)$. The condition of spinodal instability at a temperature T^* corresponds to [141]

$$\Delta(q) \equiv \det[E - \rho W(q)C(q)]_{q=q^*} = 0, \qquad (23)$$

where $\Delta(q)$ is the determinant of the matrix equation (Eq. 17), q^* represents the wave vector of maximum instability, and E is the unit diagonal matrix.

The value $q = q^*$ at which $\Delta(q^*) \to 0$ allows one to estimate the spatial scale r^* of the arising structure ($r^* = 2\pi/q^*$).

Figure 27 shows the plots of T^* and r^* versus L for the regular and random-block copolymers at the volume fraction of macromolecules $\Phi = 0.8$ corresponding to a typical polymer melt. It is seen that T^* increases in both cases with an increase in the block length; however, at small L values, the T^* values for the random-block copolymer are higher than for the regular one. In other words, structure formation in the system of copolymers with a random distribution of block lengths occurs more readily (i.e., at a higher temperature) than in the system of regular multiblock copolymers. This result, which might appear at first glance to be surprising, will be interpreted in Sect. 3.2.3.

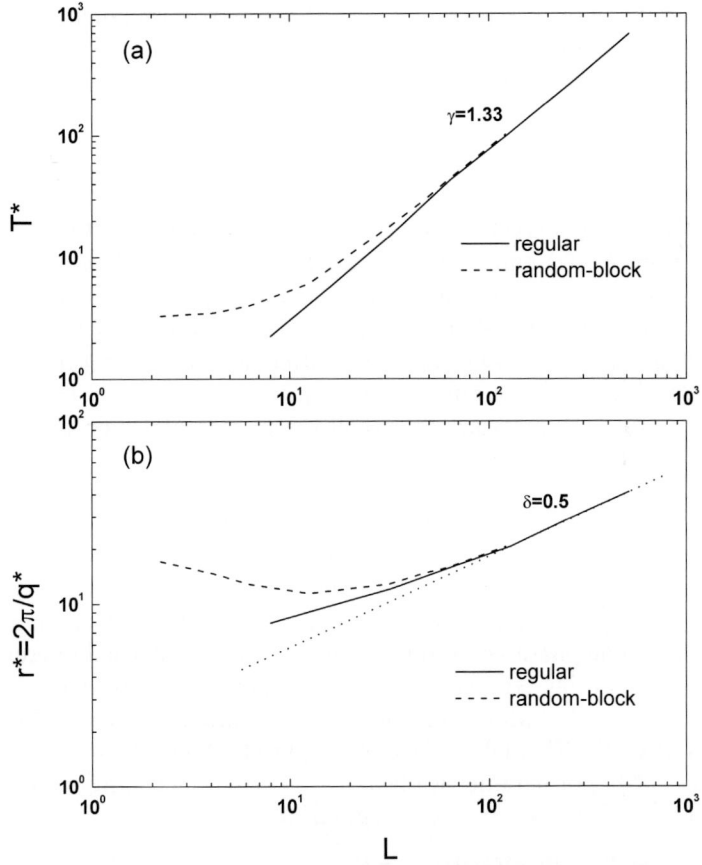

Fig. 27 The values of T^* and r^* as a function of block length L for the regular and random-block copolymers at the volume fraction of macromolecules $\Phi = 0.8$. Adapted from [153]

When the block length becomes comparable with N, distinctions between the behaviors of two copolymers practically disappear. At $L > 200$, one observes that $T^* \sim L^\gamma$ with $\gamma = 4/3$. In this case, the characteristic scale of the microdomain structure behaves as $r^* \sim L^\delta$ with $\delta = 1/2$. This dependence is caused by the fact that flexible chains in the melt have a Gaussian conformation, and the average size of any chain section of n units is proportional to $n^{1/2}$ [75]. Hence, for sufficiently large L's, the spatial scale of microinhomogeneities in the system is determined only by the block size. However, the behavior of the random-block copolymer at $L \lesssim 10^2$ is more complicated. In particular, r^* has a minimum at $L \approx 10$.

Figure 28 shows the spinodal lines for all the systems under discussion. At a low Φ ($\lesssim 0.1$), no noticeable differences are observed. This region of the phase diagram (including the critical point) corresponds to macrophase sep-

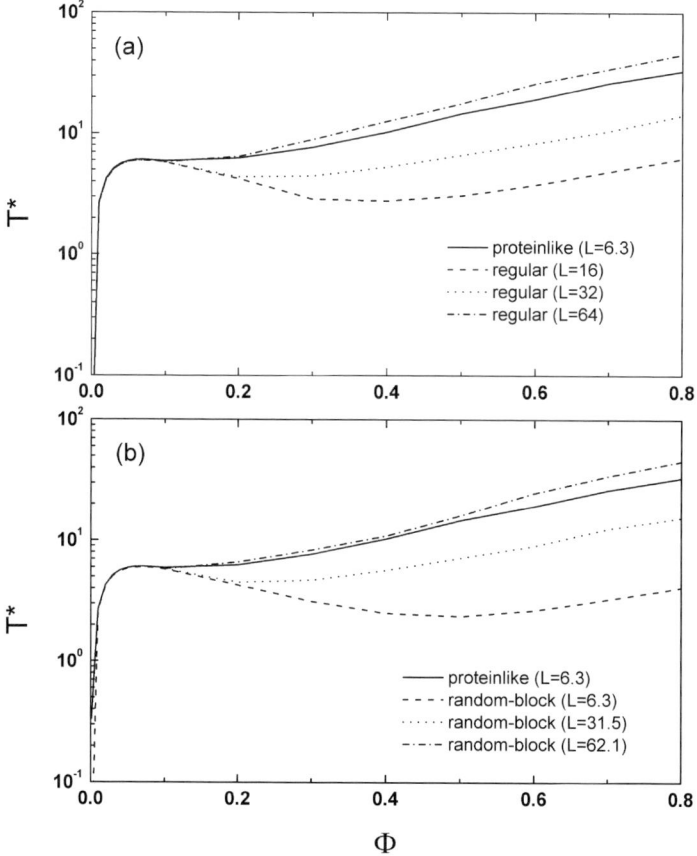

Fig. 28 Spinodals for the systems of protein-like copolymers with average block length $L = 6.3$ and of regular and random-block copolymers having different block lengths. Adapted from [153]

aration transition, MAST, i.e., to the conditions when determinant (Eq. 23) vanishes at $q^* = 0$. In this case, specific features of copolymer sequence have virtually no effect. However, there are such Φ values at which the condition $\Delta(q^*) \to 0$ is met at $q^* \neq 0$ (Fig. 29). Transition from the $q^* = 0$ regime to the $q^* \neq 0$ regime occurs at the Lifshitz point delimiting the MAST and MIST regions.

In the $q^* \neq 0$ region, the phase behavior of protein-like copolymers begins to sharply differ from that observed for the regular and random-block copolymers. In particular, Fig. 28 shows that the spinodal line for the 1024-unit protein-like copolymers with the average block length $L = 6.3$ is close to the spinodals of the regular and random-block copolymers in which the block length is roughly tenfold (!) larger [153]. At the same Ls, the transition temperatures for the protein-like copolymers are also several fold larger. This indicates that the processes of self-organization in the system of protein-like copolymers can proceed significantly more intensely than in the other copolymer systems. We note that a similar behavior is observed for intramolecular (coil-to-globule) transitions discussed above. Moreover, Fig. 29 shows that at a sufficiently high density of protein-like copolymers ($\Phi \gtrsim 0.3$), the domain spacing $r^* = 2\pi/q^*$ noticeably exceeds that observed for the random-block copolymers. The difference between the protein-like and regular copolymers is even greater.

Therefore, the PRISM calculations [153] show that the protein-like copolymers with A-A attractions are more "inclined" to self-organization than their random-block counterparts and regular multiblock copolymers with the same chemical composition and the same chain length.

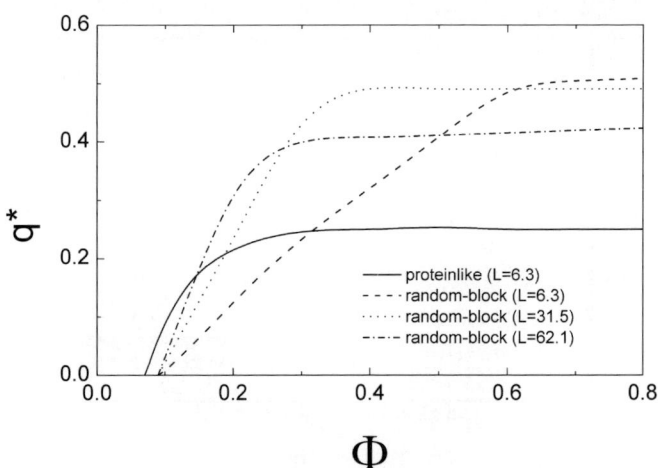

Fig. 29 Characteristic wave number q^* as a function of polymer volume fraction for the systems of protein-like copolymers with $L = 6.3$ and random-block copolymers with different block lengths. The domain spacing is defined as $r^* = 2\pi/q^*$. Adapted from [153]

3.2.2
Field-Theoretic Calculation

The RISM theory describes copolymers only in the weak-segregation limit. At much stronger interaction of monomeric units, the components should be strongly segregated, with a narrow interphase between domains. In this case, the self-consistent field (SCF) methods [134–139] can be used. They span both weak- and strong-segregation limits, as well as the intermediate regime where the other analytical approaches do not apply. Here we present some results obtained by using the real-space SCF approach [155] that is particularly well suited to screening for new types of self-assembly in copolymer melts because it requires no assumption of the mesophase symmetry.

In the SCF theory, the external mean fields acting on a polymer chain are calculated self-consistently with the composition profile. Each chain has independent statistics in average chemical potential fields, $\omega_\alpha(r)$, conjugate to the volume fraction fields, $\phi_\alpha(r)$, of monomer species α. The free energy per chain F is related to the statistical weight, $q(r,s)$, that a segment of a chain, originating from the free end of the α block and with contour length s, has at its terminus at point r. The free energy is to be minimized subject to the constraint of local incompressibility and the constraint that $q(r,s)$ satisfies a dynamical trajectory given by modified diffusion equations. Minimization of F with respect to $\omega_\alpha(r)$ and $\phi_\alpha(r)$ leads to a self-consistent set of equations [155] that are solved iteratively. The propagators $q(r,s)$, together with

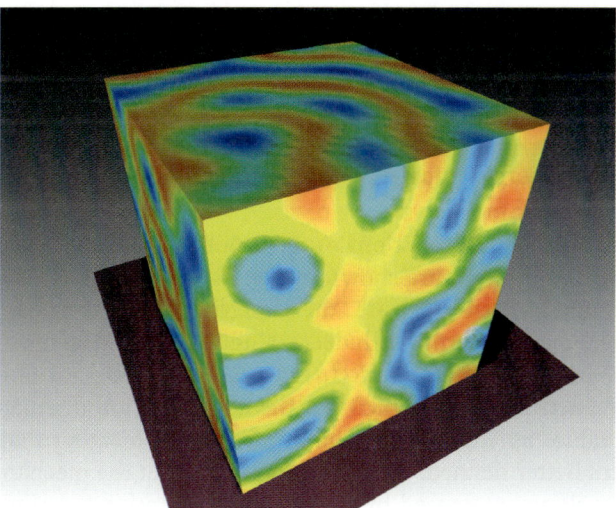

Fig. 30 Composition variations for the incompressible melt of 128-unit protein-like chains at $\chi N = 64$. Regions rich in A and B segments are shown in *red* and *blue*, respectively; intermediate regions are given in *yellow* and *green*

their conjugate propagators $q^\dagger(r,s)$, which propagate from the opposite end of the chain, are used to calculate the density fields. The calculations result in the space distribution of the volume fractions $\phi_\alpha(r)$.

As an example, we show in Fig. 30 the composition variations that are found for the incompressible melt of 128-unit protein-like chains at $\chi N = 64$, starting from a homogeneous state. The simulation box is a three-dimensional 32^3 grid with periodic boundary conditions and a side length of 6.4 (in units of R_g). Regions rich in A and B segments are shown in red and blue, respectively; intermediate regions are given in yellow and green.

As seen, the SCF theory predicts irregular bicontinuous morphology with no long-ranged order. Note that for random-block copolymers with the same average block length, no microphase separation is observed even for $\chi N = 100$.

3.2.3
Molecular Dynamics Simulation

In molecular dynamics simulations, the mean values are estimated as arithmetic averages over the configurations stored in each trajectory and over all trajectories generated for each system. With this method, thermal fluctuations are taken into account in contrast to the standard RPA model where properties are calculated over static minimum energy configurations. This leads to more reliable estimates of properties at the temperature of interest, including the strong segregation regime.

We will briefly discuss the molecular dynamics results obtained for two systems—protein-like and random-block copolymer melts— described by a Yukawa-type potential with (i) attractive A-A interactions ($\varepsilon_{AA} < 0$, $\varepsilon_{BB} = \varepsilon_{AB} = 0$) and with (ii) short-range repulsive interactions between unlike units ($\varepsilon_{AB} > 0$, $\varepsilon_{AA} = \varepsilon_{BB} = 0$). The mixtures contain a large number of different components, i.e., different chemical sequences. Each system is in a randomly mixing state at the athermal condition ($\varepsilon_{\alpha\beta} = 0$). As the attractive (repulsive) interactions increase, i.e., the temperature decreases, the systems relax to new equilibrium morphologies.

The partial scattering functions $S_A(q, \varepsilon)$ normalized by $S_A(q, 0)$ obtained at the athermal condition are shown in Fig. 31 for both systems at a few selected ε_{AA} values. When $\varepsilon_{AA} \lesssim -1$, the scattering intensity at small, nonzero wave numbers q rises sharply with decreasing ε_{AA}. Both systems show a common feature in that the peaks in $S_A(q)$ increase and their position q^*, which is directly related to the periodicity of the concentration fluctuations, shifts to lower q values as attraction becomes stronger, indicating that the structure becomes better defined. These results reflect that the mechanisms of the structure development caused by microphase separation are similar to each other. We thus see that both copolymers exhibit intermolecular aggregation as the ordering transition is approached, although we do not find a *periodic* microphase forming in both systems.

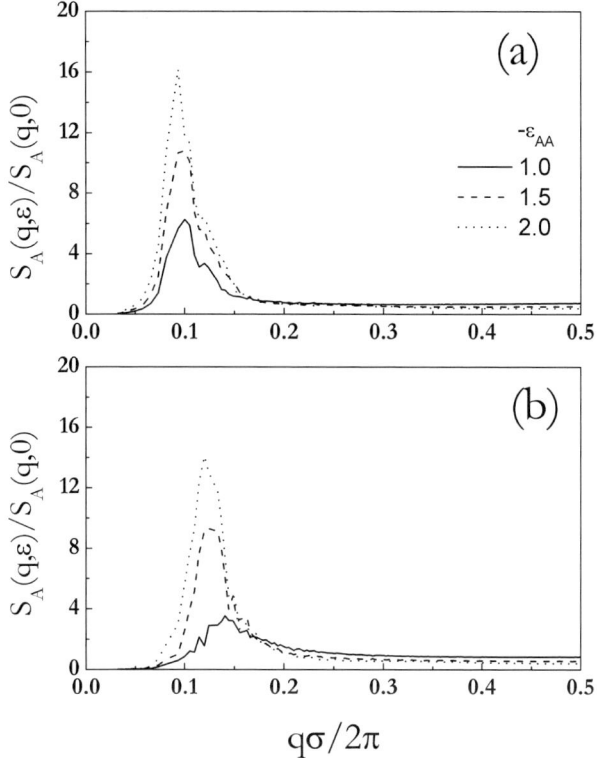

Fig. 31 Normalized partial scattering functions $S_A(q,\varepsilon)$ for the melt-like systems of **a** protein-like copolymers and **b** random-block copolymers at a few selected ε_{AA} values that characterize A-A attraction

To visualize the three-dimensional structures of these microphases more clearly, the density of A-segments was measured on a three-dimensional grid. In the A-rich domains the density of A-segments is high (~ 1), and in the B-rich domains this density is low (~ 0). The dividing surface between the A-rich domains and the B-rich domains is now represented by the isosurface where the density is midway between these values, i.e., 0.5.

When Fig. 32a is compared with Fig. 32b, it is revealed that the domain spacing in the protein-like copolymer system is longer than that in the random-block one. This observation is consistent with the results shown in Fig. 31. The same conclusion can be drawn from the molecular dynamics simulation of the systems with repulsive interactions between unlike units. As an example, we show the normalized scattering functions calculated at large incompatibility, $\varepsilon_{AB} = 1$ (Fig. 33), and the corresponding isosurfaces (Fig. 34). In Figs. 32 and 34, we observe irregular bicontinuous structures with no long-ranged order. Similar morphologies have been predicted in the SCF calculation discussed above as well as in Monte Carlo simulations of random

Fig. 32 Isosurfaces showing microphase separated systems of **a** protein-like and **b** random-block copolymers with attractive A-A interactions

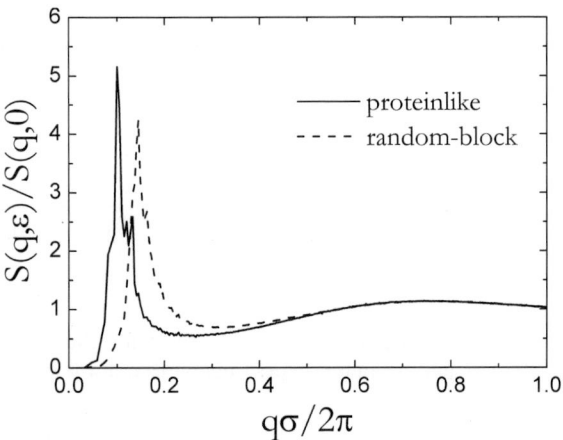

Fig. 33 Normalized scattering functions $S(q,\varepsilon)$ for the melt-like systems of protein-like and random-block copolymers with repulsive A-B interactions

copolymer melts [156, 157]. The reason is that the strong fluctuations destroy completely the stability of the ordered phases, and the disordered phase becomes thermodynamically stable everywhere. In this respect the morphology below the transition temperature resembles a disordered microstructure having an anomalously large correlation length (random wave structure) [158]. This is unique to random copolymer systems. It is clear that this structure should get more and more ordered as the randomness decreases.

What is most important for our discussion is the fact that the spatial scale r^* of the segregated structure for protein-like copolymers is appreciably larger than that for random-block copolymers with the same composition and the same average block length. Also, MIST in the protein-like copolymer system occurs at a temperature higher than that of the random-block system, which is in agreement with the prediction of the polymer RISM theory [153].

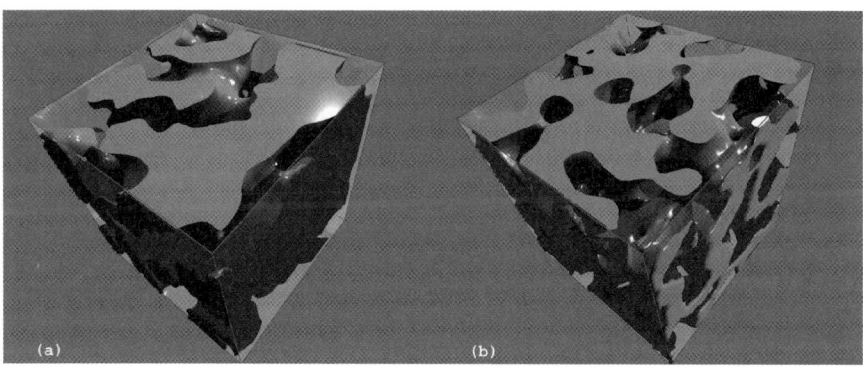

Fig. 34 Isosurfaces showing microphase separated systems of **a** protein-like and **b** random-block copolymers with repulsive A-B interactions

Obviously, these features arise from the difference between the chemical sequences of the two copolymers.

The reason behind this distinction is that the behavior of random copolymers is governed not only by the average block length L but also by the block length dispersion D_L. For correlated random copolymers, the structure formation is dominated by long blocks whose probability increases with an increase in D_L. Even at a relatively low fraction of these blocks in the chain, their effect can be decisive. As we noted while discussing Fig. 27, the largest differences between the regular and random-block copolymers manifest themselves just at relatively small L values when the chain consists of both the short and rather long chemically homogeneous sections. With an increase in L at fixed N, the width of the block length distribution decreases, and the differences gradually disappear (Fig. 27). As has been discussed in Sect. 2.2.2, the block length distribution in protein-like copolymers is described by a specific type of statistics, namely by the Lévy-flight statistics [35] with a high dispersion and a slow decrease in the block length probability. That is why the chains contain a significant fraction of long sections composed of chemically identical segments even at relatively small L values. This is precisely the major reason behind the rather unusual behavior of such copolymers in self-organization processes.

3.2.4
Evolutionary Approach

Shakhnovich and Gutin [26] (see also [28, 159, 160]) showed that it is possible to design a copolymer in such a way that it will fold into a specific conformation. To do this, they optimized the sequence, using a Monte Carlo method that randomly exchanges monomers within the sequence. In this section, we describe an extension of this approach to design a copolymer sequence that is capable of forming microphase-separated structures with the domain spacing r^* which is as large as possible.

To analyze the stability of the ordered microphases, the simplest incompressible random-phase approximation [132] can be employed. Using this approach, the critical value of the Flory–Huggins parameter, χ^*, and the corresponding spinodal temperature, $T^* = 1/\chi^*$, can be determined by the condition that the scattering intensity $S(q)$ reaches its maximum value at a nonzero wave vector q^*. Within the RPA the scattering intensity is given by [132, 142]

$$S_{RPA}^{-1}(q) = \Re(q) - 2\chi \qquad (24)$$

with

$$\Re(q) = \frac{1}{\delta w(q)} \left\{ \frac{w_{AA}(q)}{\phi_B} + \frac{w_{BB}(q)}{\phi_A} + \frac{2w_{AB}(q)}{f_A f_B} \right\}, \qquad (25)$$

where $f_A = N_A/N$, $f_B = N_B/N$, ϕ_A and ϕ_B are the volume fractions of the corresponding species, and $\delta w(q)$ is defined as $\delta w(q) = w_{AA}(q)w_{BB}(q) - f_A^{-1}f_B^{-1}w_{AB}^2(q)$. To a first approximation, one can consider macromolecules on the basis of the highly simplified unperturbed model without intramolecular excluded volume interactions. This allows us to considerably simplify the problem by calculating the matrix elements $w_{\alpha\beta}(q)$ using the Gaussian intramolecular correlation function $w_{ij}(q) = \exp\left(-q^2\sigma^2 |i-j|/6\right)$, which characterizes the distribution of segments i and j belonging to the species A and B inside a polymer.

In the model [161], Monte Carlo simulations were performed to search for point mutations that favor the microphase separation of copolymer melt. A random AB sequence is taken as an initial state, and then a procedure of the evolution (annealing) of the sequence starts. The iterative procedure consists of many mutation steps. At each Monte Carlo step, two monomers are chosen randomly and, if they happen to be of different types, an attempt is made to exchange their types (A ↔ B). This changes the copolymer sequence and the matrix elements $w_{\alpha\beta}(q)$. The resulting change in the transition temperature ΔT^* after an attempted mutation is calculated and the probability p to fix the mutation is guided by the Metropolis algorithm: p equals 1 if $\Delta T^* \geq 0$, otherwise p equals $\exp(\Delta T^*/T_s)$, where T_s is the fictitious temperature referred to as "sequence design temperature" that characterizes the tolerance to mutations in sequence space. In fact, this is the same parameter as that used in Sect. 2.2.5.

Figure 35 shows the transition temperature T^* and the domain size r^* ($= 2\pi/q^*$) as a function of T_s for different N. Note that both T^* and r^* are averaged over the ensemble of generated sequences ($\sim 10^6$). If T_s is too high, the sequences tend to become random. In contrast, when T_s is too low, the evolutionary algorithm leads to the trivial diblock sequence. In this case, one observes a typical mean-field behavior: $T^* \propto N$ and $r^* \propto N^{1/2}$ (for long

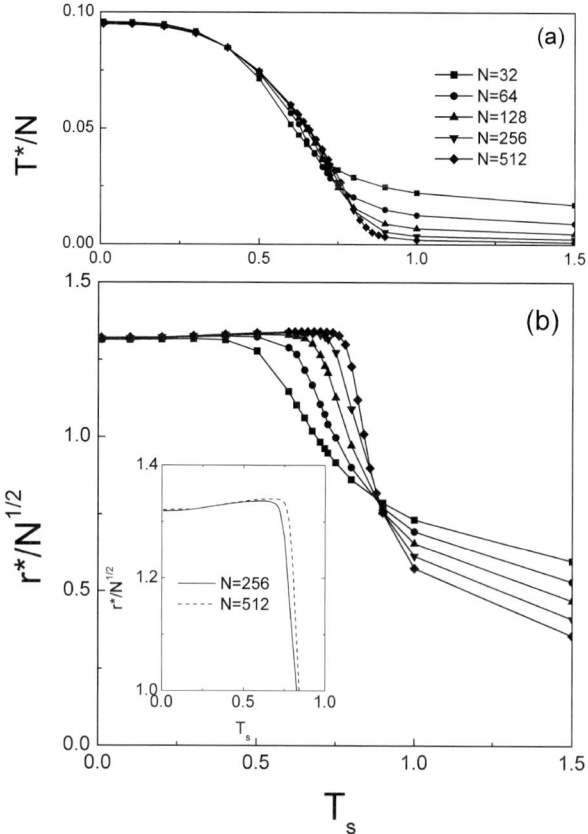

Fig. 35 a Transition temperature T^* and **b** domain size r^* ($=2\pi/q^*$) as a function of the sequence design temperature for different chain lengths N

symmetric diblock copolymers, $T^*/N = 0.096$ or $\chi^*N = 10.4$). Therefore, it is interesting to explore a range of values for T_s that yields a compromise between these two regimes. As seen, there is a certain critical design temperature ($T_s^* \approx 0.78$) near which the value of r^* has a maximum for the designed sequences with $N \geq 256$.

Let us consider the thermodynamics of the transition in sequence space from high design temperatures, where many sequences contribute more or less equally, to the lowest design temperature, where one or a few sequences dominate. Figure 36 shows the distribution of domain sizes $W(r^*)$ for the ensemble of sequences generated near T_s^*. There is a clear bimodality seen in $W(r^*)$. Therefore, the transition in sequence space occurs as a first-order-like transition. The most important observation that can be made for this transitory regime is that the sequences providing the maximum of r^* do not correspond to simple symmetric diblocks but rather they look like gradient ones. Another intriguing result is that the Jensen–Shannon divergence

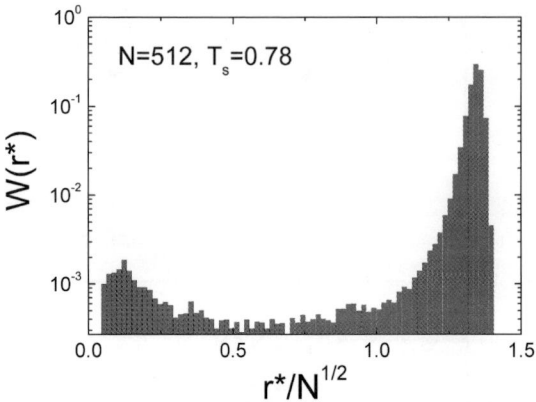

Fig. 36 Distribution of domain sizes $W(r^*)$ for the ensemble of sequences generated near T_s^*

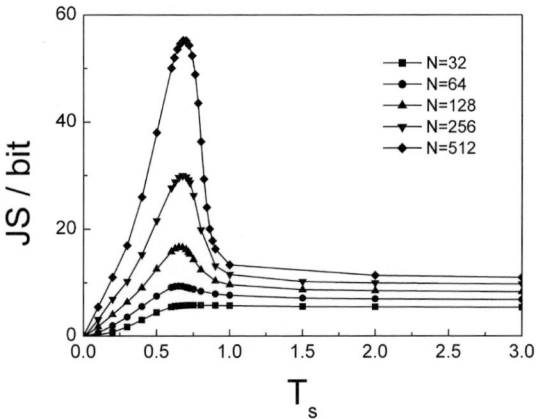

Fig. 37 The Jensen–Shannon divergence measure as a function of the sequence design temperature for different chain lengths N

measure shows a maximum in the vicinity of T_s^* (Fig. 37). This behavior is currently not completely understood, but might be a good starting point for further investigations.

3.3
Charged Hydrophobic Copolymers

Hydrophobic polyelectrolytes (HPEs) are polymers composed of covalently bonded sequences of polar (ionizable) groups which are soluble in water and hydrophobic groups which are not. This unique duality towards an aqueous environment leads to a rich spectrum of complex self-association phenomena. Such polymers can also be considered as highly simplified models of biologically important macromolecules, e.g., proteins [96, 162].

3.3.1
Solution Properties

It is well known that many of the solution properties of proteins are due to a complex interplay between short-range hydrophobic attraction, long-range Coulomb effects, and the entropic degrees of freedom [163]. The balance of these factors determines the solubility and equilibrium conformation of proteins in water. To a large extent, these properties are also dependent on the unique primary structure of the polypeptide chain. This fact motivates intensive experimental and theoretical studies directed to the design of HPEs with specific chemical sequences resembling those of biomacromolecules. In addition, enormous effort has recently been directed toward the computer simulation of model HPEs with the purposes, explicit or implicit, of imitating the conformational behavior of biological macromolecules or of obtaining a basic understanding of the mechanism of molecular aggregation. In the present section we will focus on these studies.

Conformational transitions of single-chain HPEs are today part of an important theoretical challenge because of the very specific and somewhat unusual conformations they are supposed to adopt [164]. Much attention has been paid to the study of HPEs, mainly using theoretical concepts and coarse-grained simulation models [163]. It was found that in dilute salt-free HPE solutions, the subtle and antagonist balance of electrostatic and hydrophobic interactions can lead to a large variety of conformations compared to polyelectrolytes without hydrophobic groups. In particular, simulation results demonstrate that there is a range of electrolyte concentration and hydrophobicity for which HPEs exhibit exotic but stable conformations, namely the pearl necklace and the cigar-shaped conformations [163]. By gradually decreasing the monomer hydrophobicity of a strong polyelectrolyte, it undergoes a cascade of transitions from a nearly spherical globule to a cylinder-shaped conformation, a cylinder to a pearl necklace, and a pearl necklace to a strongly extended structure, successively [165–168]. These microstructures are controlled by a balance between surface tension and electrostatic free energies.

Micka et al. [169] were the first who simulated a multichain HPE system. They studied regular copolymers with alternating neutral and charged monomers (with a charge fraction of $f = 1/3$) in a poor solvent in the presence of monovalent counterions. The paper by Micka et al. [169] nicely demonstrated that the necklace microstructures exhibit a variety of conformational transitions as a function of polymer concentration. The end-to-end distance was found to be a nonmonotonic function of concentration and showed a strong minimum in the semidilute regime.

The solution properties of charged hydrophobic-hydrophilic protein-like copolymers have been studied in the presence of both mono- and multivalent counterions using molecular dynamics simulations [170–172]. In the model, one half of the polymer units was assumed to be negatively charged (P units) while the hydrophobic H units were electrically neutral but assumed to be strongly attractive. The variation of temperature allowed for the covering of a wide range of chain states from strongly swollen polyions at high temperature to strongly collapsed polyions at low temperature. The question of primary structure has also been explored, using microscopic polymer integral equation (RISM) theory [173]. The results of the simulations [170, 171] were found to be strongly dependent on the valence of the counterions and the temperature. The effect of these factors is discussed below.

3.3.2
Designed Copolymers in the Presence of Monovalent Counterions

The main conclusion drawn from the simulations [170] is that in the presence of monovalent counterions, the charged protein-like copolymers can be soluble, even in a very poor solvent for hydrophobic units. There are three temperature regimes, which are characterized by different spatial organization of polyions and their conformational behavior.

At sufficiently high temperature, i.e., in the weak coupling regime, the chains have extended coil-like or necklace-like conformations and are distributed more or less uniformly in the solution. Such a behavior is typical for single-chain HPEs [164].

At lower temperature, due to the intrachain hydrophobic association, there is a sharp decrease in chain size: the flexible-chain polyions rearrange in globules with neutral chain sections located in the globular core and charged sections forming the envelope of the core and buffering it from solvent. The collapsed globules, however, still have a net charge and repel each other. As a result, in this intermediate temperature regime, one observes a stable solution of nonaggregating polymer globules, which are well separated from each other and form an array of colloid-like particles with partly condensed counterions [170].

Due to the presence of microscopic and mesoscopic charged particles, two different length scales and large charge ratios are involved which make the mixed system strongly asymmetric. For better understanding this type of behavior, the following computer experiment was carried out [170]. First, two globules were "stuck" together, taking into account only hydrophobic attraction while the electrostatic interactions were turned off. The obtained nearly spherical globular agglomerate was relaxed for a long time. Then, the Coulomb interaction was turned on and the system evolution was monitored in the presence of counterions. The sequence of images, which illustrates the evolution of the coalesced globules after switching on the Coulomb interaction, is shown in Fig. 38. The disintegration of the biglobular "droplet" into two separated globules is clearly seen, thereby indicating that the charged protein-like globules can indeed be stable in a poor solvent with respect to aggregation.

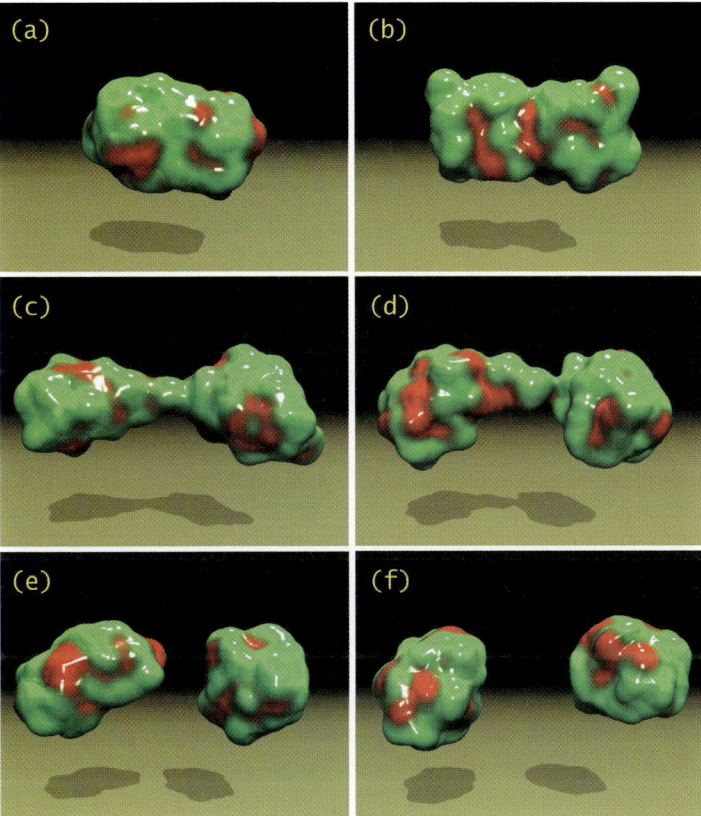

Fig. 38 a–f Sequence of images, which illustrates the evolution of the biglobular agglomerate after switching on the Coulomb interaction. For visual clarity, we present three-dimensional Connolly surfaces constructed for the polymer segments and do not depict the counterions. The repulsive Coulomb forces acting between non-compensated charges on the chains attempt to disrupt the polyion "droplet" into two "splinters", while the surface tension tries to keep it spherical. That situation is almost the same as that observed by Lord Rayleigh for highly charged liquid droplets. Depending on the droplet radius and surface tension (or temperature), the electrified droplet can become unstable, i.e., it starts to deform into an elongated ellipsoid. Such a deformation can finally lead to a fission of the droplet into two fragments of equal size and charge, if the repulsive force between the like elementary charges on the surface exceeds the forces from surface tension. Regions occupied by charged (hydrophilic) units are shown in *green* and hydrophobic regions are colored in *red*

A further decrease in temperature leads to the condensation of most of the counterions (the so-called ionomer regime) and to the aggregation of dense globules [170]. Nevertheless, they maintain their morphological integrity even in rather concentrated solutions where no large-scale aggregation is observed. Aggregated polyions form a specific supramolecular structure built up from stable individual globules connected together due to both short-range adhesive forces and effective mutual attractions mediated by counterions condensed on the strongly charged globule surfaces. In fact, the compact protein-like globules entering the aggregates behave as solid charged particles nonpenetrating each other. Interestingly, those finite size mesoglobular aggregates have an anisotropic chainlike morphology [170].

While simulations have demonstrated convincing evidence for effective attraction stemming from the counterion correlations, the physical mechanism underlying the onset of attraction is still not completely understood [163]. Also, it is not clear the role that details of the discrete charge distribution on the surface of the protein-like globules may play. In principle, this is relevant for any globular proteins with a nontrivial charge pattern on their surface. In view of this it has to be admitted that the traditional model of a homogeneously smeared charge [174] does not always seem to be correct.

3.3.3
Effect of Multivalent Counterions

When multivalent counterions ($z \geq 2$) are present in the solutions of charged protein-like macromolecules [171], the coil-to-globule transition and counterion condensation shift towards higher temperatures as compared to the $z = 1$ case [170]. Although condensed counterions of higher valence induce stronger interchain attraction and weaker repulsion, the aggregation of protein-like copolymers is not accompanied by large-scale aggregation. Instead, the formation of stable finite-size aggregates consisting of several entangled chains is observed. Charged chain segments tend to be on the surface of the aggregates, thereby screening their hydrophobic interior from the solvent. There is a penetration of the counterions into the aggregate core, although they show somewhat inhomogeneous distribution, tending to localize in outer aggregate regions. Even at low temperatures, there is no exact neutralization of negative and positive charges in the volume occupied by multichain aggregates. A certain fraction of counterions always remains in solution.

The average number of copolymer chains per aggregate, m, found in [171] is shown in Fig. 39 as a function of temperature for the systems containing counterions of different valence. It is seen that regardless of the valence of the counterion, the average aggregate size increases as the temperature is decreased. This is, of course, a quite expected result. One can conclude that for all the systems, the aggregation process becomes well pronounced in the

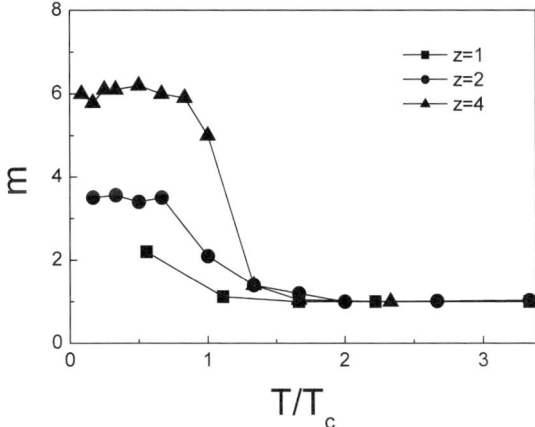

Fig. 39 Average number of 128-unit protein-like chains per aggregate as a function of reduced temperature T/T_c for counterions of different valence: ■ $z = 1$, ● $z = 2$, and ▲ $z = 4$. T_c is the temperature of counterion condensation. Adapted from [171]

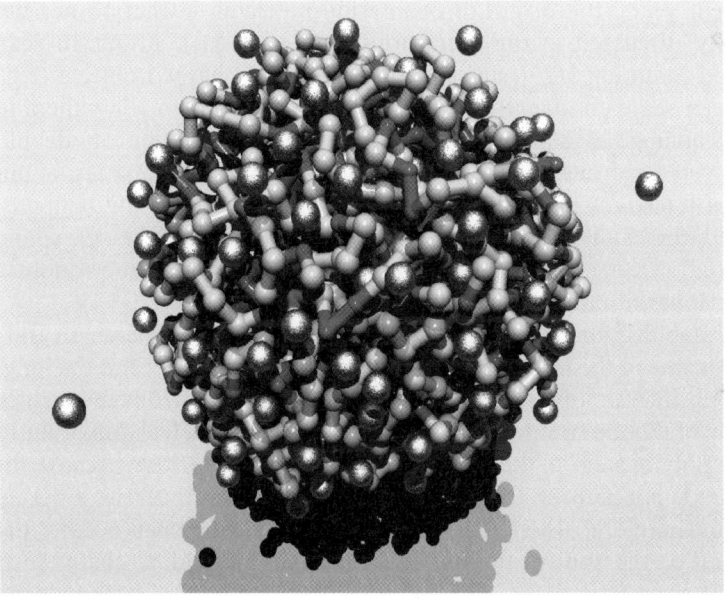

Fig. 40 Structure of a typical intermolecular aggregate formed in the presence of tetravalent counterions ($z = 4$). The aggregate is built up from seven 128-unit hydrophobic protein-like polyions, which are entangled, and 110 tetravalent counterions so that the net charge of the aggregate is − 8. It is seen that counterions condense preferably on the surface of the nearly spherical aggregate of entangled chains. The aggregate surface is covered with polar chain sections. A fraction of the counterions is floating in the immediate vicinity of the polyions. The charged chain groups are presented as *light gray spheres*, and the neutral chain groups are shown in *dark gray*. Counterions are shown as *larger spheres*

temperature region $T \leq T_c$, where T_c is the critical temperature of counterion condensation. In this case, the chains adopt compact conformations. Although the existence of separated single-chain globules is a prevalent structural motif for the $z = 1$ system [170], nevertheless, the occasional formation of intermolecular aggregates is observed at very low temperature. In the presence of di- and tetravalent ions, the average aggregate size is considerably larger. However, for all the systems studied in [171] the interchain aggregates had a finite size even at the lowest temperature; that is, no large-scale aggregation is observed. Figure 40 shows a typical snapshot of an intermolecular aggregate formed at $z = 4$.

3.3.4
Stabilization Mechanism

There are two driving forces causing the formation of compact conformations of charged copolymer chains: the short-range hydrophobic attraction and the counterion-mediated attraction between charged chain segments. In recent years, the origin of counterion-mediated attraction has been extensively discussed in the literature (see, e.g., [163]). About 10 years ago, Ray and Manning [175] suggested that two like-charged objects (e.g., rigid rods) can share condensed counterions in such a way to allow them to form what is analogous to a chemical bond. This "bridging type" model has been extensively used to explain the counterion-induced precipitation of polyelectrolytes [163, 175] and other related phenomena. When the temperature is reduced, the counterion condensation takes place that makes the counterion-mediated attraction stronger. Obviously, this effect is more pronounced for counterions with higher charge.

The stabilization of the finite-size aggregates (both single-chain and multi-chain) is due to the interplay between the attractive forces and the long-range Coulomb interactions together with the contribution from the translational entropy of mobile counterions. The point is that only a fraction of the charges on the polyions are neutralized by condensed counterions because the temperature is not zero, so that each copolymer chain still carries a net negative charge. Therefore, when the formation of the aggregates occurs, their net charge (i.e., the sum of the net charges of the individual chains in the aggregate) turns out to be nonzero for all aggregate sizes. The electrostatic repulsion due to the net-charge, whose range is set by the screening length under given conditions, is the *first factor* responsible for the stabilization of finite size aggregates. Obviously, the repulsion is longer in range than the attraction. Also, the simulations [170–172] show that for the HP copolymers having a protein-like primary structure, the significant fraction of charged chain segments is concentrated on the aggregate surface. Such a morphology assists the stabilization. On the other hand, it seems evident that this will not be the case for polyions with a purely random primary structure or for regular copolymers with an alternating distribution of charged and neutral groups along the chain.

In the counterion condensation regime, the remaining (noncondensed) counterions are partially immobilized near the charged monomer units and this results in translational entropy losses. This is the *second ingredient* of the stabilization mechanism.

For small clusters, the long-range Coulomb repulsion is more important, while for large interchain aggregates the stabilization should originate mainly from the counterion entropy decrease. Indeed, as the aggregates grow and occupy a larger fraction of the available volume in the system, a larger fraction of the free counterions will be found in the aggregates, thereby leading to an increase in the corresponding free energy connected with the translational entropy losses. Also, it should be noted that the counterion mobility inside an aggregate is reduced when the chains contract significantly and the aggregates become denser. All these arguments explain the existence of the finite-size aggregates built up from charged copolymers in a poor solvent.

The stabilization mechanism described above is practically the same as that suggested in [176]. In particular, the theory developed in this work predicts that in the dilute solution of associating polyelectrolytes, finite-size clusters having some optimum dimension should exist. Moreover, this stabilization mechanism is universal and should be valid for all attracting polyelectrolytes independent of the nature of the attraction between chain segments and the internal structure of the aggregates. To illustrate this assertion, the authors of [176] present the following qualitative arguments. When the polyions aggregate into a multichain cluster, the intermolecular potential energy of attraction per chain is a monotonically decreasing function of the aggregation number, m. It is clear that this energy should decrease up to some finite negative constant value for the macroscopic cluster in the $m \to \infty$ limit. On the other hand, the sum of the electrostatic repulsive energy and the counterion entropy decrease (per chain) is a monotonically increasing function of m. As a result, there should be a minimum of the total free energy at some finite value of m that corresponds to the formation of optimum finite-size aggregates for the given conditions. An analogous stabilization mechanism may come via the generalization of the approaches developed by Borue and Erukhimovich [177] and by Joanny and Leibler [178].

We can add a few words about the nonmonotonic behavior of the chain size observed for the low-temperature region [170, 171]. Let us compare the conformation of a polyion entering a single-chain and multichain aggregate. In the single-chain aggregate (i.e., in a dense polymer globule), every strongly collapsed polyion has a size of the order of $N^{1/3}$. In the large multichain aggregate, however, each flexible-chain macromolecule is entangled with other chains, which form a surrounding essentially similar to that characteristic of a polymer melt. In the $m \to \infty$ limit, each flexible chain should approximately behave as a Gaussian chain with the size of the order of $N^{1/2}$, i.e., its size increases. That is why the chains included into larger aggregates formed in the presence of multivalent ions have larger sizes than those observed for small single-chain aggregates [171]. The chain expansion is also clearly seen in Fig. 40.

3.3.5
Experimental Results

Experimentally, there are some hints for the existence of finite-size HPE aggregates.

Carbajal-Tinoco et al. [179] have shown that even very hydrophobic sodium poly(styrene-co-styrene sulfonane) chains close to the solubility limit in water can be treated as completely isolated, colloid-like particles. Peng and Wu [180] have studied an aqueous solution of copolymers prepared by polymerization of N-vinylcaprolactam and sodium acrylate, P(VCL-co-NaA). Under poor solvent conditions in the presence of Na^+ ions, they observed a temperature-induced coil-to-globule transition of the chain conformation, leading to a single-chain core-shell nanostructure in which negatively charged COO^- groups are concentrated in the solution-exposed part of the globule. The average hydrodynamic radius and the average monomer density gradually decrease as the solvent became poorer, but neither the apparent weight-average molar mass and the average number of aggregation change, clearly indicating that there is no intermolecular aggregation. In the presence of Ca^{+2} ions, the intermolecular aggregates became larger, but their size was always finite, showing the existence of the stable mesoglobular phase. These results are in good qualitative agreement with the findings [170] for charged protein-like copolymers. The formation of the finite size clusters has also been observed for hydrophobically modified polyelectrolyte gels [181].

It is known that, depending on the primary amino acid sequence of proteins, there are two possible scenarios of protein aggregation in solution, when the attractive hydrophobic interaction dominates over the stabilizing electrostatic repulsion as the net charge of the protein is reduced. In the first case, globular proteins are able to form stable aggregates (dimers, trimers, etc.) maintaining in general their native conformations which they have in the low-concentration limit. The self-association and dimerization of lysozime, experimentally studied for a considerable time [182–184], is a typical example of this scenario. In the second case, proteins can also aggregate, but, in order to gain the maximum potential energy and to form the most stable structure, they should refold into a conformation dissimilar from the native (single-chain) state, the aggregation scenario suggested, e.g., for the cellular prion protein PrP [185, 186], capable of converting from the PrP^c form with α-helices into the PrP^{sc} misfolded structure having β-sheets. It may be assumed that the simple model of the charged protein-like copolymers in the presence of monovalent counterions under conditions corresponding to the low-temperature regime reflects some general features of the aggregative processes observed for real proteins in the framework of the first scenario described above.

3.4
Hydrophobic-Amphiphilic Copolymers

The coil-to-globule transition in synthetic polymers, occurring under poor solvent conditions, is almost invariably accompanied by precipitation. This significantly complicates the data analysis, preventing the clear correlation between the experimental observations and the molecular parameters. At the same time, globular proteins differ from the globules of synthetic polymers in three main ways. (i) Globules formed from homopolymers and random copolymers are typically insoluble in aqueous medium, whereas many of globular proteins are water-soluble. (ii) In a poor solvent, when polymer segments attract each other strongly, synthetic globules stick together and form intermolecular clusters or aggregates even in very dilute solutions. It is the aggregation that makes them insoluble. At sufficient concentration, aggregation results in polymer precipitation. Globular proteins fold, and may form small aggregates (quaternary structure) preserving their globular native state, but even these are still water soluble. (iii) Globular proteins are mobile in solution due to their solubility. The aggregation of synthetic globules dramatically slows down their diffusion in the polymer precipitate.

It is thought that protein globules are soluble in water because of the special primary sequence: hydrophobic amino acids effectively join together and, by doing so, they are avoiding the surrounding aqueous environment. One of the major driving forces in the folding of proteins is to place each of the types of amino acids (hydrophilic and hydrophobic) in an environment appropriate for its solution properties. This can be achieved by locating the majority of hydrophilic amino acids on the globule exterior where they can bond to water molecules, while most hydrophobic amino acids are clustered in a central core where they bind to each other in an effectively water-free environment.

Having in mind this fundamental principle of protein organization, we can now discuss how to optimize copolymer sequences to obtain better aggregation stability. Since globule formation and solution stability cannot be simultaneously realized with any random sequence of monomeric units, we can formulate the following problem: is it possible to design such a sequence of an HP copolymer that provides globules protection from aggregation in solution?

Why is the understanding of the aggregation mechanism so important? First, it is needed for in vitro or computer design of new (soluble) protein-like copolymers. Second, it can help to gain insight into the stability of protein solutions. It is known that protein association leading to reduced biological activity plays a vital role in fundamental biological processes. In particular, aggregation of the human proteins (e.g., lysozyme) that can form stable amyloid fibrils is associated with a range of fatal diseases including systemic amyloidosis, Alzheimer's disease, and transmissible spongiform encephalopathy [187]. Knowing which conformational rearrangements converge on the same final fold is important for understanding the determinants of protein structure, and may enable the development of rational approaches to the inhibition of protein condensation diseases. Simple isotropic models that treat the proteins as hard spherical colloids with short-range attractive interactions explain some features of the protein phase diagram. They, however, fail to describe the properties of protein solutions quantitatively and cannot address phenomena such as protein self-assembly [174, 188].

In the past few years, a number of groups have analyzed the aggregation mechanism of heteropolymers using computer simulations employing both lattice and off-lattice representation based on the HP copolymer model. With this model, Abkevich et al. [160] proposed a simple algorithm that biases sequence sampling toward compact and water-soluble sequences. Using Monte Carlo simulations, the competition between chain folding and aggregation was studied by following the simultaneous folding of two designed copolymer chains within the framework of a lattice model [189]. It was found that aggregation is determined by partially folded intermediates formed at an early stage in the folding process. Giugliarelli et al. investigated how the interaction potentials affect the solubility and compactness of short heteropolymers on a two-dimensional lattice [190]. Maximally compact conformations were found to be destabilized in solution as the intermolecular potential varies. Timoshenko and Kuznetsov [191] simulated the formation of clusters consisting of several linear heteropolymers in dilute solutions. They found that at relatively low concentrations of heteropolymers with sufficiently strong competing interactions, such clusters ("mesoglobules") are more stable in a selective solvent as compared to single chains. Bratko and Blanch [192, 193] considered a lattice model designed to examine the competition between intramolecular interactions and intermolecular association, resulting in the formation of aggregates of misfolded chains (see also [194, 195]). Linear protein-like chains, capable of forming core-shell conformations, do not exhibit such a high tendency for aggregation as their random and random-block counterparts [40]. Nevertheless, they are not completely protected from aggregation [40], in contrast to many real protein globules. Therefore, it is instructive to look at other factors, which can be responsible for aggregation stability.

Real proteins are built up both from hydrophobic and polar amino acid residues, some of the latter can be charged. Many of the conformational and collective properties of proteins are due to a complex interplay between short-range (hydrophobic) effects and long-range (Coulomb) interactions. Electrostatic effects can also determine some of the unique solution properties of globular proteins. We have already discussed the results of simulations for charged protein-like hydrophobic-hydrophilic copolymers with a fixed charge distribution under poor solvent conditions [170, 171]. Indeed, for this model we have observed a solution of nonaggregating globular macroions. It should be kept in mind, however, that many of the nonaggregating proteins are uncharged.

Thus, we can conclude that until now, theoretical and experimental attempts to design nonaggregating heteropolymers have not given a single-valued positive result. In our opinion, the basic reason for such failures lies in the fact that in these works a study was limited only to ordinary linear heteropolymer chains. Meanwhile, it is far from clear that the desired result can be in principle accessible on the basis of simple linear models in the case of electro-neutral heteropolymers, since in this case there are no real ways of creating the insurmountable (or very high) energy barriers, preventing large-scale aggregation. Nevertheless, we should note that for very long chains, certain irregular arrangements of H and P units along the chain may help to prevent precipitation of the copolymers thus leading to formation of stable finite aggregates and globules [196] (see also [197]).

Another method leading to nonaggregating copolymers may be connected with the molecular design of their monomeric units. We have discussed an extended variant of the HP model, the HA side-chain model [97], that explicitly takes into account the amphiphilic nature of hydrophilic segments.

It is believed that, using this model and the methods of sequence design suitable for computer simulations, one can construct copolymers capable of forming nonaggregating heteropolymer globules, with the ultimate objective of learning how to manipulate the polymer chemistry and system conditions in order to preclude the aggregation processes.

Although we are primarily interested in the intermolecular effects, we will first characterize the behavior of single amphiphilic copolymers in a selective solvent.

3.4.1
Single Amphiphilic Chains

Homopolymers consisting of amphiphilic monomer units (poly-A, Fig. 24a) were simulated, using a Langevin molecular dynamics method and the "side-chain" model [97]. The simulations of the hydrophobically driven conformational transitions under the variation of solvent conditions have shown that for this model, a variety of novel structures with high complexity are possible, depending on the interaction between hydrophobic (H) and hydrophilic (P) sites. Specifically, the thermodynamically stable anisometric structures have been observed, including disk-like structures, stretched necklace-like conformations, and cylindrical-shaped conformations. Also, it was demonstrated that the chain size R_g as a function of the quality of the solvent can behave in an irregular manner, showing an *increase* when the solvent becomes poorer for hydrophobic sites. This unusual behavior is connected with the formation of strongly elongated core-shell conformations having a locally cylindrical symmetry and is consistent with existing experimental data [198]. Moreover, for the range of the chain lengths N simulated ($N \leq 1024$), the formation of such conformations can lead to the $R_g \propto N^{0.9}$ scaling under a poor solvent condition.

In order to collect information on the qualitative features of the chain conformations, we can directly look at many snapshots of the amphiphilic chain when the solvent quality is progressive worsened. Figure 41 shows a series of typical snapshots obtained for the chain with a 256-unit backbone at the strong H – P segregation.

When the repulsive interactions between monomers dominate, the chain has the usual coil-like conformation (Figs. 41a and b). As the solvent becomes poorer, we observe chain folding and this leads to the formation of specific necklace-like conformations where single "pearls" of hydrophobic groups surrounded by hydrophilic groups are connected by stretched chain sections (Figs. 41c and d). In this regime, the mobility of each monomer is quite high and monomer position fluctuations are still large. Pearls are locally in equilibrium, linked to one another by fluctuating chain sections. With worsening solvent quality the size of pearls increases and, as a result, their number decreases. Finally, for very strong attraction between H sites, the pearls coalesce and form an object that looks like a sausage, which then transforms to a cylindrical-shaped object with the cross-section increasing

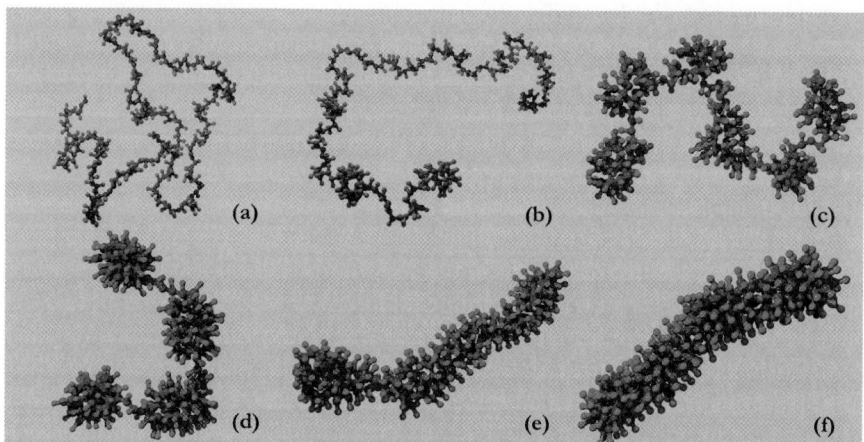

Fig. 41 Snapshot pictures illustrating typical conformations of the amphiphilic chain (poly-A) of length $N = 256$ for the strong H-P segregation, at different H – H attraction increasing from **a** to **f**. Hydrophobic beads are shown as *dark gray spheres* and hydrophilic beads are presented as *light gray spheres*. The sizes of all the spheres are schematic rather than space filling

very slowly as the H – H attraction grows (Figs. 41e and f). Such strongly nonspherical aggregates, whose state is liquid-like, are thermodynamically stable. Upon visual inspection of these aggregates, one can find that the hydrophobic chain finds itself in an irregularly folded (crumpled) state. Thus, the progressive worsening of solvent quality results in the following succession of the conformational transitions at sufficiently strong H – P segregation: swollen coil → stretched necklace-like conformation → sausage-like object → cylindrical-shaped conformation. We would like to emphasize that such transitions should be considered rather as a smooth shape evolution of the model polymer than sharp transitions. In other words, there is no well-defined temperature transition between different regimes but rather a smooth transition from the necklace-like regime to the sausage regime and further to the cylindrical regime. Therefore, it is possible to observe, e.g., a coexistence of pearls and sausage for a given solvent condition.

It is important to note that the observed compact microstructures are formed due to strong intramolecular segregation of chemically different H and P groups tending to minimize the number of H – P contacts unfavorable under poor solvent conditions for H sites: each of the hydrophobic (hydrophilic) groups has a tendency to have hydrophobic (hydrophilic) nearest neighbors and to avoid having hydrophilic (hydrophobic) nearest neighbors in a poor solvent. As a result, a core composed of mainly hydrophobic groups turns out to be surrounded by a thin dense "skin" composed of mainly hydrophilic groups. It is clear that such a situation is very similar to that characteristic of usual low-molecular-weight amphiphiles, which form micelles in a dilute solution [199, 200]. Indeed, the "side-chain" model [97] bears resemblance to a system of N small chemically connected HP surfactants. In a polar solvent, surfactants can form micelles with a dense hydrophobic core surrounded by a hydrophilic shell. Also, such a behavior is rather common for polysoaps which, due to their amphiphilic character, are able to build up intramolecular self-assembled structures in polar as well as in apolar media [201–205]. From this viewpoint, we should treat the conformational transitions found for the extended HP model (Fig. 24a) rather as an intrachain micellization than as a true coil-to-globule transition. In particular, the necklace-like conformation with hydrophobic pearls surrounded by hydrophilic groups (Figs. 41c and d) has to be considered as a string of micelles but it has nothing in common with the pearl-necklace model structure proposed by Rubinstein and co-workers [164, 206] for flexible polyelectrolytes in poor solvents where, due to the Rayleigh charge instability, highly stretched segments alternate with collapsed (micro)globules along the chain. Structurally, the intramolecular micelles observed for the necklace-like state are similar to micelles formed by free low-molecular-weight surfactants; however, unlike ordinary micelles, intramolecular micelles need no critical concentration of polysurfactants for their formation because in this case there is no loss of translational entropy. To understand many of the results of the computer simulations [97], simple theoretical arguments [207–210] that predict the formation of nonspherical micelles can be used.

The conformational transitions observed in the simulations [97] resemble in some aspects the so-called *zipping* transitions [211], the process in which two strongly attracting strands composing the polymer come in contact in such a way as to form a bound double structure, which remains swollen and does not assume compact configurations. The cylindrical-shaped conformations in which the hydrophobic backbone is in a locally collapsed state (Figs. 41e and f) look a lot like three-dimensional zipped structures.

Single hydrophobic-amphiphilic (HA) copolymers with the same HA composition but with different distribution of H and A units along the main hydrophobic chain were also simulated [212]. In particular, regular copolymers comprising H and A units in alternating sequence, regular multiblock copolymers composed of H and A blocks of equal lengths, and the quasirandom protein-like copolymers with a quenched primary structure were studied. These copolymers are schematically depicted in Fig. 24b,c, and d.

Under poor solvent conditions for hydrophobic segments, all the copolymers form compact conformations, irrespective of the primary structure. However, the morphology of these conformations dramatically depends on copolymer sequence, especially for long chains. It was found that single protein-like polyamphiphiles (Fig. 24d) can readily adopt conformations of compact spherical globules with the hydrophobic chain sections clustered at the globular core and the hydrophilic side groups forming the envelope of this core and buffering it from polar solvent. This morphology closely resembles that of micelles or globular proteins. For all the chain lengths studied, these structures are nearly spherical with small fluctuations. For the range of the hydrophobic chain lengths N simulated in this study ($N \leq 255$), the chain size R_g as a function of N behaves as $R_g \propto N^\nu$ with $\nu \approx 0.28$, the exponent expected for collapsed chains.

The globules of relatively short regular multiblock copolymers with a fixed block length of $L = 3$ (Fig. 24c) are also spherical or nearly so. On the other hand, at poor solvent conditions, the compact conformations of long regular copolymers tend to be elongated in one direction, especially for the alternating HA sequence (Fig. 24b). The hydrophobic core formed by these copolymers increases with chain length, in a manner that can be understood on the basis of a uniform core which expands with chain length in one direction, and a sharp hydrophobic/hydrophilic interface whose width is essentially constant. For sufficiently long chains, the formation of such conformations leads to the $R_g \propto N^\nu$ scaling with $0.86 \leq \nu \leq 0.89$. This scaling exponent is practically the same as that observed for poly-A chains. The fact that the scaling exponent is slightly smaller than the rod-like prediction of $\nu = 1$ is likely due to the fact that N is very close to, though not yet in, the scaling regime. Nevertheless, it is believed that in the $N \to \infty$ limit, because of strong thermal fluctuations, the locally folded chain as a whole would look like an infinitely coiled "garden hose" (or a worm-like superchain) having a finite thickness and a finite persistent length, and the limiting scaling exponent would be close to that expected for the good solvent regime, due to repulsive interactions between the outer hydrophilic groups. In other words, for very large N we expect a crossover from the $R_g \propto N$ regime to the $R_g \propto N^{0.59}$ regime.

There are experimental evidences of some facts predicted in the simulations [97, 212].

Kikuchi and Nose [198, 213] have reported on systematic experimental studies of poly(methylmethacrylate)-*graft*-polystyrene (PMMA-*g*-PS) with short branches in a selective solvent (isoamyl acetate) which is a good solvent for PS. Under given solvent conditions, this copolymer behaves as an amphiphilic copolymer, bearing a resemblance to the model considered in our simulation. At high branch density, the authors [213] have observed the formation of thermodynamically stable unimolecular rod-like micelles formed via intramolecular segregation of the PMMA backbone and PS branches, with the shrunken PMMA backbone making the rodlike core covered with PS chains. Also, it has been found that the rod is not necessarily rigid, but may be flexible in the weakly segregated state and becomes more rigid with stronger segregation upon decreasing temperature, i.e., upon the progressive worsening of solvent quality for the PMMA backbone.

Selb and Gallot [214] have demonstrated that poly(styrene)-*graft*-poly(4-vinyl-N-ethylpyridium bromide) forms unimolecular micelles in water/methanol mixtures. These experimental data can be treated as an indirect confirmation of the simulation result [97] that sufficiently long regular copolymers with amphiphilic monomer units do form intramolecular anisometric micellar structures in a poor solvent.

The presence of stable single-chain core-shell nanostructures in a solution of amphiphilic copolymers has also been observed by Wu and Qiu [215]. Using a combination of static and dynamic laser light scattering, they have found that a linear poly(N-isopropylacrylamide) chain grafted with poly(ethylene oxide) (PNIPAM-*g*-PEO) in water can undergo a coil-to-globule transition to form spherical single-chain aggregates with a collapsed PNIPAM chain backbone as the hydrophobic core and the grafted short PEO chains as the hydrophilic shell. In general, these colloid-like nanostructures are similar to those observed in the simulations [212] for protein-like amphiphilic copolymers.

In a series of papers [216, 217], Nakata and Nakagawa have studied the coil-globule transition by static light scattering measurements on poly(methyl methacrylate) in a selective solvent. They have found that the chain expansion factor, $\alpha^2 = R_g^2/R_{g\Theta}^2$, plotted against the reduced temperature, $\tau = 1 - \Theta/T$, first decreases with decreasing τ, as it should be, but then begins to increase (see, e.g., Fig. 2 presented in [217]) In the authors opinion, "the increase of

α^2 with decreasing temperature conflicts with theoretical predictions and an intuitive notion of the expansion factor." However, taking into account the amphiphilic nature (although weakly-pronounced) of poly(methyl methacrylate) and the simulation data reported for the extended HP model [97], this "anomalous" behavior can be understood. Indeed, in solvent selective to side groups, the incompatibility of chemically different groups is effectively increased when the attraction between groups composing the chain backbone becomes stronger. This is accompanied by pushing away the soluble side groups from the insoluble micellar core and by stretching the macromolecule as a whole.

Williams and co-workers [179, 218] have studied the structural changes and chain conformations of a series of hydrophobic sodium poly(styrene-*co*-styrene sulfonate)'s of various charge fractions in a poor solvent using static light scattering and small-angle X-ray scattering techniques. By varying the charged monomer fraction, f, it was possible to change the degree of hydrophobicity of this copolymer and the corresponding hydrophobic/hydrophilic interactions. From the scattering measurements, the so-called apparent radii of gyration, R_g^{app}, were determined for different values of f and concentrations. From the analysis of these results (see Tables 1 and 2 of [179]) it is seen that the values of R_g^{app} show an irregular behavior as a function of f at all the polymer concentrations studied. At large f, when the repulsive electrostatic forces dominate, the copolymer chains have an expanded conformation. With decreasing f, the intrachain short-range hydrophobic attractions begin to dominate and, as a result, the chain size decreases. However, at $f \lesssim 1/2$, an unexpected increase in R_g^{app} is distinctly observed; although, at first sight, R_g^{app} should further decrease, taking into account a monotonous growth in hydrophobic attraction. It is clear, that such an "unexpected" behavior is consistent with the simulations [97, 212] and can be explained on the basis of the discussion presented above.

3.4.2
Coil-Globule Transition Versus Aggregation

Here, we describe and compare the results of simulations for two multichain systems corresponding to alternating and protein-like HA copolymers [212]. The multichain systems consisting of 127-unit copolymers were simulated for the range of the effective interaction parameter $\widetilde{\chi}$ (which is similar to the Flory–Huggins parameter) under solvent conditions when single chains can form strongly collapsed conformations.

In Fig. 42, we show the ratio R_{gP}^2/R_{gH}^2 (R_{gH}^2 and R_{gP}^2 are the partial mean-square radii of gyration calculated separately for hydrophobic and hydrophilic beads) as a function of the interaction parameter $\widetilde{\chi}$. We see that this ratio is an increasing function of $\widetilde{\chi}$. Qualitatively the same picture is observed for isolated chains. Such behavior is due to the fact that, as the attraction

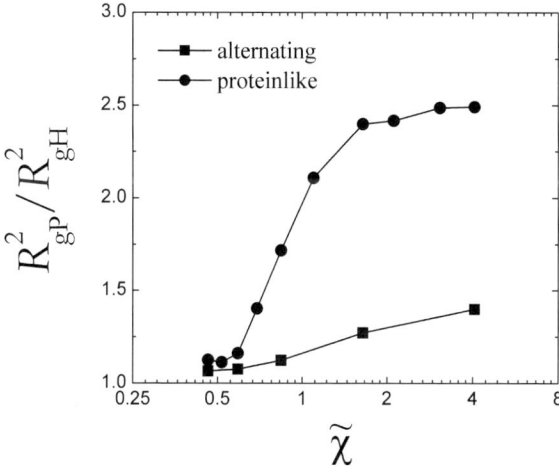

Fig. 42 Ratio R_{gP}^2/R_{gH}^2 as a function of the interaction parameter $\tilde{\chi}$ in a semi-logarithmic scale for the 127-unit ■ alternating and ● protein-like chains in the corresponding multichain systems. The parameter $\tilde{\chi}$ is similar to the Flory–Huggins interaction parameter χ and characterizes solvent quality in an integral manner. Sufficiently large values of $\tilde{\chi}$ ($\tilde{\chi} \gtrsim 1$) correspond to a poor solvent. Solvent quality becomes poorer with decreasing temperature or with increasing $\tilde{\chi}$. Adapted from [212]

between H segments increases, the value of R_{gP}^2 decreases more slowly than R_{gH}^2, thus leading to demixing of H and P segments and facilitating their intramolecular microphase separation. This trend is more pronounced for the protein-like copolymers than for the alternating copolymers, suggesting an idea that the former should be more protected against intermolecular aggregation.

A direct way to study the process of aggregation is to monitor the change in the corresponding free energy, ΔG. To this end, following the standard quasi-chemical approach, one can treat the polymer solution as a multicomponent system, where intermolecular aggregates of different size (A_M, $M > 1$) are present in equilibrium with unimers (A_1), $A_1 \rightleftarrows A_M$. These species are treated as distinct chemical components, each characterized by its own solution concentration and chemical potential. The concentrations $[A_1]$ and $[A_M]$ of the species (or their mole fractions) can be found via the integration of the center-of-mass pair correlation function; this gives an estimate for the overall association equilibrium constant K and ΔG, which is the difference in the standard Gibbs free energy between chains belonging to intermolecular aggregates and unimers. When $\Delta G < 0$, it is energetically favorable for the chains to merge. If $\Delta G > 0$, the unimers are favorable.

The calculated values of ΔG are presented in Fig. 43 as a function of the interaction parameter $\tilde{\chi}$. As seen, lowering the temperature, or equivalently,

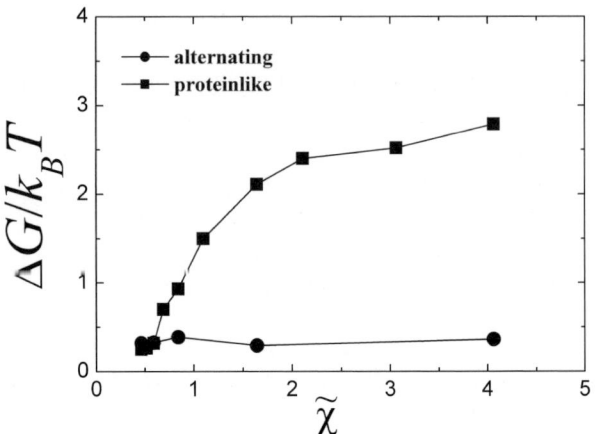

Fig. 43 Aggregation free energy ΔG as a function of the interaction parameter $\tilde{\chi}$ for the multichain systems of (■) alternating and (●) protein-like copolymers. Adapted from [212]

increasing the interaction parameter $\tilde{\chi}$, shifts the equilibrium $A_1 \rightleftarrows A_M$ ($M \geq 2$) in the system of protein-like copolymers toward the nonaggregated state, and this is reflected in Fig. 43 as an increase in ΔG.

At first sight, such behavior is somewhat counterintuitive. Actually, from a general consideration it would be possible to expect the opposite behavior, when worsening in the solvent quality facilitates the aggregation and thus reduces the aggregation free energy. However, the observed behavior becomes quite clear if one takes into account the results discussed above. An increase in the free energy is explained by the appearance of a dense hydrophilic shell around the formed spherical globules, which serves as a practically insurmountable energy barrier preventing the aggregation. When the solvent becomes poorer, chains are compressed, and this is accompanied by a strengthening of this hydrophilic "protective barrier". On the other hand, the free energy of aggregation estimated for the system of alternating copolymers weakly depends on the solvent quality and remains close to zero for all values of the energy parameter $\tilde{\chi}$. This behavior also can be understood on the basis of the data concerning the conformational structure of this copolymer. Since at the same HA composition the cylindrical globules of alternating copolymer have the larger surface-to-volume ratio, the hydrophilic shell is not so dense, and therefore, it does not ensure a sufficient protection, while for the protein-like copolymers rather high densities of the hydrophilic shell are reached.

Moreover, the thorough analysis of globular conformations shows that this layer is almost absent near the faces of the cylinder [212]. This facilitates the formation of multiglobular aggregates in the solution of regular copolymers.

These facts explain the more expressed tendency of regular copolymers toward aggregation.

For visual analysis of the simulated configurations, one can employ the technique based on the construction of isosurfaces. In this way, the global system morphology can be studied. The snapshots of the low-temperature configurations ($\tilde{\chi} \approx 4$) show that in this regime no large-scale aggregation of individual globules is observed for the multichain systems, implying that in this case the system lies in the stable one-phase region (Fig. 44). In this respect, the behavior observed for the model amphiphilic copolymers is similar to that found for charged hydrophobic/hydrophilic protein-like chains [170–172]. The simulation [212] predicts the formation of specific microphase-separated morphologies in which strongly attracting hydrophobic chain sections form a distinct population of globules which are stabilized by a dense layer of hydrophilic beads. It is clear that the driving force for the microphase separation is competing interactions, that is, the strong attraction between the hydrophobic groups and repulsive interactions associated with the hydrophilic species. One may say that the intramolecular microphase separation prevents intermolecular aggregation, thus stabilizing the solution of globules. Thus, under poor solvent conditions, one observes a stable solution of nonaggregating polymer globules which are well-separated from each other and form an array of colloid-like particles. Because of the fact that the amphiphilic globules are size- and shape-persistent objects, this allows them to maintain their morphological integrity even in concentrated enough solution.

Generally speaking, the reason for this behavior is simple. It is known that low-molecular-weight surfactants dramatically increase the stability of polymers and are widely used to prevent aggregation in polymer solutions. In the HA model, "surfactants", i.e., amphiphilic A groups, are incorporated into the polymer chain, thus ensuring the stabilizing effect. From the tempera-

Fig. 44 Snapshot pictures representing the isosurfaces generated under poor solvent conditions for the multichain systems composed of the 127-unit amphiphilic copolymers with **a** protein-like and **b** alternating distribution of H and A groups along the chain

ture dependencies of aggregation free energy (Fig. 43) one can conclude that below the collapse transition temperature there is a free energy barrier preventing the aggregation of copolymer globules. Therefore, in a macroscopic system, precipitation of a macroscopic polymer-rich phase should be suppressed. This is quite different from the solution of usual linear polymers (both homo- and heteropolymers), where the aggregation process in a poor solvent is not associated with a free energy cost and the aggregation is taking place together with the single-chain collapse transition.

The situation with the alternating copolymers is not so clear. Although in this case the large-scale aggregation is also not observed [212], such copolymers were found to be capable of forming sufficiently large intermolecular aggregates, as seen in Fig. 44.

3.5
Adsorption Selectivity

The focus here will be on the consideration of designed copolymers exhibiting selective interactions with the surfaces and interfaces. In particular, we will consider some properties of adsorption-tuned copolymers and partly cross-linked polymer envelopes that function as a molecular dispenser.

3.5.1
Adsorption-Tuned Copolymers

The adsorption of homopolymers at an impenetrable surface is a well-studied problem [219, 220]. Much less is known about copolymer adsorption (in which only one of two comonomers interacts with the surface), although the problem has been studied by several groups [221–225]. Regular copolymers with a periodic sequence of comonomers, adsorbing at a planar surface, have been studied by Moghaddam et al. [226]. In the case where the copolymer is random, the most interesting case is *quenched randomness* where the sequence of comonomers is fixed during the computation of thermodynamic quantities, which are averaged over the quenched comonomer sequences. Grosberg et al. [227] have considered copolymers with periodic quenched randomness. The case of nonperiodic quenched randomness has also been studied using a variety of techniques [228–230]. Moghaddam and Whittington [231] have used multiple Markov chain Monte Carlo methods to investigate the adsorption of a random copolymer at a homogeneous surface. Adsorption of an ideal correlated random copolymer at a liquid-liquid interface has been investigated theoretically in [232].

Zheligovskaya et al. [55] have simulated the adsorption of quasirandom adsorption-tuned copolymers (ATC). The critical adsorption energy as well as some characteristics of the adsorbed single chains (statistics of trains, loops, and tails) were studied. All these properties were compared with those of random copolymers with the same content of adsorbed segments and random-block copolymers with the same composition and the same average numbers of adsorbed and nonadsorbed blocks.

It was found that the difference in the primary structure of the chains leads to the difference in the critical adsorption energy ε^* and the characteristics of adsorbed chains. In particular, the ATC chains have the smallest (by the absolute value) adsorption energy ε^*. The random copolymers are characterized by the largest ε^*, as compared to other copolymers. This fact is simply explained by the difference in block lengths: the random copolymers have the shortest blocks. This is consistent with the analytical results [233] for regular AB copolymers, according to which the absolute value of the critical adsorption energy decreases with the increasing block length at the same fraction of adsorbed and nonadsorbed segments. At the same time, the difference in the critical adsorption energy for the random-block and ATC chains (which are characterized by the same average block lengths) can be explained only by the details of the ATC primary structure. It turns out to be that the ATC chains have significantly longer end nonadsorbed blocks, as compared to their random-block counterparts. As a result, the adsorbed segments are placed more compactly in each ATC chain. This specific feature of the ATC primary structure promotes adsorption of ATC chains. The studied characteristics of the adsorbed chains are also different for the three copolymers.

Thus, the difference in the adsorption behavior of ATC, random, and random-block chains can be rationally explained by taking into account the specific features of the primary structure of these chains.

The obtained results support the general idea of a conformation-dependent sequence design of copolymers proposed [18–20]. That is, the generated ATC sequence memorizes some features of the specific parent conformation of the adsorbed homopolymer. In particular, the position of adsorbed segments turned out to be tuned in the best way for subsequent adsorption. It is not surprising, therefore, that this memorized hidden information became apparent as soon as we considered the adsorption of ATC chains. Among the three types of AB copolymers which were studied [55], ATC chains adsorb better at a given adsorption energy. In other words, the AB chain "learns to be adsorbed" in the parent conformation, and this "experience" is used in the subsequent "life" of this copolymer.

3.5.2
Molecular Dispenser

To characterize the complexes formed between molecular dispenser described in Sect. 2.2.4 and colloidal particles, the probability $P(\sigma, T)$ of finding a complex made from the copolymer envelope and the particle of a given size, σ, was calculated as a function of temperature T [57].

After the preparation of the copolymer envelope (Fig. 10), the parent particle of size σ_p was eliminated, and another particle of a given diameter σ was introduced into the system. A new particle was placed far from the copolymer envelope; that is, initially it was not interacting with the envelope. During the stochastic motion in the bulk, the particle and envelope came into collision with each other, thus forming a copolymer-particle complex. Under the thermal agitation the complex was splitting and reassembling again. The function $P(\sigma, T)$, which is related to the energy of interaction between the copolymer envelope and particle, was calculated as an average over $\sim 10^2$ independent realization.

It was found that the selectivity of the complex formation strongly depends on the number of crosslinks in the envelope. Typical results for a moderately crosslinked envelope are shown in Fig. 45a. It is seen that the selectivity of the complex formation with the particle of a certain size is indeed reached, that is, the idea of a molecular dispenser works.

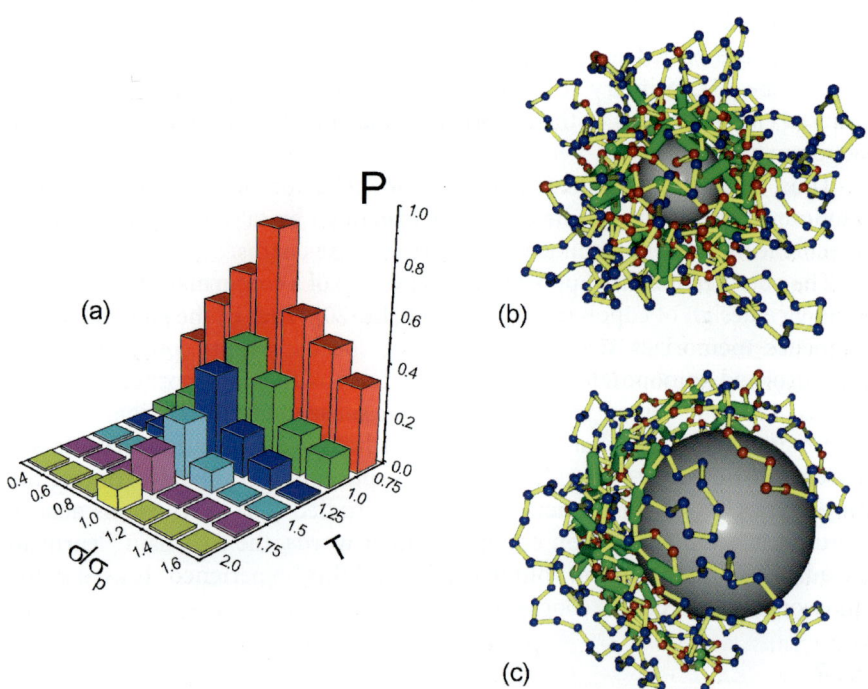

Fig. 45 a Probability of finding a complex made from a 512-unit copolymer envelope and a particle of a given size, σ, at the temperature T for the case when the copolymer envelope has 48 crosslinks. Snapshots of the complexes made from a 512-unit copolymer envelope for **b** $\sigma/\sigma_p = 0.8$ and **c** $\sigma/\sigma_p = 1.8$, where σ_p is the size of the parent particle. Adsorbed and non-adsorbed chain segments are colored in *red* and *blue*, respectively; crosslinks are shown as *green sticks*. Adapted from [57]

Generally, the structure of the polymer-particle complex can be found from the minimization of free energy that includes the polymer-particle interaction energy, entropies of nonadsorbed monomer units and units localized at the surface of the particle, and typically, for the system under consideration, the elastic deformation of crosslinked macromolecule. Such theoretical analysis, following the lines of [234, 235], can explain the specific behavior of $P(T, \sigma)$ observed for the envelopes with different numbers of crosslinks n_J [57]. According to [235], when the number of crosslinks is small enough, $n_J \ll N^{1/2}$, all junctions contribute mainly to the formation of simple loops along the chain and the polymer molecule as a whole conserves linear structure. On the other hand, when $n_J > N^{1/2}$, the macromolecule becomes really crosslinked, forming a kind of loose network. In the simulation [57], the cross-links introduced between adsorbed monomeric units stabilize a hollow-spherical structure of the copolymer envelope with mesh-like architecture and the cage structure of the central cavity. Thus, the selective property of the moderately crosslinked copolymer envelope is based on the existence of a certain mesh size, which restricts the penetration of big particles into the copolymer envelope.

The reason for the selective adsorption of a colloidal particle of parent size is explained by the typical snapshots in Figs. 45b and c. We see that the particle of parent or smaller size ($\sigma \leq \sigma_p$) is fully absorbed by the central cavity (Fig. 45b), because the corresponding fitting was ensured by the sequence design procedure (Fig. 10b). On the other hand, a particle of larger size ($\sigma > \sigma_p$) turns out to be too big for a central cavity (Fig. 45c), and thus the complex formed does not saturate all the possibilities for the attraction of red units to the surface of the particle. As to small particles, they easily penetrate inside the molecular dispenser, but the complex formed is not stable (especially at high temperature) because of the small surface of such particles. All these factors explain the peak in $P(T, \sigma)$ observed at $\sigma \approx \sigma_p$ for moderately crosslinked copolymer envelopes (Fig. 45a).

4
Conclusion

In this review, we reported on several new synthetic strategies that allow the synthesis of copolymers with a broad variation of their sequence distributions. The fundamental principle of these strategies is conformational-dependent sequence design (CDSD), which takes into account a strong coupling between the conformation and primary structure of copolymers during their synthesis. Using computer simulation techniques, we have attempted to show that rather simple methods, such as polymer-analogous reactions and normal radical copolymerization, can lead to nontrivial chemical sequences, long-range correlations, and gradient structures, if they take place under unusual physical conditions.

The results presented in this review demonstrate that the CDSD polymer-analogous transformation is a versatile approach that allows various functional copolymers such as bioinspired protein-like macromolecules (which give soluble globules with a segregated core-shell microstructure), molecular dispensers (which are able to selectively absorb nanoparticles of a given size), and adsorption-tuned copolymers to be obtained. Our discussion has focused on copolymer sequences exhibiting large-scale compositional heterogeneities and long-range statistical correlations between monomeric units. These features are intrinsically related to the CDSD scheme and they cannot be explained by the basic stochastic processes such as random sequence or Markov chain; the first has no correlations and the second only has short-range correlations. Problems associated with the evolution of copolymer sequences have been considered from the viewpoint of the emergence of information complexity in the sequences in the course of evolution.

Presently the main technique for the synthesis of copolymers are free-radical polymerization methods [11-17]. For a limited range of comonomers, anionic and cationic polymerizations are also used [236, 237].

We have reviewed results on the computer modeling of radical copolymerization under heterogeneous conditions, including the following synthetic methodologies: solution copolymerization with simultaneous globule formation, emulsion polymerization with polymerizing hydrophobic and amphiphilic monomers, copolymerization near a chemically homogeneous surface that selectively adsorbs one type of polymerizing monomer, and template copolymerization near patterned surfaces. Because the implementation of conventional radical polymerization is much easier than that of anionic or CRP processes, it is worthwhile to try to use the emerging features of the copolymer sequences (long-range correlations and gradient structures) to design new types of copolymers with sophisticated functional properties. Many of these advances are likely to lead soon to novel applications.

Also, we have discussed advances that have recently been achieved in the computer simulation and theoretical understanding of designed copolymers in solution and in bulk. The focus was on amphiphilic protein-like copolymers and on hydrophobic polyelectrolytes. Here, we have tried to demonstrate how the copolymer sequence dictates the structure and properties of polymer systems. In many cases, the presence of long-range correlations in designed copolymers can bring about dramatic changes in their physical properties with respect to the corresponding copolymers whose sequence has only minor correlation between adjacent monomeric units.

Acknowledgements Many of our colleagues have contributed to these studies. In particular, we are grateful to A.V. Berezkin, A.Yu. Grosberg, V.A. Ivanov, S.I. Kuchanov, P. Reineker, and A.N. Semenov whose ideas and/or encouragement were critical for some of the questions discussed in this work. The financial support from the Alexander von Humboldt Foundation, Program for Investment in the Future (ZIP), SFB 569, NWO, and RFBR (project # 04-03-32185) is highly appreciated.

References

1. Hamley IW (1998) The physics of block copolymers. Oxford University Press, Oxford
2. Blumenfeld LA, Tikhonov AN (1994) Biophysical thermodynamics of intracellular processes: molecular machines of the living cell. Springer, Berlin Heidelberg New York
3. Li H, Helling R, Tang C, Wingreen N (1996) Science 273:666
4. Xia Y, Levitt M (2004) Curr Opin Struct Biol 14:202
5. Shakhnovich EI (1996) Folding & Design 1:R50
6. Shakhnovich EI (1997) Curr Opin Struct Biol 7:29
7. Shakhnovich EI (1998) Folding & Design 3:45
8. Frauenfelder H, Wolynes PG, Austin RH (1999) Rev Mod Phys 71:S419
9. Pande VS, Grosberg AY, Tanaka T (2000) Rev Mod Phys 72:259
10. Konig JL (1980) Chemical microstructure of polymer chains. Wiley, New York
11. Greszta D, Madrare D, Matyjaszewski (1994) Macromolecules 27:638
12. Moad G, Solomon DH (1995) The chemistry of free radical polymerization. Elsevier, Oxford
13. Malmstrom EE, Hawker CJ (1998) Macromol Chem Phys 199:923
14. Clay P, Christie DI, Gilbert RG (1998) Controlled radical polymerization. In: Matyjaszewski K (ed) ACS Symposium Series 685. Am Chem Soc, Washington, DC, Chap. 7
15. Matyjaszewski K, Xia J (2001) Chem Rev 101:2921
16. Davis KA, Matyjaszewski K (2002) Adv Polym Sci 159:14
17. Davis KA, Matyjaszewski K (2002) Statistical, gradient, block and graft copolymers by controlled/living radical polymerizations. Springer, Berlin Heidelberg New York
18. Khokhlov AR, Khalatur PG (1998) Physica A 249:253
19. Khalatur PG, Ivanov VI, Shusharina NP, Khokhlov AR (1998) Russ Chem Bull 47:855
20. Khokhlov AR, Khalatur PG (1999) Phys Rev Lett 82:3456
21. Khokhlov AR, Grosberg AY, Khalatur PG, Ivanov VA, Govorun EN, Chertovich AV, Lazutin AA (2001) In: Broglia RA, Shakhnovich EI (eds) Proceedings of the International School of Physics *Enricon Fermi*, Course CXLV: Protein folding, evolution and design. IOS Press, Amsterdam, p 313–332
22. Khokhlov AR, Khalatur PG, Ivanov VA, Chertovich AV, Lazutin AA (2002) In: Mac Kernan D (ed) Challenges in molecular simulations (SIMU Newsletter). CECAM, Lyon, vol 4, p 79–100
23. Khokhlov AR, Khalatur PG (2004) Curr Opin Solid State Mater Sci 8:3
24. Khokhlov AR, Berezkin AV, Khalatur PG (2004) J Polym Sci A Polym Chem 42:5339
25. Khalatur PG, Berezkin AV, Khokhlov AR (2004) Recent Res Develop Chem Phys 5:339
26. Shakhnovich EI, Gutin AM (1993) Proc Natl Acad Sci USA 90:7195
27. Pande VS, Grosberg AY, Tanaka T (1994) J Phys France 4:1771
28. Pande VS, Grosberg AY, Tanaka T (1994) Proc Natl Acad Sci USA 91:12972
29. Irbäck A, Peterson C, Potthast F, Sandelin E (1998) Phys Rev E 58:R5249
30. Gupta P, Hall CK, Voegler AC (1998) Protein Sci 7:2642
31. Lau KF, Dill KA (1989) Macromolecules 22:3986
32. Hamley IW (ed) (2004) Developments in block copolymer science and technology. Wiley, New York
33. Schultz GE, Schirmer RH (1979) Principles of protein structure. Springer, Berlin Heidelberg New York
34. Carmesin I, Kremer K (1988) Macromolecules 21:2819

35. Govorun EN, Ivanov VA, Khokhlov AR, Khalatur PG, Borovinsky AL, Grosberg AY (2001) Phys Rev E 64:040903
36. Shlesinger MF, Zaslavskii GM, Frisch U (1996) Lévy flights and related topics in physics. Springer, Berlin Heidelberg New York
37. Kuchanov SI, Khokhlov AR (2003) J Chem Phys 118:4684
38. Virtanen J, Baron C, Tenhu H (2000) Macromolecules 33:336
39. Virtanen J, Tenhu H (2000) Macromolecules 33:5970
40. Virtanen J (2002) Self-assembling of thermally responsive block and graft copolymers in aqueous solutions (PhD Thesis). Department of Chemistry, University of Helsinki, Helsinki, p 42
41. Li W, Kaneko K (1992) Europhys Lett 17:655
42. Fink TMA, Ball RC (2001) Phys Rev Lett 87:198103
43. Peng C-K, Buldyrev SV, Goldberger AL, Havlin S, Sciortino F, Simon M, Stanley HE (1992) Nature (London) 356:168
44. Peng C-K, Buldyrev SV, Havlin S, Simons M, Stanley HE, Goldberger AL (1994) Phys Rev E 49:1685
45. Pande VS, Grosberg AY, Tanaka T (1994) J Chem Phys 101:8246
46. Irbäck A, Peterson C, Potthast F (1996) Proc Natl Acad Sci USA 93:9533
47. Gusev LV, Vasilevskaya VV, Makeev VY, Khalatur PG, Khokhlov AR (2003) Macromol Theory Simul 12:604
48. Yu ZG, Anh VV, Lau KS (2003) Phys Rev E 68:021913
49. Irbäck A, Peterson C, Potthast F (1996) Proc Natl Acad Sci USA 93:9533
50. Yu ZG, Anh VV, Lau KS (2001) Phys Rev E 64:031903
51. Doukhan P, Oppenheim G, Taqqu MS (eds) (2003) Theory and applications of long-range dependence. Birkhäuser, Boston
52. Khalatur PG, Berezkin AV, Khokhlov AR (2005) J Chem Phys (submitted)
53. Kuchanov SI (2000) Adv Polym Sci 152:157
54. Wesson L, Eisenberg D (1992) Protein Sci 1:227
55. Zheligovskaya EA, Khalatur PG, Khokhlov AR (1999) Phys Rev E 59:3071
56. Starovoitova NY, Khalatur PG, Khokhlov AR (2004) In: forces, growth and form in soft condensed matter: at the interface between physics and biology. In: Skjeltorp AT, Belushkin AV (eds) NATO Science Series II: Mathematics, Physics and Chemistry. Kluwer, Dordrecht, vol 160, p 253
57. Velichko YS, Khalatur PG, Khokhlov AR (2003) Macromolecules 36:5047
58. Thurmond KB, Kowalewski T, Wooley KL (1997) J Am Chem Soc 119:6656
59. Huang H, Remsen EE, Kowalewski T, Wooley KL (1999) J Am Chem Soc 121:3805
60. Sanji T, Nakatsuka Y, Ohnishi S, Sakurai H (2000) Macromolecules 33:8524
61. Stewart S, Liu G (1999) Chem Mater 11:1048
62. Discher BM, Won Y-Y, Ege DS, Lee JCM, Bates FS, Discher DE, Hammer DA (1999) Science 284:1143
63. Nardin C, Hirt T, Leukel J, Meier W (2000) Langmuir 16:1035
64. Trefonas P, West R, Miller RD (1985) J Am Chem Soc 107:2737
65. Smith JM (1972) On evolution. Edinburgh Univ Press, Edinburgh
66. Volkenstein MV (1983) General biophysics. Academic Press, New York
67. Leninger AL, Nelson DL, Cox MM (1993) Principles of biochemistry. 2nd edn, Worth Publishers, New York
68. Grosberg AY, Khokhlov AR (1997) Giant molecules: Here and there and everywhere. Academic Press, New York
69. Gatlin LL (1972) Information theory and the living system. Columbia Univ Press, New York
70. Khalatur PG, Novikov VV, Khokhlov AR (2003) Phys Rev E 67:051901

71. Chertovich AV, Govorun EN, Ivanov VA, Khalatur PG, Khokhlov AR (2004) Eur Phys J E 13:15
72. Grosberg AY (1984) Biofizika 29:569
73. Wullf G, Sarhan A (1972) Angew Chem Int Ed 11:341
74. Polowinski S (1997) Template polymerization. ChemTec Publishing, Toronto Scarborough
75. de Gennes P-G (1979) Scaling concepts in polymer physics. Cornell University Press, Ithaca New York
76. Berezkin AV, Khalatur PG, Khokhlov AR (2003) J Chem Phys 118:8049
77. Berezkin AV, Khalatur PG, Khokhlov AR, Reineker P (2004) New J Phys 6:44
78. Berezkin AV, Khalatur PG, Khokhlov AR (2005) Polymer Sci A 47:66
79. Flory PJ (1953) Principles of polymer chemistry. Cornell University Press, Ithaca New York
80. Lozinskii VI, Simenel IA, Kurskaya EA, Kulakova VK, Grinberg VY, Dubovik AS, Galaev IY, Mattiasson B, Khokhlov AR (2000) Dokl Chem 375:273
81. Lozinsky VI, Simenel IA, Kulakova VK, Kurskaya EA, Babushkina TA, Klimova TP, Burova TV, Dubovik AS, Grinberg VY, Galaev IY, Mattiasson B, Khokhlov AR (2003) Macromolecules 36:7308
82. Wahlund P-O, Galaev IY, Kazakov SA, Lozinsky VI, Mattiasson B (2002) Macromol Biosci 2:33
83. Siu M-H, Zhang G, Wu C (2002) Macromolecules 35:2723
84. Siu M-H, He C, Wu C (2003) Macromolecules 36:6588
85. Dotson NA, Galvan R, Laurence RL, M Tirrell (1996) Polymerization process modeling. Wiley, New York
86. Berezkin AV, Khalatur PG, Khokhlov AR (2005) Polymer Sci (submitted)
87. Starovoitova NY, Khalatur PG, Khokhlov AR (2003) Dokl Chem 392:242
88. Starovoitova NY, Berezkin AV, Kriksin YA, Gallyamova OV, Khalatur PG, Khokhlov AR (2005) Macromolecules 38:2419
89. Chan HS, Dill KA (1991) Annu Rev Biophys Chem 20:447
90. Peppas NA, Huang Y (2002) Pharmaceutical Res 19:578
91. Muthukumar M (1995) J Chem Phys 103:4723
92. Golumbfskie AJ, Pande VS, Chakraborty AK (1999) Proc Natl Acad Sci USA 96:11707
93. Kriksin YA, Khalatur PG, Khokhlov AR (2005) J Chem Phys 122:114703; Kriksin YA, Khalatur PG, Khokhlov AR (2005) Math Model 17:3
94. Berezkin AV, Solov'ev MA, Khalatur PG, Khokhlov AR (2004) J Chem Phys 121:6011
95. Berezkin AV, Solov'ev MA, Khalatur PG, Khokhlov AR (2005) Polymer Sci A 47:622
96. Alexandridis P, Lindman B (eds) (2000) Amphiphilic block copolymers: self-assembly and applications. Elsevier, Amsterdam
97. Vasilevskaya VV, Khalatur PG, Khokhlov AR (2003) Macromolecules 36:10103
98. Kriksin YA, Khalatur PG, Khokhlov AR (2003) Macromol Symp 201:29
99. Banavar JR, Maritan A (2003) Rev Modern Phys 75:23
100. Ivanov VA, Chertovich AV, Lazutin AA, Shusharina NP, Khalatur PG, Khokhlov AR (1999) Macromol Symp 146:259
101. van den Oever JMP, Leermakers FAM, Fleer GJ, Ivanov VA, Shusharina NP, Khokhlov AR, Khalatur PG (2002) Phys Rev E 65:041708
102. Ptitsyn OB (1992) In: Creighton TE, Freeman WH (eds) Protein folding. W.H Freeman, New York, p 253–300
103. Kriksin YA, Khalatur PG, Khokhlov AR (2002) Macromol Theory Simul 11:213
104. Nishio I, Sun S-T, Swislow G, Tanaka T (1979) Nature 281:208
105. Swislow G, Sun S-T, Nishio I, Tanaka T (1980) Phys Rev Lett 44:796
106. Chu B, Ying Q (1996) Macromolecules 29:1824

107. Wu C, Zhou S (1996) Phys Rev Lett 77:3053
108. Chu B, Ying Q, Grosberg AY (1995) Macromolecules 28:180
109. Nakata M, Nakagawa T (1997) Phys Rev E 56:3338
110. Wu C, Wang X (1998) Phys Rev Lett 80:4092
111. Zhang G, Wu C (2001) Phys Rev Lett 86:822
112. von Rague-Schleyer P, Allinger NL, Clark TC, Gasteiger J, Kollman PA, Schaefer HF (1998) Encyclopedia of computational chemistry. Wiley, New York
113. Rost B, Sander C (1995) Protein structure prediction by neural networks. In: Arbib M (ed) The Handbook of brain theory and naural networks. The MIT press, Cambridge, MA, p 772–775
114. de Gennes P-G (1975) J Phys Lett 36:L55
115. Buguin A, Brochard-Wyart F, de Gennes P-G (1996) C R Acad Sci Paris 322:741
116. Ostrovsky B, Bar-Yam Y (1994) Europhys Lett 25:409
117. Abrams CF, Lee N-K, Obukhov S (2002) Europhys Lett 59:391
118. ten Wolde PR, Chandler D (2002) Proc Natl Acad Sci USA 99:6539
119. Lee N-K, Abrams CF (2004) J Chem Phys 121:7484
120. Polson JM, Moore NE (2005) J Chem Phys 112:024905
121. Yue K, Dill KA (1992) Proc Natl Acad Sci USA 89:4163
122. Cooke IR, Williams DRM (2003) Macromolecules 36:2149
123. Bates FS, Fredrickson GH (1999) Phys Today 52:32
124. Fredrickson GH, Bates FS (1996) Annu Rev Mater Sci 26:501
125. Bates FS, Fredrickson GH (1996) Annu Rev Phys Chem 41:525
126. Khalatur PG (2000) Polymer Sci C 42:229
127. Ikkala O, ten Brinke G (2004) Chem Commun 2131
128. Hajduk DA, Harper PE, Gruner SM, Honeker CC, Kim G, Thomas EL, Fetters LJ (1994) Macromolecules 27:4063
129. Förster S, Khandpur AK, Zhao J, Bates FS, Hamley IW, Ryan AJ, Bras W (1994) Macromolecules 27:6922
130. Goldacker T, Abetz V, Stadler R, Erukhimovich I, Leibler L (1999) Nature 398:137
131. Nap R, Erukhimovich I, ten Brinke G (2004) Macromolecules 37:4296
132. Leibler L (1980) Macromolecules 13:1602
133. Helfand E, Wasserman ZR (1982) In: Goodman I (ed) Developments in block copolymers. Applied Science, London, p 99
134. Matsen MW, Schick M (1994) Phys Rev Lett 72:2660
135. Matsen MW, Schick M (1996) Curr Opin Colloid Interface Sci 1:329
136. Matsen MW, Bates FS (1996) Macromolecules 29:1091
137. Matsen MW (2001) J Phys Condens Matter 14:R21
138. Drolet F, Fredrickson GH (1999) Phys Rev Lett 83:4317
139. Fredrickson GH (2002) J Chem Phys 117:6810
140. Schweizer KS, Curro JG (1994) Adv Polym Sci 116:319
141. Schweizer KS, Curro JG (1997) Adv Chem Phys 98:1
142. David EF, Schweizer KS (1994) J Chem Phys 100:7767
143. David EF, Schweizer KS (1994) J Chem Phys 100:7784
144. Chandler D, Andersen HC (1972) J Chem Phys 57:1930
145. Chandler D, McCoy JD, Singer SJ (1986) J Chem Phys 85:5977
146. Gutin AM, Sfatos CD, Shakhnovich EI (1994) J Phys A 27:7957
147. Angerman H, ten Brinke G, Erukhimovich IY (1996) Macromolecules 29:3255
148. Angerman H, ten Brinke G, Erukhimovich IY (1996) Macromol Symp 112:199
149. Potemkin II, Panyukov SV (1998) Phys Rev E 57:6902
150. Semenov AN (1997) J Phys II (France) 7:1489
151. Semenov AN, Likhtman AE (1998) Macromolecules 31:9058

152. Semenov AN (1999) Eur Phys J B 10:497
153. Zherenkova LV, Talitskikh SK, Khalatur PG, Khokhlov AR (2002) Dokl Phys Chem 382:23
154. Zherenkova LV, Khalatur PG, Khokhlov AR (2003) Dokl Phys Chem 393:293
155. Fredrickson GH, Ganesan V, Drolet F (2002) Macromolecules 35:16
156. Swift BW, Olvera de la Cruz M (1996) Europhys Lett 35:487
157. Houdayer J, Müller M (2002) Europhys Lett 58:660
158. Dobrynin AV, Erukhimovich IY (1995) J Phys I (France) 5:365
159. Gutin AM, Abkevich VI, Shakhnovich EI (1995) Proc Natl Acad Sci USA 92:1282
160. Abkevich VI, Gutin AM, Shakhnovich EI (1996) Proc Natl Acad Sci USA 93:839
161. Gusev LV, Khalatur PG, Khokhlov AR (in preparation)
162. Branden C, Tooze J (1991) Introduction to protein structure. Garland, New York
163. Holm C, Joanny JP, Kremer K, Netz RR, Reineker P, Seidel C, Vilgis TA, Winkler RG (2004) Adv Polym Sci 166:67
164. Rubinstein M, Dobrynin AV (1999) Curr Opin Colloid Interface Sci 4:83
165. Chodanowski P, Stoll S (1999) J Chem Phys 111:6069
166. Limbach HJ, Holm C, Kremer K (2002) Europhys Lett 60:566
167. Lee N, Thirumalai D (2001) Macromolecules 34:3446
168. Chang R, Yethiraj A (2003) J Chem Phys 118:6634
169. Micka U, Holm C, Kremer K (1999) Langmuir 15:4033
170. Khalatur PG, Khokhlov AR, Mologin DA, Reineker P (2003) J Chem Phys 119:1232
171. Mologin DA, Khalatur PG, Khokhlov AR, Reineker P (2004) New J Phys 6:133
172. Khokhlov AR, Khalatur PG (2005) Curr Opin Colloid Interface Sci 10:22
173. Zherenkova LV, Khalatur PG, Khokhlov AR (2003) J Chem Phys 119:6959
174. Braun FN (2002) J Chem Phys 116:6826
175. Ray J, Manning GS (1994) Langmuir 10:2450
176. Potemkin II, Vasilevskaya VV, Khokhlov AR (1999) J Chem Phys 111:2809
177. Borue VY, Erukhimovich IY (1988) Macromolecules 21:3240
178. Joanny J-P, Leibler L (1990) J Phys (France) 51:545
179. Carbajal-Tinoco MD, Ober R, Dolbnya I, Bras W, Williams CE (2002) J Phys Chem B 106:12165
180. Peng S, Wu C (2001) J Phys Chem B 105:2331
181. Philippova OE, Andreeva AS, Khokhlov AR, Islamov AK, Kukin AI, Gordeliy VI (2003) Langmuir 19:7240
182. Sophianopoulos AJ, Holde KEV (1964) J Biol Chem 239:2516
183. Wang F, Hayter J, Wilson LJ (1996) Acta Crystallogr D 52:901
184. Booth DR et al. (1997) Nature 385:787
185. Prusiner SB (1991) Science 252:1515
186. Cohen FE, Pan K-M, Huang Z, Baldwin M, Fletterick R, Prusiner SB (1994) Science 264:530
187. Booth DR, Sundetll M, Bellotti V, Robinso CV, Hutchinson WL, Fraser PE, Hawkins PN, Dobson CM, Radfordt SE, Blaket CCF, Pepys MB (1997) Nature 385:787
188. Lomakin A, Asherie N, Benedek GB (1999) Proc Natl Acad Sci USA 96:9465
189. Broglia RA, Tiana G, Pasquali S, Roman HE, Vigezzi E (1998) Proc Natl Acad Sci USA 95:12930
190. Giugliarelli G, Micheletti C, Banavar JR, Maritan A (2000) J Chem Phys 113:5072
191. Timoshenko EG, Kuznetsov YA (2000) J Chem Phys 112:8163
192. Bratko D, Blanch HW (2001) J Chem Phys 114:561
193. Bratko D, Blanch HW (2003) J Chem Phys 118:5185
194. Gupta P, Hall CK, Voegler AC (1998) Protein Sci 7:2642
195. Toma L, Toma S (2000) Biomacromolecules 1:232

196. Semenov AN (2004) Macromolecules 31:226
197. Govorun EN, Khokhlov AR, Semenov AN (2003) Eur Phys J E 12:255
198. Kikuchi A, Nose T (1996) Macromolecules 29:6770
199. Mittal KL (ed) (1977) Micellization, solubilization, and microemulsions. Plenum Press, New York, London, vol 1 and 2
200. Smit B (1993) Computer simulations of surfactants. In: Allen MP, Tildesley DJ (eds) Computer simulation in chemical physics. Kluwer Academic Publishers, Dordrecht, p 461–472
201. Laschewsky A (1995) Adv Polym Sci 124:1 and references cited herein
202. Borisov OV, Halperin A (1995) Langmuir 11:2911
203. Borisov OV, Halperin A (1996) Europhys Lett 34:657
204. Borisov OV, Halperin A (1996) Macromolecules 29:2612
205. Zhou SQ, Chu B (2000) Adv Mater 12:545
206. Dobrynin AV, Rubinstein M, Obukhov SP (1996) Macromolecules 29:2974
207. Eriksson JC, Ljunggren S (1990) Langmuir 6:895
208. Khalatur PG, Khokhlov AR, Nyrkova IA, Semenov AN (1996) Macromol Theory Simul 5:713
209. Khalatur PG, Khokhlov AR, Nyrkova IA, Semenov AN (1996) Macromol Theory Simul 5:749
210. Borisov OV, Zhulina EB (2005) Macromolecules 38:2506
211. Baiesi M, Carlon E, Orlandini E, Stella AL (2001) Phys Rev E 63:041801
212. Vasilevskaya VV, Klochkov AA, Lazutin AA, Khalatur PG, Khokhlov AR (2004) Macromolecules 37:5444
213. Kikuchi A, Nose T (1996) Polymer 37:5889
214. Selb J, Gallot Y (1981) Makromol Chem 182:1491,1513,1775
215. Wu C, Qiu X (1998) Phys Rev Lett 80:620
216. Nakata M, Nakagawa T (1997) Phys Rev E 56:33838
217. Nakata M, Nakagawa T (1999) J Chem Phys 110:2703
218. Carbajal-Tinoco MD, Williams CE (2000) Europhys Lett 52:284
219. Eisenriegler E (1993) Polymers near surface. World Science, Singapore
220. Netz RR, Andelman D (2003) Phys Rep 380:1
221. Cosgrove T, Finch NA, Webster JRP (1990) Macromolecules 23:1334
222. Joanny J-F (1994) J Phys II (France) 4:1281
223. Sommer JU, Daoud M (1995) Europhys Lett 32:407
224. Whittington SG (1998) J Phys A Math Gen 31:3769
225. Chakraborty AK (2001) Phys Pep 342:259
226. Moghaddam MS, Vrbova T, Whittington SG (2000) J Phys A Math Gen 33:4573
227. Grosberg A, Izrailev S, Nechaev S (1994) Phys Rev E 50:1912
228. Garel T, Huse DA, Leibler S, Orland H (1989) Europhys Lett 8:9
229. Gutman L, Chakraborty AK (1994) J Chem Phys 101:10074
230. Orlandini E, Tesi MC, Whittington SG (1999) J Phys A Math Gen 32:469
231. Moghaddam MS, Whittington SG (2002) J Phys A Math Gen 35:33
232. Denesyuka NA, Erukhimovich IY (2000) J Chem Phys 113:3894
233. Zhulina EB, Skvortsov AM, Birshtein TM (1981) Vysokomol Soed A 23:304
234. Erukhimovich IY (1978) Vysokomol Soed B 20:10
235. Panyukov SV, Potemkin II (1997) J Phys I (France) 7:273
236. Hsieh HL, Quirk RP (1996) Anionic polymerization: principles and practical applications. Marcel Dekker, New York
237. Hillmyer M (1999) Curr Opin Solid State Matter Sci 4:559

Folding and Formation of Mesoglobules in Dilute Copolymer Solutions

Guangzhao Zhang[1,2] · Chi Wu[3] (✉)

[1]Laboratory of Macromolecular Colloids and Solutions,
The Hefei National Laboratory for Physical Sciences at Micro-scale, Hefei,
230026 Anhui, China

[2]Department of Chemical Physics, University of Science and Technology of China,
Hefei, 230026 Anhui, China

[3]Department of Chemistry, The Chinese University of Hong Kong, Shatin,
N.T., Hongkong, China
chiwu@cuhk.edu.hk

1	Introduction	104
2	Experimental Section	108
2.1	Preparation of Amphiphilic Copolymers	108
2.1.1	Poly(*N*-isopropylacrylamide) (PNIPAM) Homopolymer	108
2.1.2	Linear NIPAM-*co*-VP Copolymers	109
2.1.3	Grafted PNIPAM-*g*-PEO Copolymers	110
2.1.4	Segmented PNIPAM-*seg*-St Copolymers	111
2.1.5	PNIPAM-*co*-KAA and PAM-*co*-NaAA Ionomers	112
2.1.6	P(DEA-*co*-DMA) Copolymers	113
2.1.7	PNIPAM-*co*-MACA Copolymers	113
2.2	Laser Light Scattering (LLS)	114
2.3	Ultra-Sensitive Differential Scanning Calorimeter (US-DSC)	116
3	Folding of Neutral Chains in Extremely Dilute Solutions	116
3.1	Coil-to-Globule Transition of Linear PNIPAM Homopolymer Chains	117
3.2	Folding of Amphiphilic Copolymer Chains	122
3.2.1	Hydrophilically Modified PNIPAM Copolymer Chains	123
3.2.2	Hydrophobically Modified PNIPAM Copolymer Chains	138
4	Ionomers—From Intrachain Folding to Interchain Association	145
4.1	PNIPAM-*co*-KAA Copolymer Chains	145
4.2	PAM-*co*-NaAA Copolymer Chains	151
5	Formation of a Stable Mesoglobular Phase in Dilute Solutions	154
5.1	Effect of Comonomer Composition	155
5.2	Effect of Comonomer Distribution	158
5.3	Viscoelastic Effect on Formation and Stabilization of Mesoglobules	162
6	Conclusion	169
	References	173

Abstract It is known that linear homopolymer chains can undergo a coil-to-globule-to-precipitation transition when the solvent quality gradually changes from good to poor. It is also known that the observation of the coil-to-globule transition without any interchain association is extremely difficult if not impossible. On the other hand, the folding of individual amphiphilic copolymer chains, such as protein chains, in an extremely dilute solution is much easier. As the copolymer concentration increases, inevitable interchain association accompanied by intrachain folding can result in a stable mesoglobular phase (the aggregation of a limited number of chains), existing between single-chain globules and macroscopic phase separation (precipitation). In this article, we mainly review what we have accomplished in the last ten years by starting with a brief discussion of the folding of linear poly(N-isopropylacrylamide) (PNIPAM) homopolymer chains in water. Our focus is the folding of different hydrophilically or hydrophobically modified PNIPAM copolymer chains in extremely dilute solutions as well as the formation of stable mesoglobules made of amphiphilic copolymer chains in dilute solutions. The discoveries of the molten globular state and the "ordered-coil" state between the random-coil and compacted globular states will be illustrated. The effects of both the comonomer composition and distribution on the folding of individual copolymer chains into some unique core-shell nanostructures as well as the formation of the mesoglobular phase are discussed. The double roles of hydrophobic interaction in the formation and stabilization of stable mesoglobules will be explained in terms of the viscoelastic effect.

Keywords Folding · Aggregation · Mesoglobule · Copolymer

Abbreviations
Section 1
CTAB	hexadecyltriethylammonium bromide
LCST	lower critical solution temperature
LLS	laser light scattering
M_w	weight-average molar mass
M_w/M_n	polydispersity index
NMR	nuclear magnetic resonance
PNIPAM	poly(N-isopropyl-acrylamide)
PNIPAM-g-PEO	poly(N-isopropylacrylamide)-graft- poly(ethylene oxide)
$\langle R_h \rangle$	average hydrodynamic radius
τ_{crum}	relaxation time of polymer chain in the crumpled globular state
τ_{eq}	relaxation time of polymer chain in the compact globular state

Section 2
A	measured base line (Eq. 2)
A_2	second virial coefficient (Eq. 1)
AIBN	2,2′-azobis(isobutyronitrile)
C	concentration
D	translational diffusion coefficient (Eq. 4)
dn/dC	specific refractive index increment
f	constant related to internal and rotational motions (Eq. 4)
$g^{(1)}(t,q)$	first-order electric field time correlation function (Eq. 2)
$G^{(2)}(t,q)$	intensity-intensity time correlation function (Eq. 2)
$G(\Gamma)$	line-width distribution function of Γ (Eq. 3)
GPC	gel permeation chromatography

k_B	Boltzmann constant
k_d	dynamic second-order virial coefficient (Eq. 3)
KPS	potassium persulfate
MACA	2′-methacryloylaminoethylene)-3α,7α,12α-trihydroxy-5β-cholanoamide
MeOK	potassium methoxide
n	refractive index of solvent
N_A	Avogadro constant
NIPAM	N-isopropylacrylamide
NIPAM-co-VP	N-isopropylacrylamide-co-vinylpyrrolidone
PNIPAM-seg-St	segmented copolymer of N-isopropylacrylamide and styrene
PNIPAM-co-KAA	N-isopropylacrylamide-co-potassium acrylic acid
PAM-co-NaAA	acrylamide-co-sodium acrylic acid
P(DEA-co-DMA)	poly(N,N-diethylacrylamide-co-N,N-dimethylacrylamide)
PEO	poly(ethylene oxide)
q	scattering vector
$\langle R_g^2 \rangle_z^{1/2}$ (or $\langle R_g \rangle$)	z-average root mean square radius of gyration
$R_{vu}(q)$	Rayleigh ratio for unpolarized scattered light
$R_{vv}(q)$	Rayleigh ratio for vertically polarized scattered light
St	styrene
T	absolute temperature
TEMED	N,N,N′,N′-tetramethylethylenediamine
THF	tetrahydrofuran
US-DSC	Ultra-sensitive Differential Scanning Calorimeter
VP	1-vinyl-2-pyrrolidone
$\langle \Gamma \rangle$	z-average line-width
η	viscosity of solvent
θ	scattering angle
λ_0	wavelength of light in vacuum

Section 3

AFM	atomic force microscopy
C_p	partial heat capacity
$f(R_h)$	hydrodynamic radius distribution
h_{brush}	thickness of a polymer brush
ΔH	change of enthalpy
$\langle I \rangle$	average scattered light intensity
I_1/I_3	fluorescence intensity ratio
L_{shell}	shell thickness of a core-shell structure
M_c	mass of the core of a core-shell particle (Eq. 5)
M_s	mass of the shell of a core-shell particle (Eq. 5)
R	radius of the core-shell particle (Eq. 5)
R_c	radius of the core of the core-shell particle (Eq. 5)
SMFS	single-molecule force spectroscopy
T_{max}	temperature corresponding to the maximum heat capacity
α	static expansion factor
$\langle \rho \rangle$	average chain density
ρ_c	density of the core of a core-shell particle
$\langle \rho \rangle_{globule}$	average chain density in globule state
ρ_s	density of the shell of a core-shell particle
σ	grafting density

χ	Flory–Huggins interaction parameter

Section 4

d_f	fractal dimension of a cluster
DLCA	diffusion-limited cluster aggregation
HPAM	partially hydrolyzed poly(acrylamide)
$M_{w,agg}$	weight average molar mass of aggregates
N_{chain}	average number of chains inside an aggregate
RLCA	reaction-limited cluster aggregation
$\langle S \rangle_{ionic}$	average surface area per ionic group
$\langle V \rangle_{chain}$	average hydrodynamic volume of polymer chains

Section 5

a_m	length of monomer (Eq. 10)
$\langle D \rangle$	transitional diffusion coefficient of aggregates (Eq. 10)
DEA	N,N-diethylacrylamide
DMA	N,N-dimethylacrylamide
D_m	diffusion coefficient of monomer (Eq. 10)
ΔG	change of Gibbs free energy
l_o	interaction range (Eq. 10)
N_{chain}	average aggregation number
N_m	number of monomer
ΔS	change of entropy
$T_{aggregation}$	aggregation temperature
$\langle v \rangle$	mean thermal velocity (Eq. 10)
α	scaling exponent
τ_c	collision or interaction time (Eq. 10)
τ_e	time for chain entanglement (Eq. 10)
ϕ_p	average polymer concentration in aggregates (Eq. 10)

1
Introduction

Conformation and phase transition of polymer chains in solution is not only a fundamental problem in polymer research, but also directly linked to the property of a polymeric material. In the long development of polymer science, the research into properties of a polymer solution has always been an important part. Polymer researchers had been puzzled for many years by the simple question whether the conformation of a flexible linear homopolymer chain can change from a random "coil" to a thermodynamically stable single-chain "globule" when the solvent quality gradually changes from good to poor, but still remain in the one phase region. On the other hand, everyday in nature protein chains fold into stable globules though how they fold with no or only a very few mistakes still remains a mystery.

Protein chains generally contain hydrophobic, hydrophilic and/or charged amino acid residues, which can be regarded as amphiphilic copolymers in a broad definition. The coordinate and cooperative interactions, such as intra- and inter-chain hydrogen bonding, hydrophobic attraction and electrostatic interaction lead to some complicated bio-active structures [1]. Different theories were proposed to explain various properties of proteins from a biological point of view [2–5]. Recently, computer simulation was also used to construct different copolymers with hydrophobic and hydrophilic units to imitate proteins. Particularly, the coil-to-globule transition of different types of copolymer chains was simulated to demonstrate how the comonomer distribution, i.e., the sequence difference in structure, could greatly influence the folding of a single copolymer chain in dilute solution [6–8].

Khokhlov et al. [8] simulated three AB copolymer chains with an identical composition and length, but different comonomer distributions on the chain backbone. Their results showed that for the chain with a globular protein-like structure in which soluble comonomer B was incorporated on the periphery of a collapsed A chain backbone, the chain folding would be easier than that of a random copolymer without a designed sequence. Moreover, the resultant globule was stable and its chain density was higher. The simulation suggested that such a chain could "memorize" or "inherit" some special functional properties of the parent collapsed state. Timoshenko et al. [9] also showed that for a given degree of amphiphilicity, the folding of an AB copolymer chain with a segmented comonomer distribution was easier and the resultant mesoglobular phase was more stable in comparison with a random copolymer chain under the same condition.

However, it is a rather difficult experimental challenge, if not impossible, to prepare a pair of AB copolymers with a similar composition and a similar chain length, but different comonomer distributions on the chain backbone. It is known that poly(N-isopropyl-acrylamide) (PNIPAM) is a thermally sensitive polymer with a lower critical solution temperature (LCST $\sim 32\,°C$) [10–14]. This interesting thermal property has made PNIPAM a simple model for the simulation of protein de-naturation in aqueous solution even though real protein chains are much more complicated [15]. Recently, using two grafted copolymers, PNIPAM-g-PEO, respectively, prepared at temperatures below and near the LCST of PNIPAM, Tenhu et al. [16] studied the comonomer distribution dependence of the chain aggregation. Their interesting results showed that the copolymer chains prepared at different temperatures had different LCSTs, supporting the computer simulation in a general sense. However, we should not forget that chain aggregation is a complicated process, which involves intrachain contraction and interchain association. Therefore, the ultimate test and experimental challenge would be the study of the effect of comonomer distribution on the coil-to-globule transition (folding) of individual amphiphilic copolymer chains without involving any interchain association in dilute solution.

Before discussing the folding of amphiphilic copolymer chains, let us first briefly examine the past studies of the coil-to-globule transition of homopolymer chains in dilute solutions. More than three decades ago, Stockmayer [17] suggested that in dilute solutions, a flexible linear homopolymer chain can change its conformation from an expanded coil to a collapsed globule if the solvent quality gradually changes from good to poor, but still remain in the one-phase region, on the basis of Flory's mean-field theory [18]. This prediction has been extensively studied both theoretically and experimentally [19–28]. Note that most of the past studies were concentrated on polystyrene solutions because such a study requires a very high molar mass ($> 10^7$ g/mol) homopolymer with a narrow molar mass distribution ($M_w/M_n < 1.1$). Useful experimental results were obtained using static and dynamic laser light scattering (LLS) and interpreted by the existing theory. However, success was limited and no one had observed the thermodynamically stable single-chain globules in such polystyrene solutions. This was because polymer chains always started to undergo an interchain association before each chain has a chance to fully collapse into a globule.

The interchain association had frustrated researchers in this field for many years. In 1993, Grosberg et al. [29] stated that the true equilibrium single chain collapse had not yet been observed experimentally for simple uncharged homopolymers without mesogenic groups. They predicted a two-stage kinetics for the collapse of a single chain, a fast crumpling of the unknotted chain followed by a slow knotting of the collapsed polymer chain. Such two-stage kinetics were roughly observed by Chu et al. [30, 31] in the study of the folding of single polystyrene chains before macroscopic precipitation. In this study, dynamic laser light scattering (LLS) was used to monitor the change of the average hydrodynamic radius ($\langle R_h \rangle$) of individual polystyrene chains in cyclohexane after an abrupt temperature change from 35 (the Θ-temperature) to 29 °C. It was found that the hydrodynamic radius distribution contained two different species, which were attributed to single polystyrene chains and aggregates of the polystyrene chains. From the time dependence of $\langle R_h \rangle$ after the abrupt temperature change, two relaxation times τ_{crum} and τ_{eq}, respectively, for the crumpled globular state and the compact globular state, were reported.

As for thermodynamically stable single-chain globules, Grosberg et al. [29] even claimed in the same 1993 article that "Practically, it cannot be observed with modern instrumentation and the sample preparation technique at this time". Actually, we noted by that time that nearly all the past studies had been conducted in organic solvents. Considering protein folding in water, we proposed to study such a "coil-to-globule" transition by using thermally sensitive water-soluble homopolymer. This is because in water hydrophobic interaction and hydrogen bonding are much stronger than the Van der Waals interaction in organic solvents. Therefore, PNIPAM as a well-known system in the study of intelligent hydrogels became a natural choice. It has been proven that such a proposal is in the right direction. Note that PNIPAM is a wonderful thermally sensitive polymer with the following chemical structure (Scheme 1).

Its isopropyl and carbon-carbon chain backbone groups are hydrophobic, but its acrylamide group is hydrophilic. A proper balance among interac-

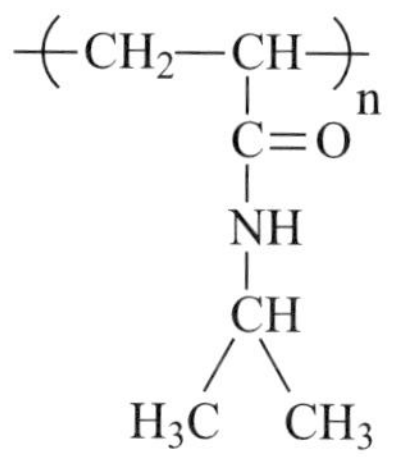

$$-(\mathrm{CH_2-CH})_n$$
$$|$$
$$\mathrm{C=O}$$
$$|$$
$$\mathrm{NH}$$
$$|$$
$$\mathrm{CH}$$
$$/\ \backslash$$
$$\mathrm{H_3C\ \ CH_3}$$

Scheme 1 Poly(N-isopropylacrylamide) (PNIPAM)

tion between these opposite groups and water leads PNIPAM to a convenient lower LCST of ∼ 32 °C. Namely, it is soluble in water at lower temperatures, but precipitates out at temperatures higher than 32 °C. Qualitatively, this delicate balance between hydrophilic and hydrophobic interaction is gradually broken as the solution temperature increases. This is because the dissolution of PNIPAM in water has a negative overall entropy change (the effect on water clusters) that is unusual, but typical for water-soluble polymers. Our recent NMR studies showed that at temperatures below the LCST, some water molecules are associated with the amide group; and during the transition, the associated water molecules dissociate.

It is well known that for a broadly distributed sample, longer chains will normally undergo the phase transition first, leading the entire solution into a thermodynamically unstable two-phase region. After overcoming some difficulties encountered in the sample preparation, we successfully prepared some narrowly distributed ($M_w/M_n < 1.1$) high-molar mass ($M_w > 10^7$ g/mol) linear PNIPAM homopolymers. Armed with these special PNIPAM samples, we were able to study the coil-to-globule transition of single PNIPAM chains in extremely dilute solutions (∼ 5 mg/mL) by using a combination of static and dynamic LLS. Finally, stable single-homopolymer-chain globules were, for the first time, observed in 1994 and the results were published in 1995 [32]. After that, a systematic study on a number fundamental problems associated with such a folding transition of homopolymer chains in water was carried out, such as the internal motions of linear coiled chains [33], the discovery of the molten globular state of a collapsed chain during the coil-to-globule transition [34], the coil-to-globule transition of linear chains grafted on a surface [35, 36], the first observation of the reversible globule-to-coil transition of linear homopolymer chains in solution [37, 38], the difference between the coil-to-globule transition of PNIPAM in normal water and in deuterated water [39], and the solvent composition induced coil-to-globule transition of PNIPAM in a mixture of methanol and water [40, 41].

After accumulating sufficient experience in the preparation of narrowly distributed long homopolymer chains and in the study of individual homopolymer chains in dilute solutions, we gradually moved into the direction of the folding and formation of the mesoglobular phase of amphiphilic copolymer chains in dilute solutions. Various copolymers with different comonomers, such as ionic or nonionic and hydrophilic or hydrophobic, were inserted or attached to the PNIPAM chain backbone with different comonomer compositions and distributions. It should be noted that there were many other research groups who studied the effects of comonomer, especially comonomer composition, on the association of amphiphilic copolymer chains in dilute solutions. However, it is our intention in this article to mainly review what has been done in our laboratory by using a combination of static and dynamic LLS to observe the folding and formation of mesoglobular phase of copolymer chains in dilute solutions. Other authors in this volume will review other theoretical and experimental studies of the association of copolymer chains. This review starts from a brief discussion of the folding of long homopolymer chains in dilute solution. Further discussion of copolymer chains will be divided according to the nature of the comonomers used in the preparation of PNIPAM copolymers. In order to facilitate our discussion, we will outline some details of our experiments in the following section.

2
Experimental Section

2.1
Preparation of Amphiphilic Copolymers

In this review, hydrophilically and hydrophobically modified poly(N-isopropylacrylamide) (PNIPAM) copolymers are mainly used to illustrate how amphiphilic copolymer chains can fold from an extended random coil to a collapsed globule in extremely dilute solutions and associate to form a stable mesoglobular phase which exists between single-chain globules and macroscopic precipitation. The copolymers used can be prepared by free-radical reaction.

2.1.1
Poly(N-isopropylacrylamide) (PNIPAM) Homopolymer

N-isopropylacrylamide (NIPAM) monomer (courtesy of Kohjin Ltd) can be purified by re-crystallization in a benzene/n-hexane mixture and azobisisobutyronitrile (AIBN) (from Aldrich, analytical grade) can also be purified by re-crystallization. In a typical free-radical polymerization, 18 g NIPAM monomer was first dissolved in 150 mL benzene with 1 mol % of AIBN added as the initiator. The solution mixture was then degassed through three cycles of freezing and thawing. Polymerization was carried out in an oil bath at 56 °C for 30 h under a positive nitrogen pressure. The solvent was removed by evaporation after the polymerization. The resultant crude polymer was further dried and then dissolved in acetone. The polymer was recovered by adding the acetone solution dropwise into n-hexane. Upon filtering and drying, a white fabric-like polymer can be obtained. The yield is normally higher than $\sim 70\%$. Details of the polymerization can be found elsewhere [42]. The resultant PNIPAM can be fractionated several times by precipitation from an extremely carefully dried acetone solution to n-hexane at $\sim 25\,^\circ\mathrm{C}$. It should be emphasized that the use of the dried solvents is one of the key factors for success in the preparation of a narrowly distributed PNIPAM sample. A careful combination of both the fractionation and filtration can lead to some narrowly distributed ($M_w/M_n < 1.1$) high molar mass PNIPAM samples ($M_w > 10^7$ g/mol). The extremely dilute solution ($\sim 10^{-6}$ g/mL) of PNIPAM in water at lower temperatures can be clarified with 0.5-μm filters. The chemical structure of PNIPAM has been listed before.

2.1.2
Linear NIPAM-co-VP Copolymers

Comonomer 1-vinyl-2-pyrrolidone (VP) comonomer can be purified by distillation at reduced pressure prior to use. Potassium persulfate (KPS) can be purified in a mixture of water and methanol. NIPAM-co-VP copolymers with different amounts of VP can be prepared at temperatures lower or higher than the LCST of PNIPAM by free radical polymerization in water with an initiator of KPS/N,N,N',N'-tetramethylethylenediamine (TEMED) redox. The resultant copolymer can be harvested by precipitation, i.e., pouring the reaction mixture into an equal volume of methanol. Each resultant copolymer can be further purified by several cycles of re-dissolution in water and precipitation in methanol to ensure a complete removal of residual monomers. The final product can be dried under reduced pressure at $40\,^\circ\mathrm{C}$.

The copolymer can be further fractionated by precipitation from acetone solution to n-hexane at room temperature. In each case, only the first fraction should be used to obtain narrowly distributed high molar mass copolymer chains for LLS measurement. ^{1}H NMR can be used to characterize the copolymer composition. The ratio of the peak areas of the methine proton of the isopropyl group in NIPAM and the two protons neighboring the carbonyl group in VP can be used to determine the VP content. The composition of each NIPAM-co-VP copolymer was found to be close to the feeding monomer ratio prior to the copolymerization. The nomenclature used hereafter for these copolymers is NIPAM-co-VP/x/y, where x and y are the copolymerization temperature ($^\circ$C) and the VP content (mol%), respectively. The solution with a concentration of as low as 3.0×10^{-6} g/mL can be clarified with a 0.45 μm Millipore Millex-LCR filter to remove dust before the LLS measurement. The resistivity of deionized water used should be close to 18 MΩ cm. The chemical structure of poly(NIPAM-co-VP) is as follows (Scheme 2).

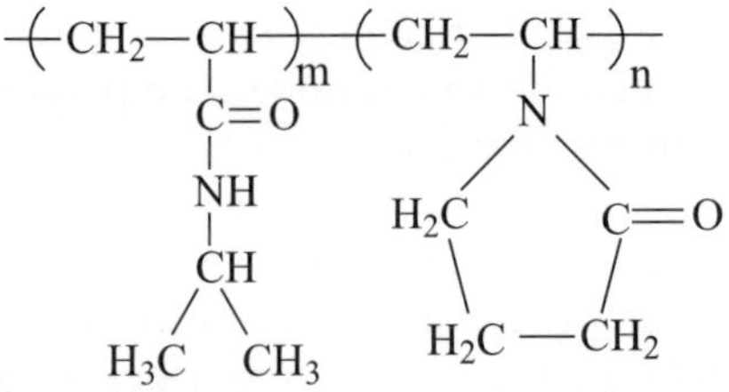

$$\mathrm{-\!\!\left(CH_2\!-\!\!\underset{\underset{\underset{\underset{H_3C\ \ CH_3}{\diagup\ \diagdown}}{CH}}{\underset{NH}{|}}}{\overset{|}{\underset{C=O}{|}}}CH\right)_{\!\!m}\!\!\left(CH_2\!-\!\!\underset{\underset{H_2C\!-\!CH_2}{\diagdown\ \diagup}}{\underset{\diagup\ \ \diagdown}{\overset{|}{N}}{\atop H_2C\ \ \ C=O}}CH\right)_{\!\!n}}$$

Scheme 2 Poly(N-isopropylacrylamide-co-1-vinyl-2-pyrrolidone) (Poly(NIPAM-co-VP))

2.1.3
Grafted PNIPAM-g-PEO Copolymers

Poly(ethylene oxide) (PEO) macromonomers end capped with a reactive methacrylate group can be synthesized by anionic ring-opening polymerization of ethylene oxide in tetrahydro-furan (THF) using potassium methoxide (MeOK) as the initiator. The weight average molar mass and polydispersity of the PEO macromonomers can be determined by gel permeation chromatography (GPC) using chloroform as the eluent and PEO standards. The molar mass of PEO macromonomer is in the range 3000–10 000 g/mol with a polydispersity index (M_w/M_n) less than 1.2. The PNIPAM grafted with PEO macromonomers can be prepared by free-radical copolymerization of different amounts of the PEO macromonomers into the PNIPAM chain backbone in water at temperatures either lower or higher than the LCST of PNIPAM. In a typical reaction, a 250-mL two-neck flask equipped with a nitrogen inlet tube and a magnetic stirrer is used. 0.03 mol NIPAM and different amounts of PEO can be added to a proper amount of deionized water to obtain 1–5 wt % solutions. The KPS/TEMED redox is normally used as the initiator. The molar ratio of KPS/TEMED is 1 : 1. The KPS and TEMED are respectively dissolved in water with concentrations of 0.009 M and 0.045 M. A few mL KPS solution is normally added into the reaction mixture. The solution should be repeatedly degassed at 20 °C and then purged with nitrogen for half an hour before the reaction. After heating the reaction mixture to 45 °C, a few mL TEMED solution is added and the reaction is carried out at this temperature for a certain time in a water bath to control the monomer conversion less than 50%. The PNIPAM-g-PEO copolymers can be purified by dialysis in a large amount of water. The final product should be dried under a reduced pressure at 40 °C. The copolymer can be further purified by several cycles of precipitation/fractionation from an acetone solution to n-hexane at 35 °C. The apparent weight average molar mass (M_w) of PNIPAM-g-PEO can be determined by laser light scattering. The copolymer composition can be estimated by ^{1}H-NMR. The chemical structure of PNIPAM-g-PEO is as follows (Scheme 3).

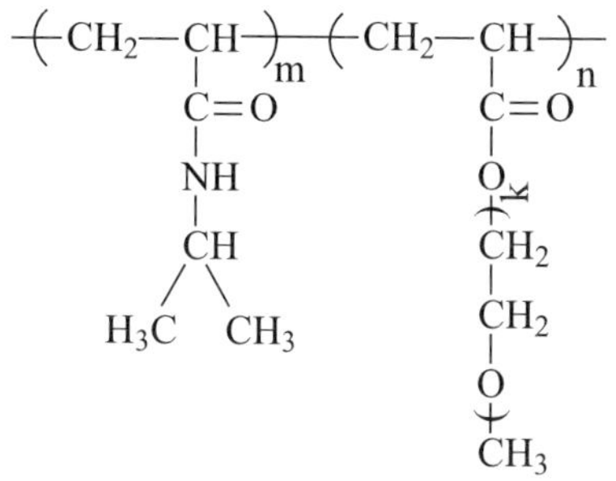

Scheme 3 Poly(N-isopropylacrylamide-*graft*-poly(ethylene oxide) (PNIPAM-*g*-PEO))

2.1.4
Segmented PNIPAM-*seg*-St Copolymers

Hydrophobically modified PNIPAM-*seg*-St segmented copolymers can be prepared by evenly inserting short styrene segments (stickers) into a PNIPAM chain backbone using the micellar polymerization. In this method, hydrophobic styrene (St) monomers is first solubilized inside small micelles made of surfactant, hexadecyltriethylammonium bromide (CTAB). KPS and TMED can be used to initiate the polymerization of hydrophilic NIPAM monomers dissolved in the continuous aqueous medium. When the free radical end of a growing PNIPAM chain enters a micelle, styrene monomers entrapped inside start to react to form a short hydrophobic segment (sticker). In this way, the coming-in-and-out of different micelles of each free-radical chain end can "connect" short styrene blocks on a PNIPAM chain.

In a typical reaction, initial concentrations of NIPAM, styrene, CTAB, KPS, and TMEDA are 0.16 M, 5.24 mM, 17.3 mM, 0.34 mM, 0.67 mM, respectively. The styrene content (3.9 mol %) of the resultant segmented PNIPAM-*seg*-St copolymer can be determined by pyrolysis gas chromatography. The average degree of polymerization between two styrene segments can be over a wide range, mainly depending on the initial NIPAM/styrene ratio. The resultant copolymer can be purified and fractionated by a number of successive dissolution-and- precipitation cycles in a mixture of extremely dried

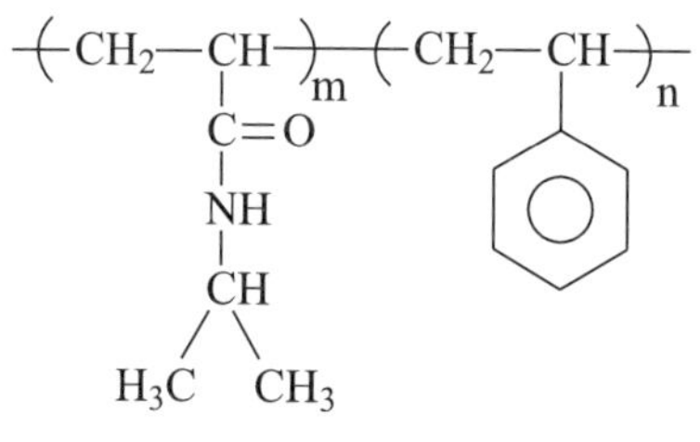

Scheme 4 Poly(N-isopropylacrylamide-*seg*-styrene) (PNIPAM-*seg*-St)

acetone and *n*-hexane at 25 °C. In each cycle, only the very first fraction obtained should be used in the next cycle. This combination of fractionation and filtration can result in narrowly distributed PNIPAM-*seg*-St chains with a high weight-average molar mass ($> 10^7$ g/mol) and a low polydispersity index ($M_w/M_n < 1.1$). The chemical structure of PNIPAM-*seg*-St is as shown in Scheme 4.

2.1.5
PNIPAM-*co*-KAA and PAM-*co*-NaAA Ionomers

PNIPAM containing a few molar percent of ionic groups on its chain backbone, ionomers, can be prepared by a free-radical copolymerization of NIPAM and other ionic comonomers, such as acrylic acid (AA), at 60 °C using AIBN as the initiator and a benzene/ethanol mixture as the reaction medium. In a typical synthesis, the reaction is conducted in a 250-mL two-neck flask equipped with a nitrogen inlet tube and a magnetic stirrer. 0.5 mmol NIPAM, proper amount of AA and 0.5 mol % AIBN are added to 50 mL solvent. The total monomer concentration is kept close to 0.5 M. After 30-min nitrogen purging, the mixture can be heated to and react at 60 °C for 1 hr in an oil bath. The monomer conversion should be controlled to be no more than 50%. After terminating the reaction, one can remove the solvent by evaporation at $T < 40$ °C under a reduced pressure. Each product of such prepared ionomers can be purified through three cycles of the acetone-to-hexane reprecipitation.

The selection of this solvent mixture is based on the following consideration: It is known that the preparation of PNIPAM homopolymer in benzene can result in higher molar mass samples than in other solvents. On the other hand, it is also known that alcohol is a relatively moderate chain transfer agent for free radical polymerization. Therefore, it is possible to control the molar mass by using a benzene/ethanol mixture with a varied composition. The average ionic content of each PNIPAM ionomer can be determined by titration using a 0.01-M potassium hydroxide solution with phenolphthalein as the indicator. A distinct color change from colorless to red can be used as an indication of the end point. Hereafter, PNIPAM ionomers neutralized with KOH are labeled as PNIPAM-*m*KAA, where "*m*" represents the average molar content of AA in the ionomer chain. On the other hand, poly(acrylamide) (PAM) ionomers can be prepared from PAM homopolymer. Acrylamide can be purified by re-crystallization. There is nothing special in such a synthesis so we will omit the details of the polymerization. The resultant PAM homopolymer can be fractionated and then hydrolyzed in NaOH and $NaCO_3$ aqueous solution. The hydrolysis is controlled by the reaction time. The hydrolysis degree can be determined by titration. The molecular parameters of these ionomer chains, such as M_w and $\langle R_g \rangle$, can be determined by LLS. Hereafter, these copolymers are denoted as PAM-*co*-*x*NaAA, where "*x*" shows the molar percent of acrylic acid groups (from the hydrolysis) on the PAM chain backbone. The chemical structure of PNIPAM-co-KAA and PAM-co-NaAA are as follows (Scheme 5).

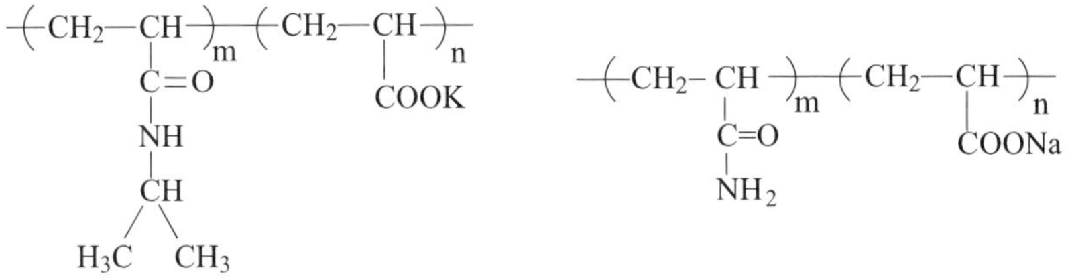

Scheme 5 Poly(N-isopropylacrylamide-co-potassium acrylate) (PNIPAM-co-KAA) and poly(N-isopropylacrylamide-co-sodium acrylate) (PAM-co-NaAA)

2.1.6
P(DEA-co-DMA) Copolymers

Poly(N,N-diethylacrylamide-co-N,N-dimethylacrylamide) P(DEA-co-DMA) copolymers with different amounts of DMA can be synthesized by free radical polymerization in THF with AIBN as the initiator (1 mol %). In a typical reaction, the solution mixture is bubbled with dry nitrogen for 30 min prior to polymerization. The temperature is then gradually raised to 68 °C in a period of 2 h and maintained for $\sim$ 18 h. Each reaction mixture was precipitated in ether or hexane after the polymerization. The copolymer composition determined by ^{1}H NMR spectra is normally close to the feed ratio of monomers prior to polymerization. The nomenclature used hereafter for these copolymers is P(DEA-co-DMA/x), where x denotes the mol % content of DMA. The chemical structure of P(DEA-co-DMA) is as shown in Scheme 6.

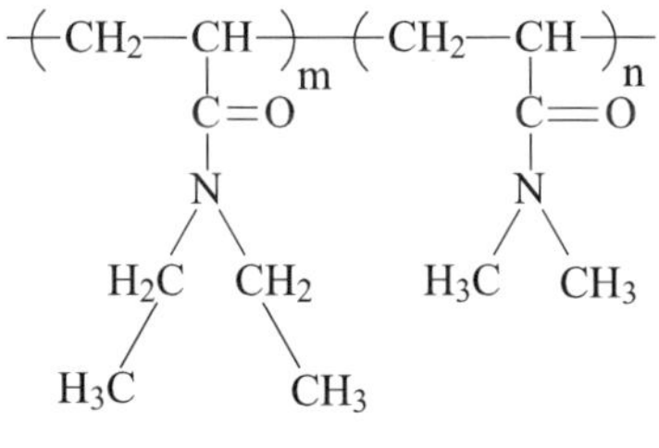

Scheme 6 Poly(N,N-diethylacrylamide-co-N-dimethylacrylamide) (P(DEA-co-DMA))

2.1.7
PNIPAM-co-MACA Copolymers

MACA is 2′-methacryloylaminoethylene)-3α,7α,12α-trihydroxy-5β-cholanoamide, a cholic acid derivative of natural bioactive amphiphilic compound. The synthetic detail of MACA is not what we would like to discuss here.

MACA as a hydrophobic comonomer can be used to modify PNIPAM. Copolymers, PNIPAM-*co*-MACA with different amounts of MACA can be synthesized by free-radical copolymerization of NIPAM and MACA in a mixture of methanol and chloroform with AIBN as the initiator. The resulting copolymers after purification can be dried in vacuum at 40 °C for 24 h. Hereafter, these copolymers are denoted as PNIPAM-*co*-*x*-MACA, where *x* denotes the molar percent of MACA. As expected, their solubility in water decreases as the MACA content or the solution temperature increases. It is also expected that the copolymer chains with a higher MACA content would have a lower LCST in comparison with PNIPAM homopolymer chains. In order to prepare a true solution, one has to dissolve these copolymers in water at low temperatures. The chemical structure of PNIPAM-*co*-MACA is as follows (Scheme 7).

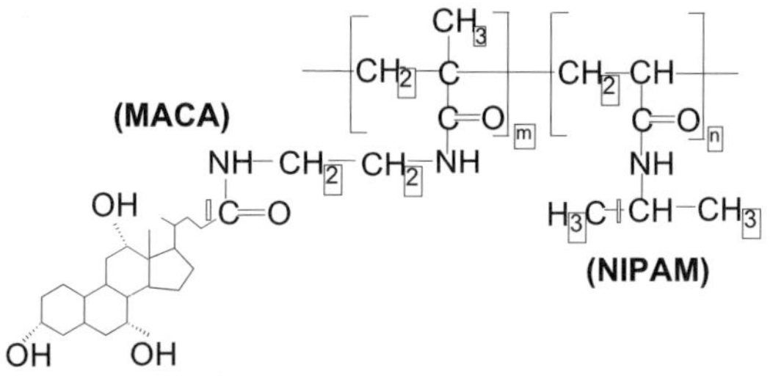

Scheme 7

2.2
Laser Light Scattering (LLS)

Laser light scattering has become a routine instrument in polymer laboratories around the world. We have used two slightly modified commercial LLS spectrometers (ALV, Germany) equipped with a multi-τ digital time correlator (ALV-5000). In our earlier experiments, we used an ALV/SP-150 spectrometer equipped with a solid-state 400-mW laser (ADLAS DPY425II, 400 mW at $\lambda_0 = 532$ nm) as the light source. Since 2002, we have used a newer version (ALV/DLS/SLS-5022F) equipped with a cylindrical 22-mW UNIPHASE He–Ne Laser ($\lambda_0 = 632$ nm) and a sensitive APD detector, in which the 22-mW red laser has the same affect as the 400-mW green laser. The incident light beam was vertically polarized with respect to the scattering plane and the intensity was regulated with a beam attenuator (Newport M-925B) so as to avoid localized heating in the light-scattering cuvette.

In static LLS [43], the angular dependence of the excess absolute time-averaged scattering intensity, known as the Rayleigh ratio $R_{vv}(q)$, is measured. For a dilute solution measured at a relatively small angle (θ), $R_{vv}(q)$ can be related to the weight average molar mass (M_w), the root mean square radius of gyration ($\langle R_g^2 \rangle_z^{1/2}$) (or simply as $\langle R_g \rangle$), the second virial coefficient (A_2), and the scattering vector (q) as:

$$\frac{KC}{R_{vv}(q)} \approx \frac{1}{M_w}(1 + \frac{1}{3}R_g^2 q^2) + 2A_2 C \tag{1}$$

where $K = 4\pi n^2 (dn/dC)^2/(N_A \lambda_0^4)$ and $q = (4\pi n/\lambda_0)\sin(\theta/2)$ with N_A, n, dn/dC, and λ_0 being the Avogadro constant, the solvent refractive index, the specific refractive index increment, and the wavelength of light in vacuum, respectively. Strictly speaking, $R_{vv}(q)$ should be $R_{vu}(q)$ because there was no analyzer in front of the detector. However, in the study of linear flexible polymer chains, depolarization is not a serious problem. Therefore, we can replace $R_{vu}(q)$ with $R_{vv}(q)$. The value of dn/dC is 0.167 mL/g in water at 25 °C, which was determined by using a special novel and precise differential refractometer [44].

In dynamic LLS [45, 46], the intensity-intensity time correlation function $G^{(2)}(t,q)$ in the self-beating mode was measured. For a Poisson distribution of the number of photons, $G^{(2)}(t,q)$ can be related to the normalized first-order electric field time correlation function $g^{(1)}(t,q)$ as [46]

$$G^{(2)}(t,q) = \langle I(0,q)I(t,q)\rangle = A\left[1 + \beta|g^{(1)}(t,q)|^2\right], \tag{2}$$

where A is the measured base line, $0 < \beta < 1$ is a parameter depending on the coherence of the detection, and t is the delay time. For a broadly distributed sample, $|g^{(1)}(t,q)|$ is related to the line-width distribution $G(\Gamma)$ by

$$|g^{(1)}(t,q)| = \langle E(t,q)E^*(0,q)\rangle = \int_0^\infty G(\Gamma)e^{-\Gamma t}d\Gamma, \tag{3}$$

Using the Laplace inversion program, CONTIN, in the correlator, we were able to calculate $G(\Gamma)$ from $G^{(2)}(t,q)$ on the basis of Eqs. 3 and 4. For a pure diffusive relaxation, Γ is related to the translational diffusion coefficient D as [47]

$$\Gamma = Dq^2(1 + k_d C)(1 + f R_q^2 q^2), \tag{4}$$

where k_d is the dynamic second-order virial coefficient, containing both thermodynamic and hydrodynamic contributions, and f is a constant related to internal and rotational motions of scattering objects. $D = \Gamma/q^2$ if the polymer solution is sufficiently dilute and $qR_g \ll 1$. D can be further converted into the hydrodynamic radius R_h using the Stokes–Einstein equation: $D = k_B T/6\pi\eta R_h$, where k_B, T, and η are the Boltzmann constant, the absolute temperature, and the solvent viscosity, respectively. For narrowly distributed samples, the cumulant analysis of $G^{(2)}(t)$ can result in an accurate average line-width $\langle\Gamma\rangle$.

It should be note that the coherent factor β in dynamic LLS should be as high as possible. The ALV instrument can reach ~ 0.95, a rather high value for a LLS spectrometer to be used for both static and dynamic LLS simultaneously. This is one of the reasons why we are able to carry out dynamic LLS in an extremely dilute solution, but still have a sufficient signal-to-noise ratio. In addition, with some modifications, one can remove straight lights from the incident bean to make a LLS spectrometer capable of measuring both static and dynamic LLS continuously in the small angle range that is particularly useful in the measurement of ultra-long polymer chains. This is because in static LLS the condition of $qR_g < 1$ is essentially required to determine the precise value of R_g; whereas in dynamic LLS the extrapolation of $q \to 0$ and the interference of the internal motions associated with the long polymer chain in dynamic LLS can be avoided. In addition, in this accessible small angle range the scattered intensity of long polymer chains is much stronger than that at high scattering angles, so that we are able to study an extremely dilute solution. The typical long-term temperature stability inside our LLS sample holder was less than $\pm 0.02\,°\text{C}$.

2.3
Ultra-Sensitive Differential Scanning Calorimeter (US-DSC)

The energy involved in the folding and association of copolymer chains in solutions can be measured by a micro-calorimeter (MicroCal Inc). We used US-DSC at an external pressure of ~ 180 kPa. The cell volume is only 0.157 mL. The heating rate can be varied and the instrument response time is normally a few seconds. All the DSC data should be corrected for instrument response time and can be analyzed using the software in the calorimeter. Note that the concentration used in DSC is normally not lower than 10^{-3} g/mL, much higher than that used in LLS (10^{-6}–10^{-4} g/mL).

3
Folding of Neutral Chains in Extremely Dilute Solutions

In comparison with copolymers, linear homopolymer chains are much simpler because there is no complication of comonomer composition, configuration and distribution on the chain backbone. However, the folding of linear homopolymer chains has been an extremely difficult problem in polymer science. Whether linear homopolymer chains can fold from an expanded random coil conformation in good solvents to a collapsed tiny globule had puzzled polymer researchers for many years and became a classic question since the 1960s. In 1995, Zhou et al. [32] finally showed, for the first time, that linear PNIPAM homopolymer chains can undergo the predicted coil-to-globule transition to form stable single-chain globules in water in the one-phase region without any interference of interchain association.

3.1
Coil-to-Globule Transition of Linear PNIPAM Homopolymer Chains

Figure 1 shows the typical angular dependence of $KC/R_{vv}(q)$ of PNIPAM homopolymer chains in the coiled and the fully collapsed globular states, respectively. The decrease of $\langle R_g \rangle$ from 127 nm to ~ 18 nm, i.e., the decreases of the slope of the lines in Fig. 1 on the basis of Eq. 1, clearly indicates the chain collapse. The transition from the coil state to the globule state can also be directly viewed from the change of the hydrodynamic radius distribution $f(R_h)$ in the inset of Fig. 1. It is worth noting that the respective extrapolations of $[KC/R_{vv}(q)]_{q \to 0}$ lead to the same intercept, indicating that there is no change in M_w on the basis of Eq. 1. The narrowly distributed $f(R_h)$ in the globule state also indicates no interchain aggregation. Moreover, the average scattered light intensity ($\langle I \rangle$) in the globule state (not shown) was independent of time over three days, which indicates the globules were stable, because $\langle I \rangle \propto M_w \propto nM^2$ on the basis of Eq. 1, very sensitive to interchain association.

The plot in Fig. 1 should be the most important and ultimate test whether the folding of individual polymer chains is free of interchain association. Unfortunately, many past studies of the coil-to-globule transition did not present such a plot. It is always questionable whether the solutions studied were truly in the "one-phase" region or in the meta-stable "two-phase" region. Note that in the one-phase region, solvents can be further divided as "good", "theta" and "poor" solvents, depending on the relative strength of the solvent-polymer and solvent-solvent interaction. While in the "two-phase" region, polymer solution forms, in principle, a concentrated layer and a dilute

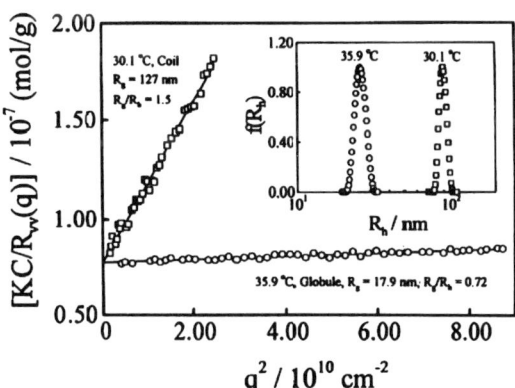

Fig. 1 Typical angular dependence of $KC/R_{vv}(q)$ of PNIPAM in water at two different temperatures, where the weight-average molar mass (M_w) and concentration (C) of PNIPAM are 1.3×10^7 g/mol and 6.7×10^{-7} g/mL, respectively. The *insert* shows the corresponding hydrodynamic radius distributions $f(R_h)$ of the PNIPAM chains respectively in the coiled and the globular states [38]

layer. In real experiments, two macroscopic layers might not be observable, especially in aqueous polymer solutions, because similar densities of polymer chains and water prevent macroscopic phase separation. Instead, only microphase separation, i.e., the formation of the mesoglobular phase, is observed, in which a limited number of polymer chains come together to form stable polymeric colloidal particles floating inside the solution. Within a reasonable experimental time scale, these particles or mesoglobules would not be able to further aggregate to form macroscopic precipitation because there is no sufficient sedimentation force to drag them down.

The earlier results concerning the folding of homopolymer chains have already been reviewed [48], which is schematically summarized in Fig. 2. The essential message of Fig. 2 is that when the solvent quality changes from good to poor, a linear and coiled homopolymer chain first shrink into a crumpet state without some additional knotting and then it passes through a molten globule state before it finally reaches its collapsed globular state. The reverse process (the dissolution or the "globule-to-coil" transition) of the collapsed chain follows a different route; namely, there exists a hysteresis between the folding and unfolding processes.

First, let us look at indirect evidence that the collapsed chain is in a crumpet state instead of a highly knotted state. Figure 3 shows the dissolving kinetics (in terms of the change of $\langle R_h \rangle$) of fully collapsed single PNIPAM chain globules, where t is the standing time after the solution temperature was quenched from 33.02 °C to 30.02 °C. Experimentally, the PNIPAM solution was prepared at 30.02 °C and its $\langle R_h \rangle$ was measured. The solution temperature was then increased to 33.02 °C and aged for more than $\sim 10^3$ s

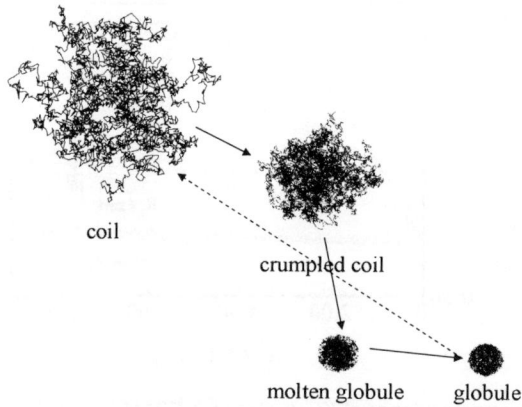

Fig. 2 Schematic of four thermodynamically stable states (random coil, crumpet coil, molten globule and collapsed globule) of a homopolymer chain in the coil-to-globule and the globule-to-coil transitions. There exists a hysteresis between the two transitions around the Θ-temperature (~ 30.6 °C) of the PNIPAM solution [37]

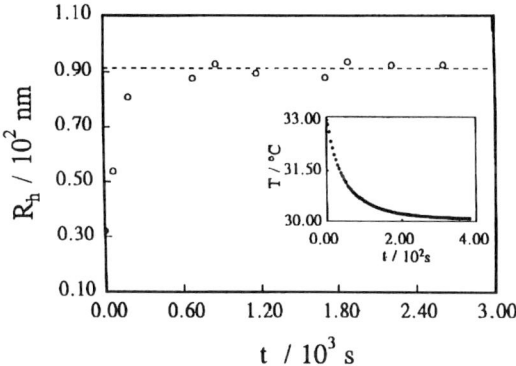

Fig. 3 Dissolving kinetics (in terms of average hydrodynamic radius R_h) of collapsed single-chain PNIPAM globules, where t is the standing time after the solution temperature was quenched from 33.02 to 30.02 °C and the *dashed line* represents a stable average value of R_h of individual PNIPAM random coils at 30.02 °C. The weight-average molar mass (M_w) of the PNIPAM sample used is 1.08×10^7 g/mol with a polydispersity index (M_w/M_n) less than 1.1 [32]

so that individual PNIPAM chains have a sufficient time to collapse into individual single-chain globules. Note that the longest aging time was 3 days and there was no change in $\langle R_h \rangle$ after ~ 800 s. The solution temperature was suddenly cooled down back to 30.02 °C and both $\langle R_h \rangle$ and t were immediately recorded after the temperature change.

The inset in Fig. 3 shows how fast the solution temperature could reach its equilibrium value, wherein a very special LLS cuvette made of a thin wall (~ 0.4 mm) glass tube was used for all kinetic studies. Figure 3 shows that the change from the globular state to the extended random-coiled state, i.e., the dissolving kinetics, was too fast to be detected in our LLS setup. In other words, before the solution reached its temperature equilibrium at 30.02 °C, individual collapsed PNIPAM globules already expanded into individual extended random coil chains. This fast dissolving time ($< \sim 300$ s) indicates that there should be no additional and extensive chain knotting inside these highly collapsed single-chain globules because it has been known that it takes one week to dissolve such a long-chain polymer (chains are entangled in bulk) even in a good solvent.

Figure 4 shows a plot of the static expansion factor (α) as a function of the relative temperature Θ/T, where α is defined as $R_g(T)/R_g(\Theta)$ and r is the number of residues that may be one monomer unit or a number of repeat units. When $T < \Theta$ (water is a good solvent for PNIPAM), the data points are reasonably fitted by the line with $r = 10^5$ calculated on the basis of Flory–Huggins theory [15]. Similar results have also been observed for linear polystyrene in cyclohexane [25, 49]. The theory works well in the good-solvent region wherein the interaction parameter (χ) is expected to be

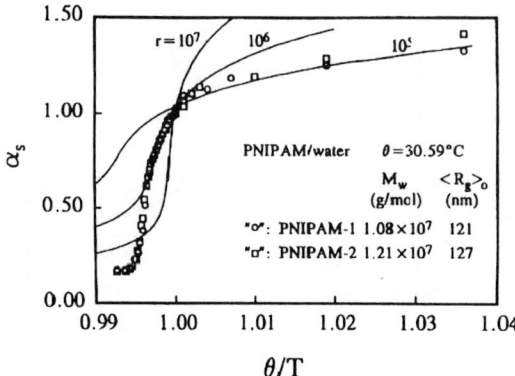

Fig. 4 Plot of static expansion factor (α_s) as a function of relative temperature Θ/T, where α is defined as $R_g(T)/R_g(\Theta)$; symbols are our measured results; and the *lines* are calculated data with three different values of r. If choosing $M = 113$ (molar mass of monomer NIPAM), we have $r \sim 10^5$ for both the PNIPAM samples used [32]

a weak function of temperature [15]. On the other hand, for temperatures higher than Θ, the measured α drops much faster than the line with $r = 10^5$. Up to now, we still do not have a clear understanding of this discrepancy. Apparently, the results can be partially fitted by the line with $r = 10^6$. The remarkable point is that α decreases to a much lower plateau than the predicted value [50, 51]. However, this does indicate that for PNIPAM in water, χ might be a function of temperature or concentration because the local concentration within the volume occupied by the chain increases ~ 500 times during the chain contraction.

Recently, Tanaka et al. [52] suggested that PNIPAM in water can be treated as a "copolymer" because they think that the PNIPAM chain backbone is hydrolyzed in a segmental fashion. They also suggested that the hydrolyzing degree or extent would dramatically decrease when temperatures are higher than the LCST. This suggestion seems to agree well with the two-state theory (hydrolysis and nonhydrolysis) used by Halperin et al. [53] In any case, the short-coming of the concentration-independent χ is obvious. Our results also showed that the average chain density $\langle\rho\rangle$ estimated from $M_w/[N_A(4/3)\pi\langle R_h\rangle^3]$ increases from 0.0025 g/cm^3 (coil) to 0.34 g/cm^3 (globule), close to ~ 0.4 g/cm^3 predicted on the basis of a space-filling model [54]. This means that the inside of single-chain PNIPAM globules is not as dry as what we originally thought and each globule still contains ~ 70–80% of water inside its hydrodynamic volume.

It has been known that $\langle R_g\rangle/\langle R_h\rangle$ can better reflect the chain conformation. Figure 5 shows the temperature dependence of $\langle R_g\rangle/\langle R_h\rangle$ in both the heating and cooling processes. During the heating, in the range 20–30.6 °C (Θ-temperature), $\langle R_g\rangle/\langle R_h\rangle$ remains nearly a constant (~ 1.5), revealing that the PNIPAM chains keep the random coil conformation as long as $T < \Theta$. The

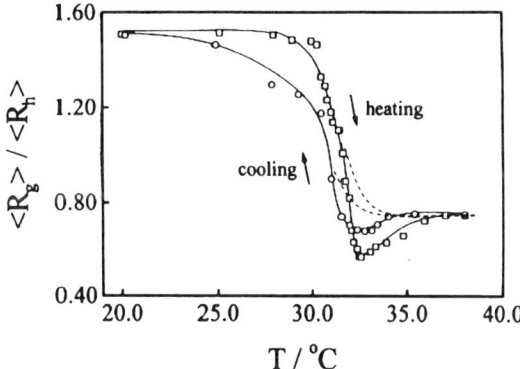

Fig. 5 Temperature dependence of the ratio of average radius of gyration to average hydrodynamic radius ($\langle R_g \rangle / \langle R_h \rangle$) in both heating and the cooling processes, where $M_w = 1.3 \times 10^7$ g/mol and $M_w/M_n < 1.05$ [38]

change of $\langle R_g \rangle / \langle R_h \rangle$ between ~ 1.5 at 20 °C to ~ 0.77 at 38 °C agrees well with the values respectively predicted for a random coil and a solid uniform sphere, clearly indicating the collapse of the PNIPAM chains. In the temperature range 30.6–38 °C, $\langle R_g \rangle / \langle R_h \rangle$ dips into a low value of ~ 0.56 before it comes back to ~ 0.77. This temperature range can be roughly divided into two sub-stages. The first one is from the Θ-temperature to 31.6 °C at which $\langle R_g \rangle = \langle R_h \rangle$; and the second one is from 31.6 to 38 °C in which $\langle R_g \rangle / \langle R_h \rangle$ reaches the minimum value.

The decrease of $\langle R_g \rangle / \langle R_h \rangle$ in the first stage reflects the conformation change from an extended random coil to a *crumpled* coil. If the folding and unfolding were an all-or-none process, the changes of $\langle R_h \rangle / \langle R_g \rangle$ in the second stage would, respectively, follow the two dashed lines in Fig. 5. However, the unexpected minimum, which has later been confirmed by many experiments related to the coil-to-globule transition, reveals that there exists another physical state between the fully collapsed globule and the unfolded random coil. This is identified as the molten globular state. In this molten globular state, the chain density is not evenly distributed inside each single-chain globule. Presumably, the surface of the collapsed single-chain globule contains many small loops (i.e., a rough surface with a lower density) because polymer chains are not infinitely flexible. We can imagine that on the one hand, these small loops are nondraining and they make $\langle R_h \rangle$ larger, and on the other hand, they have much less effect on $\langle R_g \rangle$ because their masses are relatively low. In other words, $\langle R_g \rangle$ decreases relatively faster than $\langle R_h \rangle$. This explains why the ratio of $\langle R_g \rangle / \langle R_h \rangle$ could be smaller than $(3/5)^{1/2}$ predicted for a uniform hard sphere. It is not difficult to imagine that stress must be built up within these small loops when they become smaller and smaller so that the shrinking of these small loops slows down. This is why $\langle R_h \rangle$ decreases slightly, but there is no change in $\langle R_g \rangle$, when $T > 32.4$ °C (not shown).

Figure 5 shows that in the cooling process, $\langle R_g \rangle / \langle R_h \rangle$ reaches ~ 1.5 only after $T < 25\,°C$. There is a clear hysteresis, especially around the Θ-temperature. It reveals that even water becomes a good solvent in the temperature range 25–30.6 °C, individual single-chain globules are still not completely dissolved into the randomly coiled conformation. This hysteresis implies that some additional intrachain interaction (presumably, intrachain hydrogen bonding) is formed when the chain is in its fully collapsed globular state because of a much smaller inter-segment distance and such intrachain interaction persists in the globule-to-coil process until water becomes a very good solvent. The results also revealed that the decrease of $\langle R_g \rangle / \langle R_h \rangle$ in the left side of the minimum point is because the decrease of $\langle R_g \rangle$ is relatively faster, while the increase of $\langle R_g \rangle / \langle R_h \rangle$ at the right side is mainly due to the decrease of $\langle R_h \rangle$.

The transition between a random coil and a crumpet coil can be respectively described by the existing Flory and Birshtein–Pryamitsyn theories [18, 55]. However, a quantitative description of the molten globular state still remains a challenging problem. The deviation of the existing theory from the experimental results could be, at least partially, because the molten and fully collapsed globules have different chain density distributions in comparison with the coils. On the other hand, the Flory–Huggins interaction parameter χ for PNIPAM in water might be a strong function of temperature and/or polymer concentration. The quantitative theory of polymer chain conformation in a poor solvent remains an interesting problem, which is generally related to some basic problems of semi-dilute and concentrated solutions because the local chain density inside the globule is high even though the overall concentration is very low. Moreover, to our knowledge, no one knows the dynamics of a single chain in the crumpet or the collapsed conformation.

3.2
Folding of Amphiphilic Copolymer Chains

The copolymerization of a few molar percent of water-soluble neutral or ionic comonomer on PNIPAM generally makes it more hydrophilic so that its LCST in water decreases. In the synthesis, comonomers can be inserted into or grafted on the PNIPAM chain backbone in a random or a more segmented fashion by using different chain conformations of PNIPAM in water at different temperatures. It was expected that at temperatures higher than its LCST, hydrophilic comonomers would segregate on the periphery of the collapsed PNIPAM chain backbone, while at lower temperatures, the copolymerization would lead to a more random distribution of comonomers on the PNIPAM chain backbone because both of them are hydrophilic. Therefore, by alternating the reaction temperature, we were able to incorporate hydrophilic comonomers into the PNIPAM chain backbone with different comonomer distributions. In the following discussion, we will start with some simpler cases of noncharged neutral copolymer chains and discuss how the comonomer distribution and composition can affect the folding of individual copolymer chains before moving to charged anionic copolymer chains. This is because long-range electrostatic interaction in water is always troublesome in theory.

3.2.1
Hydrophilically Modified PNIPAM Copolymer Chains

For different applications, water-soluble neutral and ionic comonomers can be incorporated into or attached to the PNIPAM chain backbone to form amphiphilic PNIPAM copolymers via free-radical copolymerization. In this section, we will use the folding of neutral PNIPAM amphiphilic copolymer chains in extremely dilute solutions ($\sim \mu$g/mL) to illustrate a general feature of the folding of hydrophilically modified copolymer chains.

Linear NIPAM-co-VP copolymers: As discussed in the Experimental Section, hydrophilic comonomer, vinyl pyrrolidone (VP), can be purposely copolymerized into PNIPAM at two different temperatures, 30 °C and 60 °C, respectively, below and above the LCST of PNIPAM homopolymer. At each temperature, the copolymers with two different VP/NIPAM ratios (5 and 10 mol %) were prepared. A proper fractionation of resultant copolymers led to narrowly-distributed long NIPAM-co-VP copolymer chains with a similar length and VP/NIPAM ratio, but different comonomer distributions.

Figure 6 shows the typical temperature dependence of both $\langle R_g \rangle$ and $\langle R_h \rangle$ of two copolymers synthesized at two different temperatures. As expected,

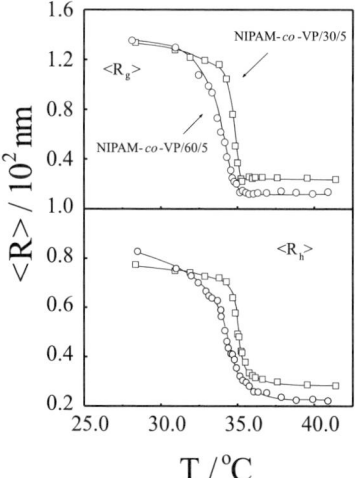

Fig. 6 Temperature dependence of z-average root-mean square radius of gyration ($\langle R_g \rangle$) and average hydrodynamic radius ($\langle R_h \rangle$) of copolymers NIPAM-co-VP/60/5 and NIPAM-co-VP/30/5 in water, where the weight average molar masses are 2.9×10^6 and 4.2×10^6 g/mol, respectively [56]

both $\langle R_g \rangle$ and $\langle R_h \rangle$ decrease sharply during the coil-to-globule transition, revealing the chain collapse at higher temperatures. Note that in each case, the average size of the collapsed chains remains nearly a constant even when the temperature (40 °C) is much higher than the LCST of PNIPAM and water becomes a very poor solvent for PNIPAM, indicating that such formed single-chain globules are very stable. In contrast, previous studies of PNIPAM homopolymer showed that stable single-chain globules could only be observed within a limited temperature range [32]. The formation of such stable single-chain globules can be attributed to the existence of hydrophilic comonomer VP. As expected, the copolymer chains with hydrophilic comonomer VP have a higher transition temperature than PNIPAM homopolymer. The detail values of the coil-to-globule transition temperature and the average hydrodynamic radii in the collapsed state for two pairs of NIPAM-co-VP copolymers with different VP contents can be found elsewhere [56].

It is not surprising to see that the chains with a higher hydrophilic comonomer VP content have a higher transition temperature. However, it is rather interesting to see that for each pair of the copolymers with a similar VP content, the copolymer prepared at 60 °C has a lower transition temperature than its counterpart prepared at 30 °C. In order to check this shift in the transition temperature, we also measured the partial heat capacity (C_p) of these copolymers in solution using a micro-calorimeter. Figure 7 shows that for the two copolymers prepared at 60 °C, the temperatures at which the maximum

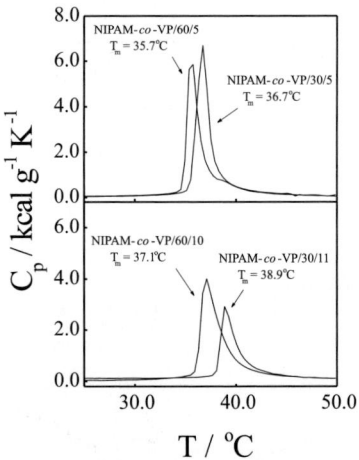

Fig. 7 Temperature dependence of partial heat capacity (C_p) of two pairs of NIPAM-co-VP copolymers in water. The weight average molar masses of NIPAM-co-VP/60/5, NIPAM-co-VP/30/5, NIPAM-co-VP/60/10 and NIPAM-co-VP/30/10 are 2.9×10^6, 4.2×10^6, 5.6×10^6 and 7.9×10^6 g/mol, respectively. The polymer concentration is 10^{-3} g/mL. The temperature was increased with a rate of 1.5 °C/min and pressure was maintained at 180 kPa [56]

heat capacity (T_{max}) occurs are indeed lower. Such a difference between the transition temperatures can only be attributed to different comonomer distributions on the PNIPAM chain backbone because the copolymers in each pair have a similar chain length and composition.

As we mentioned earlier, at lower temperatures, water is a good solvent for PNIPAM and the PNIPAM segments formed during the copolymerization and that exist as a random coil. In this way, NIPAM and VP are copolymerized in a relatively more random fashion to form a statistical copolymer. In contrast, water at 60 °C becomes such a poor solvent that the PNIPAM chain backbone collapses and the hydrophilic VP comonomer can only be incorporated on its periphery, leading to a segregation of VP, or in other words, a segmented comonomer distribution. Therefore, the average length of the PNIPAM segment between two neighboring VP segments is longer in comparison with a statistical copolymer chain with a similar VP/NIPAM ratio, as schematically shown in Fig. 8.

The lower transition temperature also indicates that the folding of the copolymer chains prepared at higher temperatures is much easier, or in a sense, these chains could "memorize" the parent collapsed globular state in which they were formed. As we discussed earlier, the conformational change can be better viewed in terms of the ratio of $\langle R_g \rangle / \langle R_h \rangle$. For a random coil and a uniform nondraining sphere, $\langle R_g \rangle / \langle R_h \rangle \sim 1.5$ and ~ 0.774, respec-

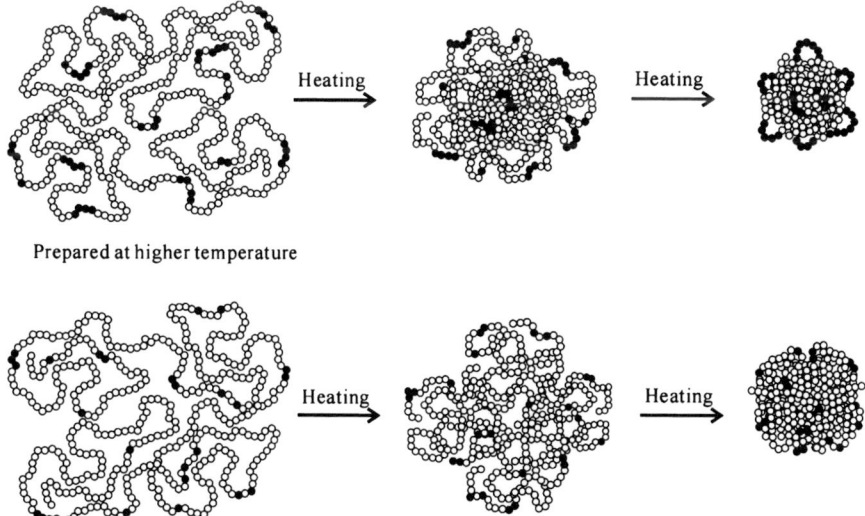

Fig. 8 Schematic of different chain conformations and the coil-to-globule transition of NIPAM-*co*-VP copolymers prepared at two temperatures, respectively, lower and higher than the lower critical solution temperature of PNIPAM homopolymer [56]

tively. In Fig. 9, the decrease of $\langle R_g\rangle/\langle R_h\rangle$ from ~ 1.65 to ~ 0.6–0.8 clearly reveals the coil-to-globule transition of individual copolymer chains. Just like PNIPAM homopolymer in water, before fully collapsing into a uniform globule, the copolymer chains with an extended randomly coiled conformation first collapses into a nonuniform structure with a value of $\langle R_g\rangle/\langle R_h\rangle$ much smaller than 0.774, i.e., the dip of $\langle R_g\rangle/\langle R_h\rangle$ at ~ 35–$36\,°C$. It should be noted that even in the fully collapsed state at higher temperatures, $\langle R_g\rangle/\langle R_h\rangle$ of single-chain globules made of NIPAM-co-VP/60/5 is still smaller than 0.774. This suggests that individual single-copolymer chain globules have a nonuniform chain density distribution inside, as previously observed by Khokhlov et al. [8] in a computer simulation. It is also schematically shown in Fig. 8.

As expected, the collapse of longer PNIPAM segments on the copolymer chain prepared at 60 °C forces the hydrophilic VP segments to stay on the periphery, leading to a "core-shell" structure with a denser PNIPAM core and a swollen VP shell presumably made of small VP loops. On the other hand, the copolymer chains prepared at 30 °C should have a more random comonomer distribution and hydrophilic comonomer VP would not be segregated together to form short VP segments, in other words, the average length of the PNIPAM segments is much shorter. The collapse of these shorter PNIPAM segments at high temperatures inevitably pulls hydrophilic comonomer VP inside, resulting in less compact, but more uniform, globules with a ratio of $\langle R_g\rangle/\langle R_h\rangle$ similar to uniform latex particles. The comparison of the coil-to-globule transition of a pair of such copolymer chains is schematically shown in Fig. 8.

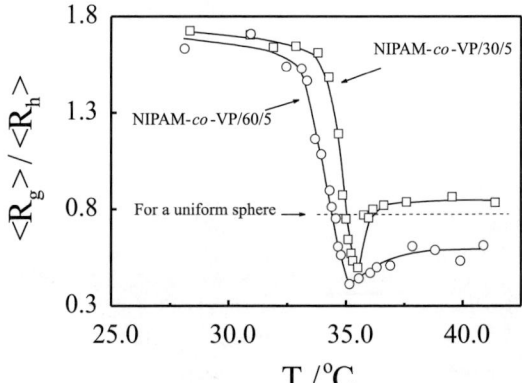

Fig. 9 Temperature dependence of the ratio of average radius of gyration to average hydrodynamic radius ($\langle R_g\rangle/\langle R_h\rangle$) of copolymer NIPAM-co-VP chains prepared at two different temperatures, respectively, lower and higher than the lower critical solution temperature of PNIPAM homopolymer. The weight average molar masses of NIPAM-co-VP/60/5 and NIPAM-co-VP/30/5 are 2.9×10^6 and 4.2×10^6 g/mol, respectively [56]

The structural difference between the collapsed globules made of the chains prepared at different temperatures can also be evidenced in their size and density. Note that in Fig. 6 in the fully collapsed state at $\sim 40\,^\circ$C, NIPAM-co-VP/60/5 globules have a smaller size than NIPAM-co-VP/30/5 globules even though they have a similar $\langle R_g \rangle$ and $\langle R_h \rangle$ in the random coiled state at lower temperatures. Therefore, the former have higher average chain densities ($\langle \rho \rangle_{\text{globule}}$). The values of $\langle \rho \rangle_{\text{globule}}$ for NIPAM-co-VP/60/5 and NIPAM-co-VP/30/5 globules are 9.6×10^{-2} and 7.0×10^{-2} g/cm^3, respectively. Such a difference in $\langle \rho \rangle_{\text{globule}}$ also indirectly reflects that the copolymer chains prepared at higher temperatures can "memorize" their parental collapsed globule state and fold back to a more compact structure, confirming the computer simulation and prediction [8, 9]. In comparison, $\langle \rho \rangle_{\text{globule}}$ of single-chain globules made of amphiphilic copolymer chains is 3–4 times lower than that made of homopolymer chains [32]. This is reasonable because the incorporation of a few percent of hydrophilic comonomer VP into PNIPAM swells the globules and retards the chain packing.

Grafted PNIPAM-g-PEO copolymers: There has been considerable interest in the study of grafted amphiphilic copolymers in selected solvents because they can form stable aggregates with a core-shell structure in solution [57–62]. For example, amphiphilic copolymer chains consisting of the hydrophobic backbone and hydrophilic branches can form stable core-shell colloidal particles in water (a selective solvent) because the hydrophobic chain backbone tends to aggregate to form a hydrophobic core, while the hydrophilic branches grafted on them are forced to stay on the periphery to form a hydrophilic corona (shell) [63–66]. Core-shell particles formed in this way are sterically stabilized and have an average size normally in the range 10–100 nm, depending on the formation condition and the copolymer structure. This has provided a new method to prepare stable surfactant-free polymeric nanoparticles.

By extending the study of the folding of linear PNIPAM homopolymer chains in water, Qiu et al. [67] prepared a series of amphiphilic PNIPAM-g-PEO copolymer chains by grafting different amounts of short PEO chains on the PNIPAM chain backbone as described in the Experimental Section. In cold water, both the PNIPAM chain backbone and the grafted PEO branches are hydrophilic so that PNIPAM-g-PEO copolymer is soluble in water as individual coiled chains. At temperatures higher than its LCST, the PNIPAM chain backbone becomes hydrophobic, but PEO branches remain hydrophilic. Therefore, a change in the solution temperature about 1–2 degrees can greatly alternate the degree of amphiphilicity of this kind of copolymer to induce the chain folding and to form stable single-chain globules with a hydrophobic PNIPAM core and a soluble hydrophilic PEO shell. Such a core-shell nanostructure can be switched on and off simply by a very small temperature variation.

Figure 10 shows a schematic of the formation of a single chain core-shell nanostructure. It is known that the short grafted PEO chains have an average

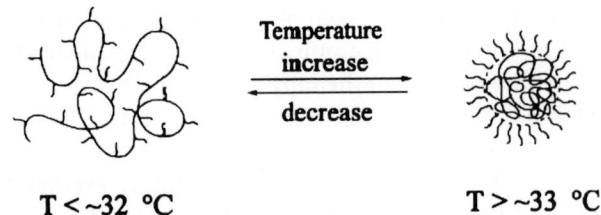

Fig. 10 Schematic of formation of a single chain core-shell nanostructure through the coil-to-globule transition of the PNIPAM-g-PEO copolymer chain backbone [67]

hydrodynamic radius of ~ 3 nm. For the PNIPAM-g-PEO copolymer chains ($M_w = 7.29 \times 10^6$ g/mol) are grafted, on average, ~ 70 short PEO chains per PNIPAM chain backbone, its $\langle R_g \rangle$ changes from 155 nm to 21 nm and the ratio of $\langle R_g \rangle / \langle R_h \rangle$ decreases from 1.5 to 0.74 during the chain folding. Assuming that these short PEO chains grafted on the periphery have a similar chain conformation as those free in water, one can estimate the shell thickness (L_{shell}) and the radius of the core to be ~ 6 nm and ~ 23 nm, respectively. Further, using the molar mass of the PNIPAM chain backbone, one can also estimate the average chain density of the core to be ~ 0.25 g/cm^3, significantly lower than the average chain density (~ 1 g/cm^3) of conventional polymeric latex particles. In other words, the PNIPAM core still contains $\sim 75\%$ of water even in its fully collapsed state. Therefore, this core-shell nanostructure might be used as a carrier to load a large amount of drugs or catalysts because of its large free volume. Another advantage of using it as a carrier is that such a single-chain core-shell nanostructure can be quickly switched on and off by a small temperature variation of only 1–2 °C.

Figure 11 shows the releasing and encapsulation of pyrene (an imitation of drugs/catalysts) by these PNIPAM-g-PEO copolymer chains in water in

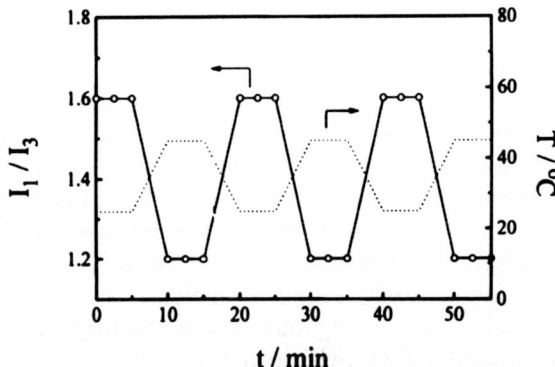

Fig. 11 Temperature dependence of fluorescence intensity ratio (I_1/I_3) of pyrene as an imitated drug/catalyst in deionized water in the presence of PNIPAM-g-PEO chains, where the concentration of pyrene is 2×10^{-7} M [67]

terms of its temperature-dependent fluorescence intensity ratio (I_1/I_3), where the pyrene concentration is $\sim 2 \times 10^{-7}$ M. It has been known that I_1/I_3, the highest energy vibrational band I_1 (373 nm) to the third highest energy vibrational band I_3 (385 nm), is sensitive to micro-environmental polarity [17]; namely, in pure water, $I_1/I_3 \sim 1.8$, while in a hydrophobic domain, I_1/I_3 could be as low as 1.2. The change of I_1/I_3 between ~ 1.6 and ~ 1.2 clearly shows that the folding and unfolding of these copolymer chains can release and encapsulate pyrene when the temperature is alternated between 25 to 45 °C. It should be noted that the change of I_1/I_3 would be a much sharper step function if one could change the temperature instantly.

Recently, Tenhu et al. [16, 58] studied the influence of the PEO content on the aggregation of such copolymer chains and found that the chains prepared at different temperatures had different values of LCST. As expected, the PEO content has a great effect on the phase transition. Following their method, Chen et al. [59] synthesized four PNIPAM-g-PEO copolymers with different PEO contents, but a similar chain length, at 45 °C. Their objective was to study the effect of the PEO content on the folding and unfolding of such amphiphilic copolymer chains in dilute solutions by using a combination of static and dynamic LLS as well as US-DSC. Surprisingly, they found that the PNIPAM-g-PEO copolymer chains with a higher PEO content underwent two transitions. One sharp transition at ~ 33 °C is related to the collapse of the PNIPAM chain backbone. Another broad transition in the range 35–45 °C disappeared when the PEO content was lower, which was attributed to the stretching and collapsing of short PEO chains grafted on the PNIPAM chain backbone.

Figures 12 and 13 summarize the temperature dependence of $\langle R_g \rangle$ and $\langle R_h \rangle$ of PNIPAM-g-PEO copolymer chains in water during one heating-and-

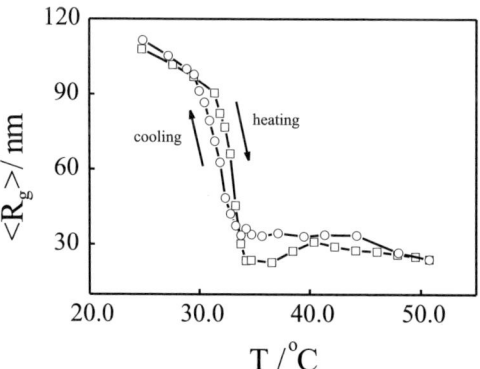

Fig. 12 Temperature dependence of z-average root-mean-square radius of gyration ($\langle R_g \rangle$) of copolymer PNIPAM-g-PEO chains in water during heating and cooling, where the weight-average molar mass (M_w) is 7.2×10^6 g/mol, the molar number ratio of NIPAM monomers to PEO macromonomers is 111, and on average, there are 392 short PEO chains grafted on each PNIPAM chain [69]

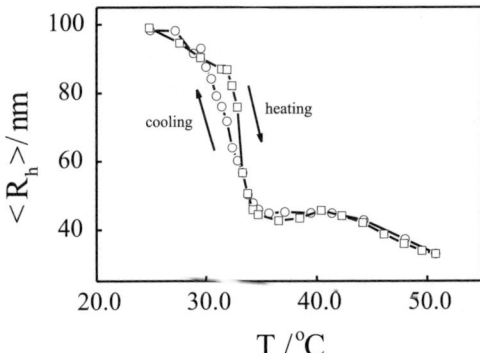

Fig. 13 Temperature dependence of average hydrodynamic radius ($\langle R_h \rangle$) of copolymer PNIPAM-g-PEO chains in water during heating and cooling, where the weight-average molar mass (M_w) is 7.2×10^6 g/mol, the molar number ratio of NIPAM monomers to PEO macromonomers is 111, and on average, there are 392 short PEO chains grafted on each PNIPAM chain [69]

cooling cycle. In the heating process, both $\langle R_g \rangle$ and $\langle R_h \rangle$ drop sharply in the range 31–33 °C, reflecting the expected coil-to-globule transition of the PNIPAM chain backbone. Further increase of the temperature first results in slight increases of both $\langle R_g \rangle$ and $\langle R_h \rangle$ and is then followed by a slow decrease over a broad temperature range. In comparison with PNIPAM homopolymer chains, such additional changes of both $\langle R_g \rangle$ and $\langle R_h \rangle$ must be related to short PEO chains grafted on the chain backbone. To explain the results (Figs. 12 and 13) we have to look at what has happened during the shrinking of the chain backbone.

As we discussed earlier, at lower temperatures, each copolymer chain in water exists as an extended random coil. The heating makes the PNIPAM chain backbone insoluble in water so that it undergoes the coil-to-globule transition. In this process, short hydrophilic PEO chains are forced to stay on the periphery of the PNIPAM core to form a core-shell nanostructure, such as schematically shown in Fig. 14a. The estimated average surface area per PEO chain at ~ 33 °C is ~ 10 nm². Further shrinking of the PNIPAM core increases the chain density on the periphery. As expected, the repulsion among different PEO chains forces them to stretch, as schematically shown in Fig. 14b. During this stage, the stretching of short PEO chains in the shell and the collapse of the PNIPAM chain backbone in the core have opposite effects on the measured $\langle R_g \rangle$ and $\langle R_h \rangle$ when the PEO content is high. The slight increases of both $\langle R_g \rangle$ and $\langle R_h \rangle$ in the range 35–40 °C indicates that the stretching of short PEO chains dominates slightly. As for the slow decreases of both $\langle R_g \rangle$ and $\langle R_h \rangle$ in the high temperature range ($T > 40$ °C), this could be explained by two possible scenarios as follows.

One is that the shrinking of a long PNIPAM chain backbone in the core overrides the stretching of short PEO chains in the shell. The other is the

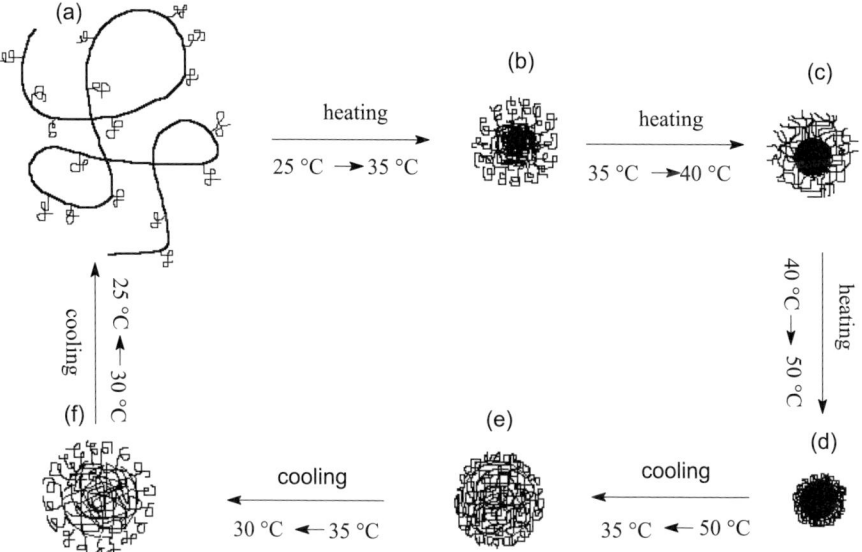

Fig. 14 Schematic of the coil-to-globule-to-coil transition of a copolymer PNIPAM-g-PEO chain with a higher PEO content during a heating-and-cooling cycle [69]

n-clustering-induced collapse of short PEO chains because we know that polymer chains in bulk or in a very concentrated solution adopt a random coil conformation, as with the Θ-temperature [70, 71]. In addition, the solvent quality of water for PEO decreases as the solution temperature increases. In order to differentiate these two scenarios, the temperature dependence of $\langle R_g \rangle / \langle R_h \rangle$ is plotted in Fig. 15 to reflect the chain density distribution. The fact that $\langle R_g \rangle / \langle R_h \rangle \sim 1.1$ at lower temperatures, instead of ~ 1.5 (an expected value for linear coil chains), reflects its branching structure because short PEO chains have a length similar to the PNIPAM segments between two neighboring grafting points. The decrease of $\langle R_g \rangle / \langle R_h \rangle$ from ~ 1.0 to ~ 0.5 clearly reveals a change of the chain conformation.

As discussed before, the lower value of $\langle R_g \rangle / \langle R_h \rangle$ confirms that the collapsed chain has a core-shell nanostructure and the collapsed PNIPAM core is denser than the swollen PEO shell. In comparison with a uniform sphere with the same $\langle R_h \rangle$, the denser core leads to a smaller $\langle R_g \rangle$. The increase of $\langle R_g \rangle / \langle R_h \rangle$ in the range 35–40 °C reflects the core-shell structure and becomes more uniform in density because the PEO chains in the shell are forced to overlap each other when the PNIPAM core continues to shrink (Fig. 14c). In Fig. 15, it is the further increase of $\langle R_g \rangle / \langle R_h \rangle$ in the range 40–50 °C that differentiates the two scenarios. Namely, if the first one was correct, $\langle R_g \rangle / \langle R_h \rangle$ should decrease because the core becomes denser. On the other hand, in the second scenario, the collapse of the PEO chains increases the chain density of the shell so that the core-shell nanostructure becomes more uniform, as

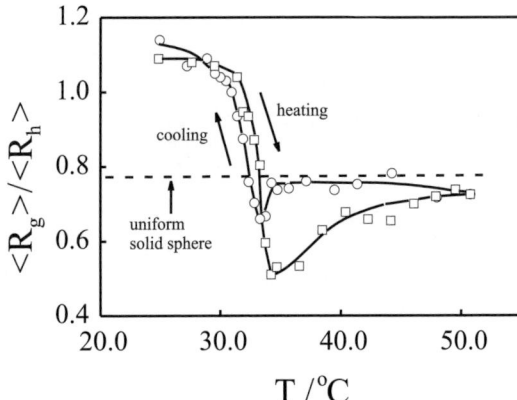

Fig. 15 Temperature dependence of ratio of average radius of gyration ($\langle R_g \rangle$) to average hydrodynamic radius ($\langle R_h \rangle$) of copolymer PNIPAM-*g*-PEO chains in water during heating and cooling. The weight-average molar mass (M_w) is 7.2×10^6 g/mol, the molar number ratio of NIPAM monomers to PEO macromonomers is 111, and there are 392 short PEO chains grafted on each PNIPAM chain [69]

schematically shown in Fig. 14d. This explains why $\langle R_g \rangle / \langle R_h \rangle$ gradually approaches 0.774 as the solution temperature increases.

A combination of Figs. 12–15 shows that the reversible globule-to-coil transition does not follow the coil-to-globule path. There exists a hysteresis between the heating and cooling processes. In the cooling process, both $\langle R_g \rangle$ and $\langle R_h \rangle$ have no peak in the range 35–50 °C and $\langle R_g \rangle / \langle R_h \rangle$ remains nearly a constant, indicating a uniform swelling, as schematically shown in Fig. 5e, very different from the shrinking process. The small dip of $\langle R_g \rangle / \langle R_h \rangle$ at $\sim$ 33 °C indicates that the collapsed PEO chains in the shell are finally swollen back into individual coils on the periphery and the PNIPAM chain backbone in the core has not reached its fully swollen state, as schematically shown in Fig. 14f.

On the other hand, when the PEO content is low, the temperature dependence of $\langle R_g \rangle$, $\langle R_h \rangle$ and $\langle R_g \rangle / \langle R_h \rangle$ has only one transition related to the coil-to-globule transition of the PNIPAM backbone. The transition temperature is close to the LCST ($\sim$ 32 °C) of PNIPAM homopolymer. Moreover, the PEO content has nearly no effect on such a transition. This is expected because when the PEO content is low there is no strong interaction among the PEO chains in the shell. Such a single transition temperature indirectly supports our discussion of the second transition; namely, it is related to the repulsion-induced stretching followed by the clustering-induced collapse of short PEO chains in the shell because the chain density increases as the PNIPAM core shrinks. It was also found that at higher temperatures, the copolymer chains with a low PEO content have a smaller $\langle R_g \rangle$ even though they are longer. The difference further reflects the repulsion-induced stretching of short PEO chains in the shell when the PEO content is higher.

The n-clustering-induced stretch and collapse of short PEO chains can be better viewed when they are grafted on a thermally sensitive PNIPAM spherical microgel [34, 70]. In the temperature range 25–35 °C, the microgel can shrink ~ 3 times in its diameter, i.e., its surface area can decrease ~ 10 times, providing a convenient way to continuously increase the grafting density because the average number of the PEO chains grafted on each microgel is fixed. The microgels were prepared by dispersion polymerization in aqueous solution at 70 °C, similar to the preparation of amphiphilic PNIPAM-g-PEO copolymer chains, but with a proper amount of crosslinking agent, N,N'-methylenebis(acryl-amide). The core was collapsed and cross-linked at the reaction temperature. The resultant microgels were successively purified by ultra-centrifugation at 40 °C to remove all remaining small molecules. The average grafting density determined by NMR was ~ 370 PEO chains per microgel, or in other words, each PEO chain occupies ~ 320 nm^2 at 25 °C.

Figure 16 shows that in the heating-and-cooling cycle, the shrinking-and-swelling of the microgel with short PEO chains grafted on its periphery is completely reversible. It indicates that there was no hysteresis in the process. This is apparently inconsistent with some previous predictions for the grafted PEO chains [71] and also different from the hysteresis observed in the same process for ultra-long PNIPAM homopolymer chains and PNIPAM-g-PEO chains [37]. Note that intrachain association of PEO unlikely exists in this case. On the other hand, the collapse of a long chain has a much higher entropy penalty than the shrinking of a microgel. The fact that no hysteresis was observed for the grafted PEO chains can also be attributed to the much shorter chain length ($\sim 10^4$ g/mol).

As the temperature increases, the PNIPAM core shrinks and the grafting density increases. It is expected that the repulsion would force the PEO-chains on its surface to stretch, i.e., the thickness of the PEO shell (h_{brush})

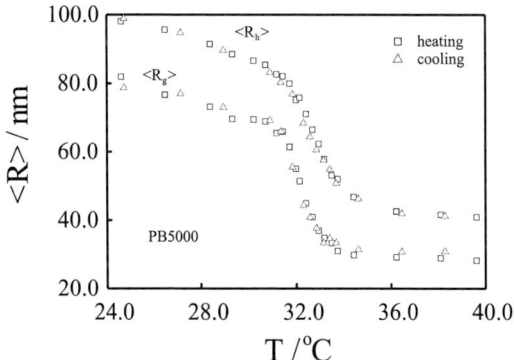

Fig. 16 Temperature dependence of average hydrodynamic radius ($\langle R_h \rangle$) and average radius of gyration ($\langle R_g \rangle$) of PNIPAM microgels grafted with linear PEO chains in the heating-and-cooling cycle, where the dispersion concentration is $\sim 1.0 \times 10^{-5}$ g/mL [70]

increases, as the grafting density increases. In this way, the PNIPAM core becomes denser, but the PEO shell becomes thicker. Note that the core has ~ 90% of mass so that the shrinking of the core has a more profound effect on $\langle R_g \rangle$, while the stretching of the grafted PEO chains in the shell slows down the decrease of $\langle R_h \rangle$, according to their own definitions. This is why $\langle R_g \rangle / \langle R_h \rangle$ decreases in the range 23–33 °C [34]. When the temperature is higher than ~ 33 °C, $\langle R_g \rangle / \langle R_h \rangle$ starts to increase and finally approaches 0.71 at ~40 °C [71]. As shown in Fig. 16, when $T > \sim 37$ °C, the decrease of $\langle R_g \rangle$ nearly stops, but $\langle R_h \rangle$ still decreases, reflecting that the increase of $\langle R_g \rangle / \langle R_h \rangle$ is attributed to the collapse of the PEO shell. This can be better viewed from the change of the PEO shell thickness.

Figure 17 shows the grafting density dependence of the average thickness of the PEO shell ($\langle h \rangle_{\text{brush}}$), i.e., the average height of the grafted PEO chains, where $\langle h \rangle_{\text{brush}}$ was obtained by two completely different methods. One is from the difference between the average radii of the microgels with and without the grafted PEO chains; namely, $\langle h \rangle_{\text{brush}} = \langle R_h \rangle_{\text{microgel+PEO}} - \langle R_h \rangle_{\text{microgel}}$. The other is indirectly from the ratio of $\langle R_g \rangle / \langle R_h \rangle$, involving a combination of static and dynamic LLS results. The principle of the second method is outlined as follows. The microgel grafted with the PEO chains can be viewed as a core-shell particle, in which the masses of the core (M_c) and the shell (M_s) are

$$M_c = \frac{4}{3}\pi \rho_c R_c^3 \quad \text{and} \quad M_s = \frac{4}{3}\pi \rho_s \left(R^3 - R_c^3\right) \tag{5}$$

where R_c and R are the radii of the PNIPAM core and the entire microgel; ρ_c and ρ_s are the densities of the core and the shell, respectively. Note that for the first approximation, the shell and the core are assumed to be uniform.

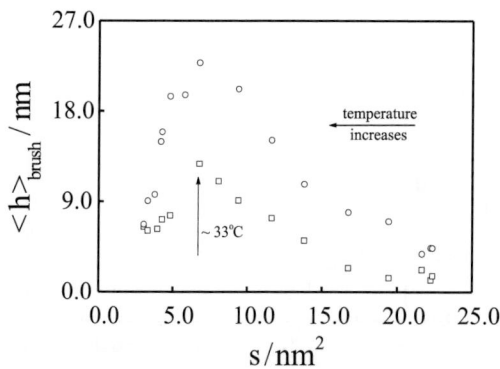

Fig. 17 Dependence of the average PEO brush height ($\langle h \rangle_{\text{brush}}$) on the surface area per grafted chain (s), where $\langle h \rangle_{\text{brush}}$ was calculated by two different methods. The squares represent $\langle h \rangle_{\text{brush}}$ from the difference between the average radii of the PNIPAM microgels with and without the grafted PEO chains; and the circles from the ratio of $\langle R_g \rangle / \langle R_h \rangle$ [70]

In reality, they are not uniform. However, this assumption will not affect our discussion. Also note that no mixing between the grafted PEO layer and the PNIPAM core is assumed. At higher temperatures, the hydrophobic PNIPAM core should not mix with the hydrophilic PEO shell. Even at lower temperatures, it would be more favorable for the PEO chains to stay on the periphery than to penetrate inside the PNIPAM gel network. Substituting Eq. 5 into the definition of R_g leads to

$$R_g^2 = \frac{\int_v \rho(r)r^2 dv}{\int_v \rho(r)dv} = \frac{\int_0^{R_c} 4\pi\rho_c r^4 dr + \int_{R_c}^R 4\pi\rho_s r^4 dr}{\int_0^{R_c} 4\pi\rho_c r^2 dr + \int_{R_c}^R 4\pi\rho_s r^2 dr}$$

$$= \frac{3\left[M_c R_c^2 R^3 - (M_c + M_s)R_c^5 + M_s R^5\right]}{5(M_c + M_s)(R^3 - R_c^3)}. \tag{6}$$

Letting the mass ratio $M_c/M_s = A$ and the radius ratio $R_c/R = x$, one can rewrite Eq. 2 as:

$$R_g^2 = \frac{3R^2\left[Ax^2 - (1+A)x^5 + 1\right]}{5\left(1 + A/(1-x^3)\right)}. \tag{7}$$

Replacing R with R_h, Eq. 7 can be re-arranged as

$$\frac{R_g}{R_h} = \left\{\frac{3\left[Ax^2 - (1+A)x^5 + 1\right]}{5(1+A)(1-x^3)}\right\}^{1/2}, \tag{8}$$

where A is a constant for a given PNIPAM microgel grafted with a fixed number of PEO chains. From each measured R_g/R_h in Fig. 16, one can numerically find a corresponding x, and then R_c. In this way, we can determine $\langle h \rangle_{\text{brush}}$ from $R - R_c$.

Figure 17 shows that before the temperature reaches its lower critical solution temperature (LCST, $\sim 33\,°C$), $\langle h \rangle_{\text{brush}}$ increases while the surface area per PEO chain (s) decreases (i.e., the increase of the temperature). The absolute values of $\langle h \rangle_{\text{brush}}$ obtained from the two different methods do not quantitatively agree with each other due to experimental uncertainties. Such stretching of the PEO chains is expected because they are forced to approach each other as s decreases. It should be stated that the shell thickness is much smaller than the average radius of the microgels (~ 100 nm) so that the grafted PEO chains can be treated as a quasi-planar brush even if the microgel surface is curved. It is helpful to note that the PEO shell at 23 °C is slightly thicker than the hydrodynamic diameter (~ 4 nm) of the PEO macromonomers free in water [72], indicating that the grafted chains are slightly elongated even at the room temperature. One of the possible explanations of the unexpected collapse of the PEO chains in the temperature range of $T > \sim 33\,°C$ might be related to the formation of an equilibrium between one dilute phase made of stretched chains and one dense phase made of collapsed chains [70, 73].

In general, the hydration has been attributed to the dissolution of PEO in water [74]. It has been shown that the interchain interaction among individual PEO chains is so strong that it is rather difficult to completely dissolve bulk PEO into individual chains in pure water [75, 76]. Nevertheless, water is peculiar due to its structured conformation and could be viewed as a polymeric solvent [77]. A theoretical study of PEO in aqueous solution showed that the pressure could also induce the collapse of the PEO chains [78]. It is known that polymer chains in bulk adopt a random coil conformation. It is our opinion that as the grafting density increases, the grafted chains are pushed together, the PEO chains are gradually dehydrated and most of the water molecules are gradually excluded. Therefore, as the grafted PEO layer becomes "drier" and "drier", it approaches the bulk state, the stretched PEO chains have to collapse at some point. It is similar to the report that in a bad or polymeric solvent, polymer brushes would gradually collapse as more solvent molecules leave the grafted layer and the configuration entropy of the solvent becomes more important [79].

Figure 18 shows that during the stretching, i.e., before reaching $\sim 33\,°C$, the brush height ($\langle h \rangle_{\text{brush}}$) can be scaled to the grafting density (σ) as $\langle h \rangle_{\text{brush}} \propto N\sigma^{1.0 \pm 0.2}$, where N is the degree of polymerization. The theoretical studies and experimental results in the past showed $\langle h \rangle_{\text{brush}} \propto N\sigma^{\beta}$ with $\beta = 1/3$ for polymer brush in a good solvent [80, 81]. Note that $\sigma^{-1} = s$, which is the average surface area occupied per grafted chain and $\langle h \rangle_{\text{brush}} \sigma^{-1}$ represents the average hydrodynamic volume per grafted PEO chain. Therefore, the scaling of $\langle h \rangle_{\text{brush}} \propto \sigma^{1.0 \pm 0.2}$ suggests that the average hydrodynamic volume per grafted PEO chain is close to a constant, or in other words, the

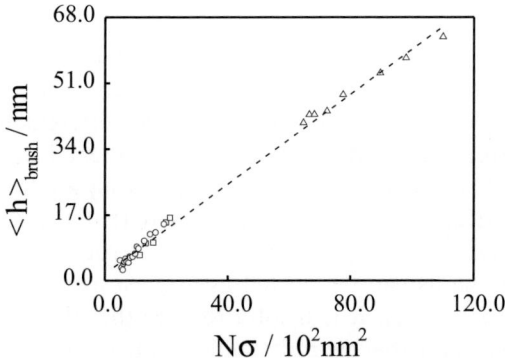

Fig. 18 Grafting density (σ) dependence of average PEO brush height ($\langle h \rangle_{\text{brush}}$) during the PEO chain stretching. The *line* represents a least square fitting of $\langle h \rangle_{\text{brush}} \propto N\sigma^{1.0 \pm 0.2}$, where N is the degree of polymerization of the grafted chain [70]

average hydrodynamic volume per grafted PEO chain is incompressible during the chain stretching. Further studies of other types of grafted polymer chains are needed to differentiate whether this incompressibility is only related to PEO, or in general, to high grafting chain density.

In order to justify the discussion about the second PEO-related transition in Figs. 12 and 13, the partial heat capacity (C_p) of four PNIPAM-g-PEO copolymers with different PEO contents in aqueous solutions was measured using a micro-calorimeter. Figure 19 shows that in the heating process, the peaks located at $\sim 33\,^{\circ}$C are similar and independent of the PEO content, which can be attributed to the PNIPAM chain backbone. The DSC results agree well with the LLS ones. The second broad peak in the range 40–50 $^{\circ}$C gradually disappears when the PEO content becomes lower. Noted that in the US-DSC measurement, both the intra-chain contraction and the inter-chain association are involved because a much higher concentration (10^{-3} g/mL) had to be used to give a sufficiently high signal-to-noise ratio. By investigating both PNIPAM hydrogels and PNIPAM linear chains in water, Laszlo et al. [82] and Ding et al. [83] found two exothermal peaks in the temperature range ~ 30–$33\,^{\circ}$C in the cooling process when the scanning rate was low and only one endothermic peak in the heating at any scanning rate. They attributed this additional peak to the kinetic effect. In Fig. 19, two peaks, respectively, located at ~ 32 and $\sim 50\,^{\circ}$C appear in both the heating and cooling.

Around the first transition peak in Fig. 19, the hysteresis is obvious. Such a hysteresis has been attributed to the formation of some additional intra-chain hydrogen bonding in the collapsed state [34]. The second broad peak reveals that the stretching of the grafted PEO chains and further collapse of the PNIPAM chains in the core occur over a wider temperature range. Figure 19 further reveals that the heating and cooling involve a certain amount of endothermic and exothermic heat, respectively. One can define the area

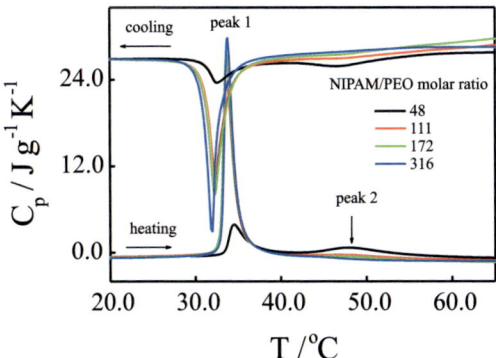

Fig. 19 Temperature dependence of partial heat capacity (C_p) of copolymer chains with different PEO contents in water, where the heating rate is 1.0 $^{\circ}$C/min [69]

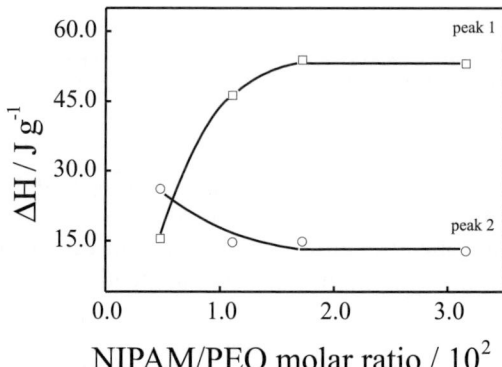

Fig. 20 PEO content dependence of endothermic heats, respectively, for both the transitions in Fig. 19 [69]

under each peak as the total endothermic or exothermic heat (the change of enthalpy ΔH) for each transition.

Figure 20 shows the PEO content dependence of the enthalpy change (ΔH) for the two transitions in Fig. 19. For the copolymer chains with a higher PEO content, the average length of the PNIPAM segment between two grafted neighboring PEO chains is short. This is why the peak at $\sim 33\,°\mathrm{C}$ related to the PNIPAM segments becomes smaller. On the other hand, when the PEO content decreases, the PNIPAM segments become longer and the PEO shell becomes less crowded so that the peak related to the stretching and collapsing of the grafted PEO chains disappear during the heating process.

3.2.2
Hydrophobically Modified PNIPAM Copolymer Chains

The presence of only a few molar percent of hydrophobic monomer units (stickers) in a hydrophilic polymer chain backbone can sufficiently trigger chain association in water or in a selective solvent to form large insoluble clusters or even a gel, depending on solvency and polymer concentration. The associating copolymers are widely used in industry as viscosity modifiers, colloidal stabilizers and surface-active agents [84–86]. Such copolymer chains can adopt a more complicated conformation in dilute solutions than a corresponding homopolymer. It has been predicted and observed that in a solvent selectively poor for the stickers, the chain can self-fold into a multi-flower nanostructure; namely, a number of neighboring stickers undergo intrachain association to form a string of micelle-like flowers along the chain backbone [87–90]. It has also been suggested that the self-folding could lead to a single flower-like core-shell nanostructure with all stickers condensed in the center and short hydrophilic chain segments between stickers swollen like a flower petal (chain loop) [86–88]. It is expected that such a chain folding may provide a simple model for the study of more complicated problems, such as protein folding and DNA packing.

However, due to experimental difficulties, especially in the sample preparation, few studies have been reported clarifying this point. In most cases, interchain association often occurs with intrachain self-folding, which spoils the study and leads to multi-chain micelles coexisting with single-chain micelles [91, 92]. To our knowledge, only Kikuchi and Nose [93] experimentally studied intrachain micelle formation. They used polystyrene-g-poly(methyl methacrylate) to observe the formation of thermodynamically stable single-chain multi-flower structures and revealed that the resultant structure was rigid and on average consisted of ca. five flower-like micelles on each chain, resembling a string of closely packed pearls. Experimentally, the formation of a single flower-like single-chain nanostructure was only reported two years ago by Zhang et al. [94] after successfully preparing long and narrowly distributed PNIPAM-seg-St copolymer chains ($M_w = 1.33 \times 10^7$ g/mol and $M_w/M_n < 1.1$) by following the micelle polymerization method developed by Candau et al. [95], and Dowling et al. [96]. On average, each chain had ~ 230 hydrophobic St stickers and each St sticker had ~ 20 styrene monomer units. Note that it is extremely difficult, if not impossible, to characterize the structure (comonomer distribution) of such prepared chains. This still remains a challenging problem in polymer characterization, especially for protein-like amphiphilic chains.

Recently, Cui et al. [97] used single-molecule force spectroscopy (SMFS) developed on the basis of atomic force microscopy (AFM) to measure the length distribution of the PNIPAM segment between two neighboring hydrophobic St stickers. They adsorbed long PNIPAM-seg-St copolymer chains on a flat PS surface, as schematically shown in Fig. 21. The flat PS substrate was prepared from a sheet extruder, which was sufficiently smooth for SMFS. The PS segments are so short that the surface can be considered very smooth; i.e., each PS segment was laid flat on the substrate. Before the adsorption, the PS substrate was thoroughly rinsed with ethanol (99.5%) and purified water (> 18 MΩ cm) and then confirmed as a blank sample by SMFS. No obvious force signal was detected during over 1000 cycles of AFM tip's approach and retraction. 0.2 mL PNIPAM-seg-PS aqueous solution was then deposited on the PS substrate and left for approximately 12 h to form a thin layer. Afterward, the sample was rinsed with purified water for 1 min to remove loosely adsorbed PNIPAM-seg-PS before being measured.

A homemade SMFS with a silicon nitride cantilever (Park, Sunnyvale, CA) was used. Each tip was calibrated by using a standard sample. The spring constants of these cantilevers were in the range 0.010–0.012 N/m. By moving the piezo tube, one could bring the sample into contact with the AFM tip so that some polymer chains were physically adsorbed onto the tip, resulting in a number of "bridges". As the distance between the tip and the substrate increased, the chains were stretched and the elastic force deflected

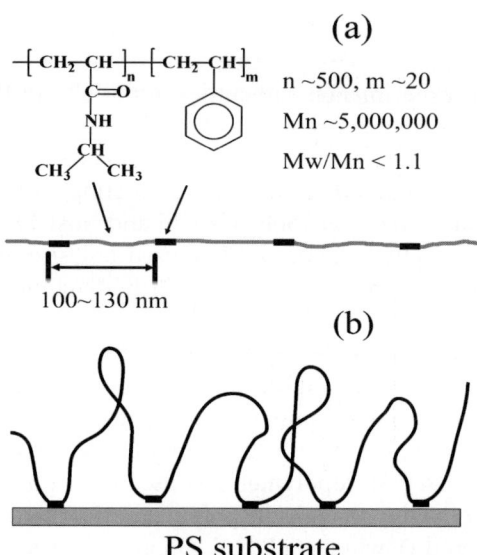

Fig. 21 a Schematic of a linear segmented poly(N-isopropylacrylamide-seg-styrene) (PNIPAM-seg-St) coplymer chain prepared by micelle copolymerization, in which short PS segments are relatively evenly distributed on the chain backbone. **b** Schematic of the adsorption of a linear PNIPAM-seg-PS chain on a hydrophobic PS substrate in water [97]

the cantilever. A recorded deflection-piezo path curve was converted into a force extension curve. It has been shown that the adhesion force between the tip and the adsorbed chain can be up to a few nano-Newtons in magnitude. Using SMFS, the weak interaction force between the copolymer chain (sticker) and the substrate can be measured. The stretching velocity used by Cui et al. [97] was in the range 520–4600 nm/s. Prior to the force measurement, a drop of purified water, acting as a buffer, was injected between the substrate and the cantilever holder, whereupon both the substrate and the cantilever were immersed in water. The force measurements were performed at 22 °C at which long PNIPAM segments are hydrophilic and soluble in water with a random coil conformation. The details of instrumentation can be found elsewhere [98, 99].

As expected, the adsorption of two insoluble short hydrophobic PS segments onto the PS substrate resulted in many PNIPAM "loops". Figure 22a shows that each force curve exhibits a saw-tooth pattern. To find such a pattern and how it is related to the chain structure, Cui et al. [97] analyzed the distribution of the distance between each two adjacent peaks (teeth) in the force curve, as shown in Fig. 22b. The Gaussian fitting of the histogram led to an average distance of $\sim$ 114 nm. This value is very close to the average length of the "repeat unit" (one long PNIPAM segment plus one short PS segment) in the copolymer chain on the basis of the synthesis condition

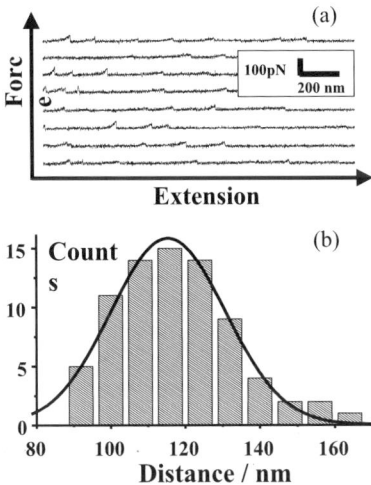

Fig. 22 a Measured force curves of linear segmented poly(N-isopropylacrylamide-*seg*-styrene) (PNIPAM-*seg*-St) copolymer chains adsorbed on a hydrophobic PS substrate in water. **b** Statistics of the distance between two adjacent peaks in the measured force curves [97]

and NMR characterization. It confirms that styrene comonomer molecules are indeed uniformly distributed on the PNIPAM chain backbone as short segments.

In an LLS study, Zhang et al. [94] dissolved this well-characterized PNIPAM-*seg*-St copolymer in deionized water for 10 days to ensure complete dissolution. The final concentration used was $7.2\,\mathrm{E}-7$ g/mL and it was clarified with a 0.5 µm Millipore Millex-LCR filter to remove dust. Note that in their experiment, the scattering volume ($\sim 10\,\mu$L) still contained 10^5–10^6 copolymer chains so that the number of density fluctuations was not a problem even in such a dilute solution. Their original objective was to determine whether such a copolymer chain could self-fold into the predicted single-flower-like core-shell nanostructure.

Figure 23 shows that the slope of $KC/R_{vv}(q)$ vs. q^2 sharply decreases as the temperature increases. It is known in LLS that the slope of each line is related to $\langle R_g \rangle$ of polymer chains. The decrease of $\langle R_g \rangle$ indicates shrinking of the chains. Such a shrinking can be viewed directly in terms of the shift of the hydrodynamic radius distribution $f(R_h)$ from ~ 124 nm to ~ 31 nm, as shown in the inset of Fig. 23, where $f(R_h)$ was calculated from the digital time correlation function measured in dynamic LLS. On the other hand, the extrapolation to $q \to 0$ leads to M_w since the copolymer concentration is extremely low. The same intercept clearly reveals no change in M_w, i.e., no interchain association. Therefore, the shrinking of the chains with increasing temperature is purely a single chain (intrachain) process. The plot of $qR_{vv}(q)/KC$ versus $q\langle R_g \rangle$ reveals the increase of the chain segment density at higher temperatures.

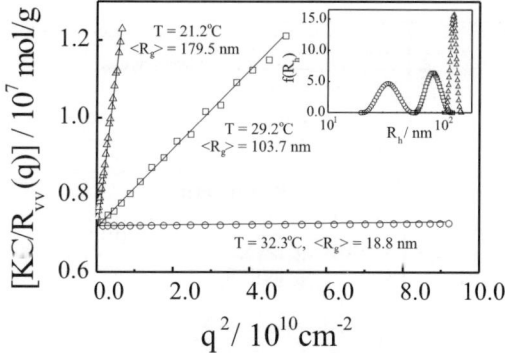

Fig. 23 Angular dependence of Rayleigh ratio ($R_{vv}(q)$) of segmented PNIPAM-*seg*-St copolymer chains in water measured from static LLS, where K is a constant, q is the scattering vector and polymer concentration (C) was 7.2×10^{-7} g/mL. The *inset* shows the temperature dependence of the hydrodynamic radius distribution $f(R_h)$ determined from dynamic LLS [94]

Figure 24 shows that both $\langle R_g \rangle$ and $\langle R_h \rangle$ decrease as the temperature increases. Each data point was obtained only after the solution had reached thermodynamically equilibrium and the measured value was stable. Note that in each curve there exists a small kink at $\sim$ 29.4 °C and that $\langle R_g \rangle / \langle R_h \rangle$ remains constant at $\sim$ 1.15 in the range 29–30.6 °C, representing an additional transition prior to the collapse of the PNIPAM chain segments. The decreases of both $\langle R_g \rangle$ and $\langle R_h \rangle$ after the kink become faster. As shown before, the coil-to-globule transition of PNIPAM homopolymer chains do not present such a "kink" [32–34, 37–40]. The sharp decrease of $\langle R_g \rangle / \langle R_h \rangle$ from $\sim$ 1.5 to $\sim$ 0.6 in the inset confirms the coil-to-globule transition of individual copolymer chains. However, a careful examination of Fig. 24 raises a number of questions.

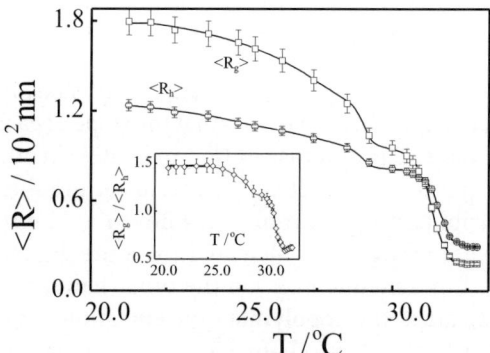

Fig. 24 Temperature dependence of z-average root-mean square radius of gyration $\langle R_g \rangle$ and average hydrodynamic radius $\langle R_h \rangle$ of PNIPAM-*seg*-St copolymer chains in water; the *inset* shows the temperature dependence of the ratio of $\langle R_g \rangle / \langle R_h \rangle$ [94]

The first question concerns the association of the stickers: do they associate into a string of micelle-like flowers or as one core-shell-like flower? If a string of flowers were formed, the overall chain structure would become more rigid and, hence, $\langle R_g \rangle / \langle R_h \rangle$ should increase. In addition, these micelle-like flowers would collapse and pack together at higher temperatures to form a uniform globule. In contrast, if only one micelle-like flower was formed, the hydrophobic stickers would condense in the center so that the resultant globule should have a core denser than the shell. According to the definitions of $\langle R_g \rangle$ and $\langle R_h \rangle$, a hard sphere with a denser core has a smaller $\langle R_g \rangle$ than a uniform sphere with the same size, but the density distribution has no effect on $\langle R_h \rangle$. In Fig. 22, the ratio $\langle R_g \rangle / \langle R_h \rangle$ decreases from ~ 1.5 to ~ 1.1 in the range 25–29.8 °C, and at higher temperatures, becomes much smaller than 0.774. As discussed earlier, the lower value of $\langle R_g \rangle / \langle R_h \rangle$ indicates a core-shell structure. They also measured the ratios of $\langle R_g \rangle / \langle R_h \rangle$ for surfactant-free polystyrene nanoparticles before and after grafting a layer of linear polymer chains. The decrease of $\langle R_g \rangle / \langle R_h \rangle$ from ~ 0.8 to ~ 0.6 was due to the hydrodynamic draining of the grafted layer [100]. As discussed before, the lower values of $\langle R_g \rangle / \langle R_h \rangle$ reflects a core-shell structure with a dense core, presumably, made of the hydrophobic styrene stickers.

The second question is related to the existence of an additional transition at ~ 29.4 °C. As expected, hydrophobic styrene stickers tend to associate in water. At lower temperatures, water is such a good solvent for PNIPAM that the copolymer chain adopts a random coil conformation. The movements of the stickers are also random and not correlated to each other because the PNIPAM segments randomly fluctuate in solution. As the temperature increases in the range 25–30.6 °C, the solvency of water for PNIPAM gradually decreases and the hydrophobic stickers tend to gather towards the chain center and to move in a more correlated fashion. Since water is not a poor solvent yet, the PNIPAM chain backbone is still in its swollen and coiled state, reflected in the fact that the decrease of $\langle R \rangle$ in Fig. 24 is only $\sim 30\%$. In order to distinguish such a chain conformation from a normal random coil, it was named an "ordered coil".

Figure 25 schematically shows such a conformation transition from the random coil to the ordered coil. As the temperature approaches ~ 30.6 °C (the Flory Θ-point), the stickers start to gather towards the center and each PNIPAM chain segment between two neighboring stickers forms a flower petal (loop). The overall chain conformation becomes flower-like. Further increase of the temperature leads to the condensation of all the stickers to form a core and the shrinking of the PNIPAM loops in the shell, resulting in a collapsed core-shell nanostructure. To confirm the existence of two such different transitions, Zhang et al. [94] carried out a thermal analysis of the copolymer aqueous solution using a Micro-calorimeter (MicroCal Inc, USA) with a heating rate of 0.5 °C/min and a response time of 5.6 s.

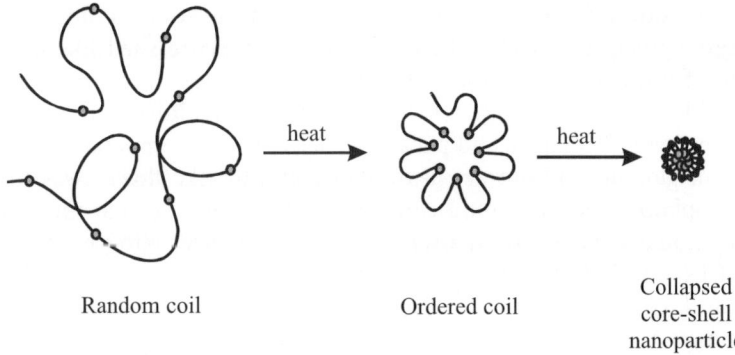

Random coil   Ordered coil   Collapsed core-shell nanoparticle

Fig. 25 Schematic of transitions of chain conformation of a segmented copolymer chain with stickers in dilute solution from a random coil to an "ordered coil" and then from an "ordered coil" to a collapsed core-shell globule as the solvency of water for the PNIPAM chain backbone decreases [94]

Figure 26 shows that for the PNIPAM homopolymer, the endotherm of the transition is slightly skewed towards the high temperature side, whereas the endotherm of a solution of PNIPAM-*seg*-St is abnormally skewed towards the low temperature side, and surprisingly, the transition occurs at a higher temperature. This is because copolymerizing hydrophobic comonomers into PNIPAM should decrease its shrinking temperature. Figure 26 shows that for a random NIPAM and styrene copolymer with a comonomer composition similar to the PNIPAM-*seg*-St copolymer used, the endotherm of the PNIPAM-*co*-St solution has the normal skew and is indeed shifted to lower temperatures in comparison with the PNIPAM homopolymer solution. Considering the kink in Fig. 26, one has to attribute the abnormal skew to a transition, which involves a small amount of energy, prior to the shrinking of PNIPAM.

The inset in Fig. 26 shows that the endotherm related to the collapse of the segmented copolymer can indeed be de-convoluted into two peaks. A small peak appears $\sim 1.2\,^\circ$C before the homopolymer peak. It is ascribed to the transition from the random coil to the ordered coil, while the large peak is related to the collapse of the PNIPAM chain backbone. Note that in comparison with the transition temperatures obtained in LLS, all the peak temperatures in DSC were ca. 1 $^\circ$C higher. This is because the transitions in DSC were slightly behind the temperature scanning, while the measurement in LLS could be considered as an infinitely slow scanning. Therefore, the small peak in DSC corresponds well with the kink in LLS. Also note that the collapse temperature of the short PNIPAM chain segments, the flower petals, is $\sim 0.6\,^\circ$C higher than that of long PNIPAM homopolymer chains. This could be ascribed to the chain length dependence of the phase transition temperature and also to the fact that the collapse of those small PNIPAM petals (loops) has to overcome an additional internal stress.

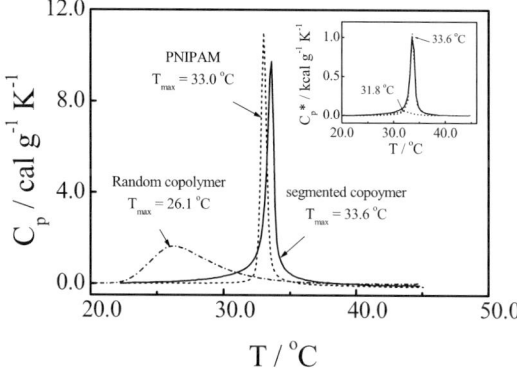

Fig. 26 Temperature dependence of partial heat capacity (C_p) of PNIPAM homopolymer, PNIPAM-co-St (4.1 mol %) random copolymer and NIPAM-seg-St (3.9 mol %) segmented copolymer in water; the polymer concentration was 1.0 g/L and the heating rate was 0.5 °C/min. The *inset* shows the de-convolution of stand partial heat capacity (C_p^*) of PNIPAM-seg-St [94]

4
Ionomers—From Intrachain Folding to Interchain Association

By definition, ionomers are copolymers with a few molar percent ionic groups inserted on their chain backbones. The solubility of ionomers in water mainly depends on the chemical nature of the chain backbone. For example, polystyrene ionomers are insoluble in water and poly(acrylamide) ionomers are completely soluble in water in the temperature range 0–100 °C. Water-insoluble ionomers have been extensively studied and reviewed and these reviews can be easily found in the literature. Hereafter we will discuss only ionomers soluble in water.

4.1
PNIPAM-co-KAA Copolymer Chains

PNIPAM has been anionically and cationically modified for different applications. Normally, such prepared copolymers are still thermally sensitive, just like PNIPAM, in water as long as its ionic content is not too high. PNIPAM ionomers generally have a higher LCST than PNIPAM homopolymers. In principle, they could be considered as one special example of hydrophilically modified PNIPAM chains as we described in Sect. 3.2.1. However, we like to review it as a special group of amphiphilic copolymers. This is because hydrophilic ionic groups have a much longer electrostatic interaction range in pure water than hydrophilic neutral groups. In other words, each ionic group can stabilize a larger surface area and play a stronger stabilization role when the PNIPAM chain backbone becomes hydrophobic at higher temperatures.

Using PNIPAM ionomers as a bridge, we will shift our discussion from the folding of individual copolymer chains in extremely dilute solutions to the formation of the mesoglobular phase of hydrophilically and hydrophobically modified copolymer chains in dilute solutions. The synthesis of PNIPAM-*co*-*x*KAA ionomers has been described before.

Figure 27 shows that for four different PNIPAM-*co*-0.8KAA concentrations, $R_{vv}(\theta)/KC$ of each solution remains nearly a constant (~ 4–5×10^6 g/mol) in the range 25–32 °C, very close to the average molar mass of individual PNIPAM-0.8KAA chains, indicating that there was no interchain aggregation [101]. When the temperature is raised to ~ 32.5–33 °C, slightly higher than the LCST of PNIPAM, $R_{vv}(\theta)/KC$ abruptly increases when the solution concentration is higher than $\sim 10^{-5}$ g/mL, clearly indicating interchain association. Further increase of the solution temperature to ~ 34–35 °C, leads $R_{vv}(\theta)/KC$ to a plateau, indicating that interchain association stops. From the ratio of $[R_{vv}(\theta)/KC]_{T=45\,°C}/[R_{vv}(\theta)/KC]_{T=25\,°C}$, the average number of the chains inside each aggregate (N_{chain}) can be estimated. The values of N_{chain} are ~ 17, ~ 8 and ~ 4, respectively, for $C = 5.0 \times 10^{-4}$, 1.0×10^{-4} and 9.5×10^{-6} g/mL. For the lowest concentration (4.7×10^{-6} g/mL), $R_{vv}(q)/KC$ is independent of temperature. Note that $R_{vv}(q)$ is proportional to the square of the molar mass of the scattering objects, i.e., a dimer scatters four times more light than an unimer. A trace amount of interchain association can lead to a large increase in $R_{vv}(\theta)/KC$. The temperature independence of $Rvv(q)/KC$ indicates no interchain association in such an extremely dilute solution even at temperatures as high as 45 °C. This is very different from the folding of individual PNIPAM homopolymer chains. First, single-homopolymer chain globules are not stable at very high temperature. Second, it is much more difficult to induce the coil-

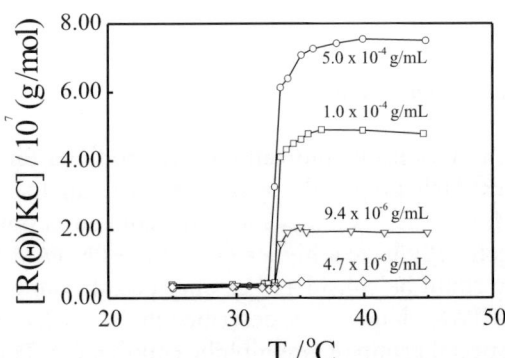

Fig. 27 Temperature and concentration dependence of the excess scattering intensity $R_{vv}(\theta)/KC$ of PNIPAM-0.8KAA in deionized water, where K is a constant, and $R_{vv}(\theta)/KC$ approximately equals the weight-average molar mass (M_w) because the solution is very dilute [101]

to-globule transition of homopolymer chains in a real experiment without any interchain association. We will come back to discuss this point later.

Figure 28 shows that in the most dilute solution, $\langle R_h \rangle$ decreases as the solution temperature increases, reflecting the intrachain coil-to-globule transition because we know from Fig. 27 that in this solution there is no interchain association in the heating process. This is similar to the collapsing process observed for a neutral PNIPAM homopolymer chain in an extremely dilute solution [32–34, 37–41]. However, we have to note that the ionomer chains used here are broadly distributed with a polydispersity index of $M_w/M_n = 1.6$–1.7 in comparison with those ($M_w/M_n < 1.1$) used in the study of the folding of individual PNIPAM homopolymer chains, which makes the experiment much easier. This is due to the stabilization role of ionic groups. The decrease of $\langle R_h \rangle$ can be divided into three stages. In the low temperature range (25–32 °C), water progressively changes from a good solvent to a poor solvent, resulting in slight contraction of the PNIPAM chain with a slightly smaller $\langle R_h \rangle$. Around the phase transition temperature (32–35 °C), the PNIPAM chain backbone undergoes the coil-to-globule transition so that $\langle R_h \rangle$ rapidly decreases. Finally, at temperatures higher than ~ 35 °C, the PNIPAM chain backbone reaches its fully collapsed globule state so that further increase of the solution temperature has little effect on $\langle R_h \rangle$.

In the solution with a higher concentration, interchain association accompanies intrachain contraction. When interchain association is dominant, $\langle R_h \rangle$ increases as the temperature increases, leading to a peak in the temperature range of 32–35 °C. At higher temperatures, $R_{vv}(q)/KC$ stops to increase at ~ 34 °C, as shown in Fig. 27, and reflects the end of interchain association. Therefore, the decrease of $\langle R_h \rangle$ at temperatures higher than ~ 34 °C is related to further collapse of the PNIPAM chain backbones inside each aggregate. Besides $\langle R_g \rangle$ and $\langle R_h \rangle$, a combination of static and dynamic LLS results can also lead to other microscopic parameters of these stable interchain aggre-

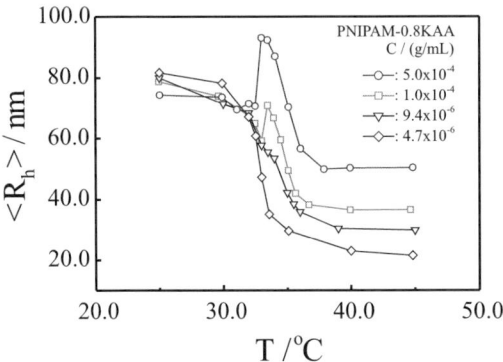

Fig. 28 Temperature and concentration dependence of the average hydrodynamic radius $\langle R_h \rangle$ of PNIPAM-0.8KAA in deionized water [101]

gates at $\sim 34\,°C$, such as the weight average molar mass ($M_{w,\,\mathrm{agg}}$), the average number of copolymer chains inside each aggregate ($\langle N \rangle_{\mathrm{agg}}$), the average surface area per ionic group ($\langle S \rangle_{\mathrm{ionic}}$), the average hydrodynamic volume of each polymer chain inside ($\langle V \rangle_{\mathrm{chain}}$) and the average chain density ($\langle \rho \rangle$) defined as $M_{w,\,\mathrm{agg}}/(4\pi \langle R_{\mathrm{h}} \rangle^3 N_A/3)$. Table 1 summarizes these microscopic parameters.

It is worth noting that at 34 °C the aggregates formed in different solutions have a similar $\langle V \rangle_{\mathrm{chain}}$ and $\langle \rho \rangle$. It indicates that at a given temperature the average degree of the shrinking of the polymer chains is similar in spite of the fact that each interchain aggregate contains different numbers of chains, in other words, interchain association has nearly no effect on intrachain contraction. Conventional wisdom tells us that interchain association should retard intrachain contraction. However, the results in the table above clearly reveal that intrachain contraction (the coil-to-globule transition) and interchain association are two *independent* and *competing* processes. This is important information.

In the fully collapsed state (45 °C), $\langle R_{\mathrm{g}} \rangle/\langle R_{\mathrm{h}} \rangle$ is in the range 0.73–0.84, indicating that the aggregates behave like uniform solid spheres without any draining. Note that this is different from single-chain core-shell structures made of PNIPAM-g-PEO copolymers. The average density $\langle \rho \rangle$ of the aggregates decreases as the aggregation number $\langle N \rangle_{\mathrm{agg}}$ decreases. It can be attributed to the imperfect packing of the copolymer chains inside each aggregate when only a few chains are packed inside each aggregate because the chain backbone is not infinitely flexible and the intrachain wrapping inside each aggregate results in a larger free volume in comparison with interchain wrapping [102]. When $\langle N \rangle_{\mathrm{agg}} > 8$, $\langle \rho \rangle \sim 0.34\,\mathrm{g/cm^3}$, similar to the value of the aggregates made of neutral PNIPAM homopolymer chains [103, 104]. For single-chain globules formed in the extremely dilute solution, $\langle \rho \rangle \sim 0.2\,\mathrm{g/cm^3}$, slightly lower than that made of a single PNIPAM homopolymer chain [32, 33], indicating that the ionic groups are on the periphery of the aggregate and have nearly no effect on the coil-to-globule transition of the PNIPAM chain backbone. The slightly lower density can be attributed to the ionic groups on the periphery, which leads to a slightly larger $\langle R_{\mathrm{h}} \rangle$ because of the hydration of ionic groups.

Table 1 Light-scattering results of PNIPAM-0.8KAA aggregates, respectively, formed at 34 and 45 °C

			34 °C				45 °C		
$\dfrac{C}{\mathrm{g/mol}}$	$\dfrac{M_{w,\mathrm{agg}}}{10^7\,\mathrm{g/mol}}$	$\langle N \rangle$	$\dfrac{\langle R_{\mathrm{h}} \rangle}{\mathrm{nm}}$	$\dfrac{\langle S \rangle_{\mathrm{ionic}}}{\mathrm{nm}^2}$	$\dfrac{\langle V \rangle_{\mathrm{chain}}}{10^5\,\mathrm{nm}^3}$	$\langle N \rangle_{\mathrm{agg}}$	$\dfrac{\langle R_{\mathrm{h}} \rangle}{\mathrm{nm}}$	$\dfrac{\langle R_{\mathrm{g}} \rangle}{\langle R_{\mathrm{h}} \rangle}$	$\dfrac{\langle \rho \rangle}{\mathrm{g/mol}^3}$
5.0×10^{-4}	7.68	17	87	17.8	1.62	17	87	0.78	0.34
1.0×10^{-4}	3.88	8	68	20.9	1.64	8	68	0.73	0.34
9.4×10^{-6}	1.86	4	54	26.3	1.64	4	54	0.75	0.28
4.7×10^{-6}	4.97	1	35	49.1	1.80	1	35	0.84	0.20

We can reasonably paint the following picture. During the coil-to-globule transition, the PNIPAM chain backbone, more precisely, the segments between two neighboring ionic groups, collapse and associate with each other, while all the ionic groups stay on the periphery of each resultant aggregate to act as stabilizers. As the chain association proceeds, the average number of ionic groups on each resultant particle should be proportional to the mass of the aggregate (M_{agg}) or the cubic of the size ($\langle R_h \rangle^3$) if we assume that the aggregates have a uniform chain density. On the other hand, the average surface area of the aggregate is only proportional to the square of the size ($\langle R_h \rangle^2$) or $M_{agg}^{2/3}$. Therefore, the surface area per ionic group is reversibly proportional to $\langle R_h \rangle$, i.e., $\langle R_h \rangle^{-1}$ or $M_{agg}^{-1/3}$. This means that $\langle S \rangle_{ionic}$ should decrease as the chain association proceeds. At the same time, the intrachain coil-to-globule transition also leads to the decrease of $\langle S \rangle_{ionic}$.

Figure 29 shows a sharp decrease of $\langle S \rangle_{ionic}$ in the temperature range of 32.5–33 °C, which exactly corresponds to the interchain association (the increase of $R_{vv}(\theta)/KC$ in Fig. 25 and the peak position of $\langle R_h \rangle$ in Fig. 26). Logically, there is a minimum value of $\langle S \rangle_{ionic}$, at which the surface of each aggregate is "fully covered" by ionic groups so that further chain association becomes impossible because of long-range electrostatic repulsion. However, the intrachain coil-to-globule transition inside the aggregates continues. This is exactly why $\langle R_h \rangle$ in Fig. 26 first increases and then decreases, but $\langle S \rangle_{ionic}$ in Fig. 29 only decreases. It is worth noting that for aggregates with different sizes formed in different solutions, $\langle S \rangle_{ionic}$ approaches a similar value at high temperatures. This is reasonable because for a given colloidal system, or more precisely, an interface (here this is water/collapsed PNIPAM), the surface area occupied per stabilizer should be a constant, independent of the particle size, if we consider that the process is thermodynamically controlled. This point has already been experimentally demonstrated with different systems, such as micro-emulsion [105–107] and the formation of various surfactant-free

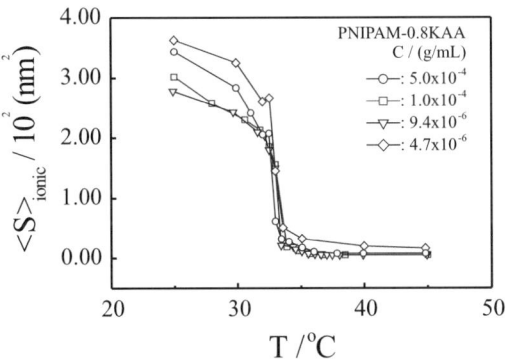

Fig. 29 Temperature dependence of the average surface area $\langle S \rangle_{ionic}$ per ionic group on the PNIPAM-0.8KAA aggregates, where $\langle S \rangle_{ionic}$ is defined as $4\pi \langle R_h \rangle^2 / \langle N \rangle_{ionic}$ [101]

polymeric nanoparticles [108–112]. Therefore, $\langle S \rangle_{ionic}$ is a fundamental parameter governing the final size of the chain aggregates.

As expected, increasing the ionic content generally resulted in smaller aggregates [101], but the competition between intrachain contraction and interchain association have a similar effect in this aspect, because for PNIPAM ionomers with a higher ionic content, it requires a much lower degree of aggregation to reach the same minimum value of $\langle S \rangle_{ionic}$. For PNIPAM-4.5KAA in an extremely dilute solution ($C = 5.0 \times 10^{-6}$ g/mL), only intrachain contraction (the coil-to-globule transition) occurs. Figure 30 shows that $\langle R_h \rangle$ returns to the starting point when the solution was cooled to 25 °C, indicating that the solution returns to its initial state in which only individual ionomer chains exist. On the other hand, the heating rate independence of $\langle R_h \rangle$ at 45 °C implies that the solution at 45 °C reaches a thermodynamic equilibrium state. However, near the coil-to-globule transition temperature (32–34 °C), the values of $\langle R_h \rangle$ in the heating and cooling processes are different at the same temperature. The hysteresis is similar to the folding of PNIPAM homopolymer chains. It can be attributed to some intrachain hydrogen bonding formed in the collapsed globular state, which persists around the Θ-temperature in the cooling process because water is still a very good solvent. This is why $\langle R_h \rangle$ is smaller. When the temperature is lower than 25 °C, water becomes such a good solvent that all intrachain association is destroyed and $\langle R_h \rangle$ returns to its initial value before the heating.

In principle, intrachain contraction should occur before interchain association as long as the solution is sufficiently dilute. The limitation is the sensitivity of our detection in laser light scattering. Note that in Fig. 27,

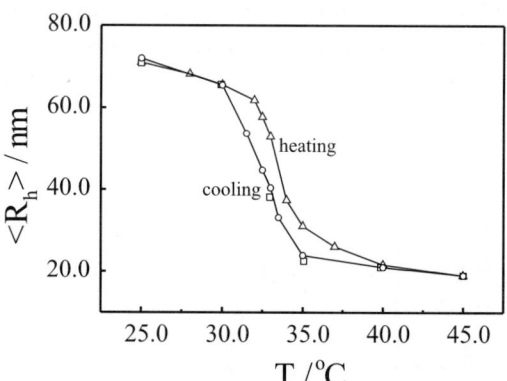

Fig. 30 Temperature dependence of $\langle R_h \rangle$ of a dilute PNIPAM-4.5KAA solution ($C = 5.0 \times 10^{-6}$ g/mL) respectively in slow heating and cooling processes, □ represents that the solution was jumped from 25 to 45 °C by a single step and then slowly cooled to each measurement temperature; △ and ○ represent that the solution was slowly heated to each measurement temperature from 25 °C to 45 °C and then slowly cooled to each measurement temperature. Every data point was obtained after the system reached its equilibrium [101]

temperature-induced intrachain contraction and interchain association simultaneously occurs when the concentration is slightly higher. However, the processes are so fast that one cannot distinguish the two processes even in dilute solutions. In the lowest concentration, only intrachain contraction occurs and ionic groups on the periphery stabilize single-chain collapsed globules. It would be ideal to find a system where under proper experimental conditions the evolution from intrachain contraction to interchain association could be observed. The following is an example.

4.2
PAM-*co*-NaAA Copolymer Chains

It is known that certain metal ions like Ca^{++} can specifically interact with carboxylic groups. If carboxylic groups are attached to a polymer chain backbone, such as partially hydrolyzed poly(acrylamide) (HPAM), the COO^-/Ca^{++} interaction could lead to intrachain contraction and interchain association through the polyion/metal "complexation" [113]. Flory and Osterheld [114] showed, as early as 1954, that Ca^{++} ions could change the chain conformation. Ohmine et al. [115] and Ben Jar et al. [116] studied the effects of monovalent and divalent cations on the collapsing of HPAM. Moreover, a few mechanistic models have been used to describe the aggregation of polyelectrolyte chains [117–119]. Michaeli [120] interpreted the aggregation as a function of the ionization degree and of the inert monovalent electrolyte concentration in terms of a stoichiomeric complex between divalent cations and anionic groups.

The aggregation kinetics have also been extensively studied [121]. The recent observation of fractal structures of the polymer clusters has sparked a renewed interest in aggregation kinetics [122]. Two distinct aggregation kinetic processes have been proposed and investigated. One is the diffusion-limited cluster aggregation (DLCA) controlled by the time taken for two clusters to collide via Brownian diffusion [123, 124]; and the other is the reaction-limited cluster aggregation (RLCA) in which the probability of forming a bond upon collision of two clusters is so high that the aggregation rate is chemically limited by its reaction rate. The RLCA has been observed in several colloid systems and modeled by computer simulation [125–128]. In general, the fractal dimension d_f is defined as: $M \sim R^{d_f}$, where M is the molar mass and R is the cluster size [124]. In RLCA, $d_f \sim 1.55$ and ~ 2, respectively, in 2-dimensional and hierarchical 3-dimensional simulations. The experimental values of d_f for the clusters formed in RLCA were $\sim 2.1 \pm 0.1$. Ball et al. [129] pointed out that in RLCA, the slightly larger experimental d_f values were due to the cluster's polydispersity.

Peng et al. [130] used the complexation of the HPAM chains in $CaCl_2$ aqueous solution to investigate the complexation-induced transition from intrachain contraction to interchain association over a wide range of the hydrolysis degrees and Ca^{++} concentrations as well as the structure of the HPAM/Ca^{++}

complexes. They hydrolyzed PAM homopolymer in an aqueous solution (10% NaOH + 10% NaCO$_3$) at 60 °C [131]. The hydrolysis degree (HD%) is controlled by the reaction time. The hydrolysis degree can be determined by titration with a 0.10 N HCl standard solution [132]. The complexation was induced by adding dropwise a proper amount of dust-free CaCl$_2$ aqueous solution into ~ 2 mL of dust-free HPAM aqueous solution. The initial concentration of the HPAM solution was kept at 1.00×10^{-5} g/mL. All the HPAM solutions used in LLS were clarified with a 0.5 μm Millipore filter and the CaCl$_2$ aqueous solution was clarified with a 0.1 μm Whatman filter (Anotop 25) to remove dust.

Figure 31 shows the kinetics of the complexation between HPAM/Ca^{++} in terms of the change of $\langle R_h \rangle$ for five different HPAM samples in a given CaCl$_2$ aqueous solution. For each HPAM sample, $\langle R_h \rangle$ approaches a plateau $\langle R_h \rangle_{max}$. As expected, $\langle R_h \rangle_{max}$ decreases with the hydrolysis degree. Note that there exists an initial stage in which $\langle R_h \rangle$ decreases, that reveals intrachain complexation in which individual chains contract before interchain association. For HPAM5, interchain complexation is dominant and initial intrachain contraction is too short to be observed, while for HPAM1, intrachain complexation becomes so dominant that there is no increase of $\langle R_h \rangle$. As for HPAM2, HPAM3 and HPAM4, the transition from *intrachain* to *interchain* is fairly clear. Figure 31 also reveals that for a given Ca^{++} concentration, the complexation-induced interchain association is directly related to the hydrolysis degree. This is because the carboxylic groups on the chain backbone act as "stickers" to "glue" different chains together, similar to the results of the PMA/Ca^{++} system reported by Yuko et al. [133].

Figure 32 shows the Ca^{++} concentration dependence of $\langle R_h \rangle_{max}$ for five different HPAM samples. The inset shows an enlargement of the low [Ca^{++}] range in which $\langle R_h \rangle_{max}$ first decreases as [Ca^{++}] increases, indicating the complexation-induced intrachain contraction. Further increase of Ca^{++} con-

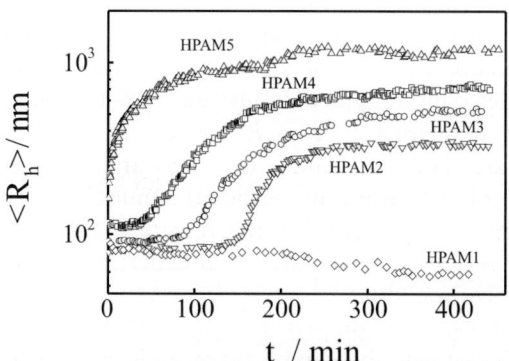

Fig. 31 Time dependence of average hydrodynamic radius ($\langle R_h \rangle$) of the HPAM/Ca^{++} complexes for complexation of HPAM chains in 0.08 M CaCl$_2$ aqueous solution [130]

centrations leads to a sharp increase of $\langle R_h \rangle$, revealing the complexation-induced transition from intrachain contraction to interchain association. The results in Figs. 31 and 32 reveal that the complexation between the HPAM chains in $CaCl_2$ aqueous solution can be dominated by either intrachain or interchain interaction, depending on both the hydrolysis degree and Ca^{++} concentration. The complexation can be viewed as follows. Each HPAM is a long coil chain with hundreds of "stickers" ($-COO^-$). Two "stickers" and one Ca^{++} ion can be driven thermodynamically together to form one $(-COO)_2Ca$ complex point. These interchain "points" result in the clustering of the HPAM chains and they further collide with each other or with individual HPAM chains, leading to larger clusters. Finally, when either Ca^{++} ions or $-COO^-$ groups are consumed, the complexation stops. In the process, clusters with different sizes were formed. Note that the *intrachain* $-COO^-$ groups are closer than those *interchain* $-COO^-$ groups in a dilute solution. For the HPAM chains with a low hydrolysis degree in a low Ca^{++} concentration, intrachain complexation is expected to be easier; while for a higher hydrolysis degree and a higher Ca^{++} concentration, interchain complexation becomes dominant. In the middle range of $[Ca^{++}]$ and $[-COO^-]$, individual HPAM chains first undergo intrachain complexation through the neighboring carboxylic acid groups on the same chain before interchain complexation becomes apparent.

Figure 33 shows double logarithmic plots of M_w versus $\langle R_h \rangle$ of HPAM/Ca^{++} complexes for a given HPAM sample but different Ca^{++} concentrations. Figure 34 shows double logarithmic plots of M_w versus $\langle R_h \rangle$ for a given Ca^{++} concentration but different HPAM samples, where the values of M_w were calculated from the measured Rayleigh ratio on the basis of Eq. 1. Note that the ratio of $\langle R_g \rangle / \langle R_h \rangle$ nearly remains a constant of ~ 1.35 in the measurable range of $\langle R_g \rangle$. Figures 33 and 34 clearly reveal that M_w can be scaled to $\langle R_h \rangle$ as $M_w \propto \langle R \rangle^{2.11 \pm 0.04}$ for different Ca^{++} concentrations and different HPAM

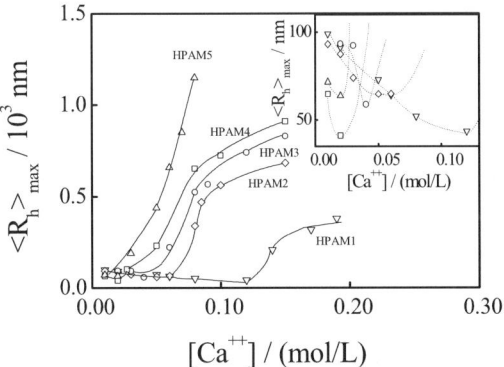

Fig. 32 Ca^{++} concentration dependence of the maximum average hydrodynamic radius $\langle R_h \rangle_{max}$, where $\langle R_h \rangle_{max}$ is the plateau value as shown in Fig. 31 [130]

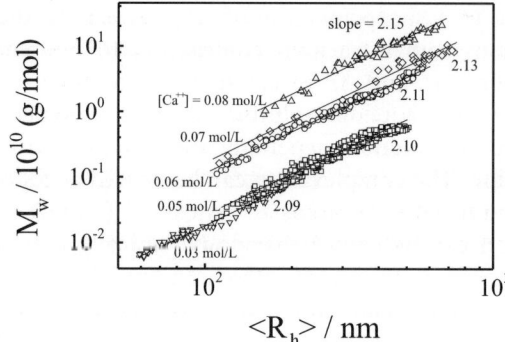

Fig. 33 Double logarithmic plots of weight average molar mass (M_w) vs. average hydrodynamic radius $\langle R_h \rangle$ for the HPAM5 chains in the presence of different amounts of Ca^{++} ions [130]

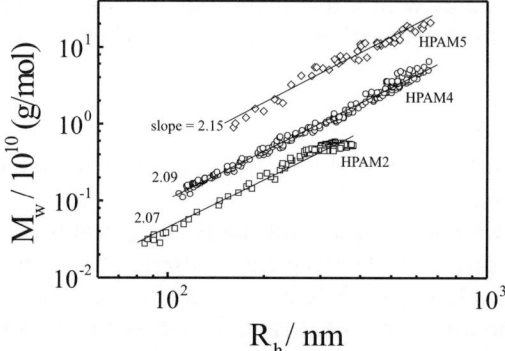

Fig. 34 Double logarithmic plots of weight average molar mass (M_w) vs. hydrodynamic radius $\langle R_h \rangle$ for different HPAM samples in 0.08 M $CaCl_2$ aqueous solution [130]

samples. This suggests that the HPAM/Ca^{++} complexes have a fractal structure with a dimension of $d_f = 2.11 \pm 0.04$, which is in good agreement with the value predicated for RLCA [134, 135].

5
Formation of a Stable Mesoglobular Phase in Dilute Solutions

The above discussion clearly shows that the increase of anionic copolymer concentration or the strength of chain interaction can lead to a transition from intrachain contraction to interchain association. It can result in a mesoglobular phase in which a limited number of copolymer chains are associated together to form polymeric colloidal particles stable between microscopic single-chain globular phase and macroscopic phase separation (precipitation). It is not a surprise to see the formation of such stable mesoglobules because ionic groups on the periphery have strong electrostatic repulsion to prevent further aggregation. Note that such long-range electrostatic interaction is difficult to handle in theory. In literature, the formation of stable mesoglobules made of neutral copolymer chains without any stabilizer has also been predicted [136–138]. It also should be noted that the effect of comonomer composition on chain association has not been systematically investigated. This is partially because it requires some challenging synthesis to control the comonomer composition.

5.1
Effect of Comonomer Composition

Recently, Siu et al. [139] studied the effect of comonomer composition on the formation of the mesoglobular phase of amphiphilic copolymer chains in dilute solutions. The copolymer used was made of monomers, N,N-diethylacrylamide (DEA) and N,N-dimethylacrylamide (DMA). Like PNIPAM, PDEA is also a thermally sensitive polymer with a similar LCST, but PDMA remains water-soluble in the temperature range ($< 60\,°C$) studied. At room temperature, copolymers made of DMA and DEA are hydrophilic, but become amphiphilic at temperatures higher than $\sim 32\,°C$. Before the association study, each P(DEA-co-DMA) copolymer was characterized by laser light scattering to determine its weight average molar mass (M_w) and its chain size ($\langle R_g \rangle$ and $\langle R_h \rangle$). The copolymer solutions (6.0×10^{-4} g/mL) were clarified with a 0.45 μm Millipore Millex-LCR filter to remove dust before the LLS measurement.

Figures 35 and 36 reveal that for each copolymer studied, the average aggregation number (N_{agg}) and average hydrodynamic radius ($\langle R_h \rangle$) of polymer clusters made of collapsed and associated P(DEA-co-DMA/x) chains increase and approach corresponding constants after a certain time, indicating the for-

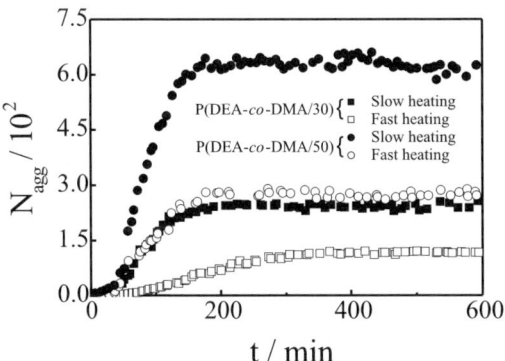

Fig. 35 Time dependence of average aggregation number (N_{agg}) of P(DEA-co-DMA) mesoglobules formed under different heating rates [139]

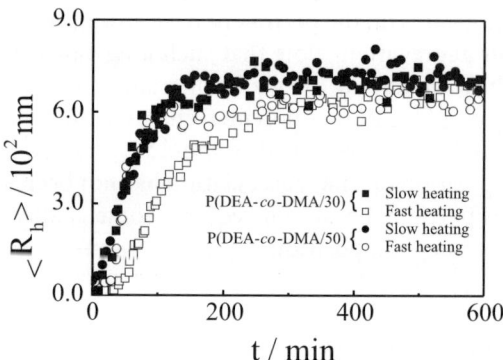

Fig. 36 Time dependence of average hydrodynamic radius ($\langle R_h \rangle$) of P(DEA-co-DMA) mesoglobules formed under different heating rates [139]

mation of stable mesoglobules, where N_{agg} was calculated from the ratio of the weight-average molar masses of the interchain aggregates and individual copolymer chains measured in static LLS. Note that for larger particles, the following Guinier plot instead of the Zimm plot (Eq. 1) has to be used.

$$\left(\frac{KC}{R_{vv}(q)}\right)_{C\to 0} \cong \frac{1}{M_w}\exp(-\frac{1}{3}R_g^2 q^2) \quad \text{for} \quad R_g q > 1. \tag{9}$$

It should be stated that aggregates formed in this way were so stable that no change in the scattering intensity was observed over months. It is also helpful to note that the stabilization in water was reached without the addition of any ion or surfactant. The stabilization is due to the concentration of hydrophilic DMA segments on the periphery of the aggregates during the microphase separation. It has been found that $\langle R_h \rangle$ is nearly independent of q in the low scattering angle range. The resultant stable mesoglobules are narrowly distributed with a relative width less than 0.05, as shown in Fig. 37. This is understandable because the association is an average process.

A combination of Figs. 35 and 36 shows that for a given heating process, N_{agg} increases with the DMA content. On the other hand, for a given copolymer, the fast heating results in a much smaller N_{agg} with a slightly larger size with $\langle R_g \rangle / \langle R_h \rangle \sim 1$, indicating that they must have a loose structure. In comparison, the aggregates formed in the slow heating have a ratio of $\langle R_g \rangle / \langle R_h \rangle \sim 0.8$. Further, it shows that at the very initial stage of the microphase transition, N_{agg} remains a constant, but $\langle R_h \rangle$ slightly decreases (not so obvious in Fig. 36) in the fast heating process, reflecting that intrachain contraction appears before interchain association. As expected, intrachain contraction must force the hydrophilic DMA segments to stay on the periphery to minimize the surface energy and slow down interchain association, resulting in a slower kinetics and smaller mesoglobules presumably consisting of many loosely packed small single- or pauci-chain collapsed globules. It

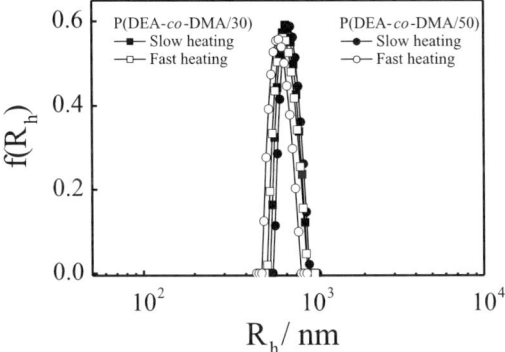

Fig. 37 Typical hydrodynamic radius distributions ($f(R_h)$) of resultant P(DEA-*co*-DMA) mesoglobules formed under different heating rates [139]

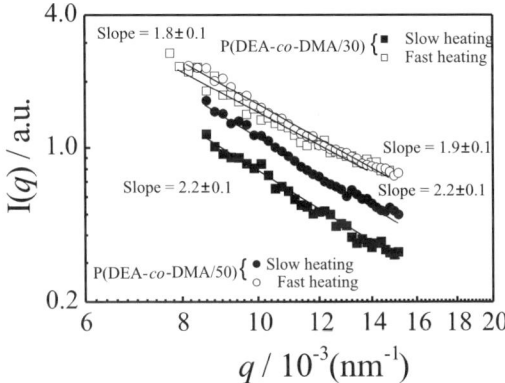

Fig. 38 Scattering vector (q) dependence of scattered light intensity (I) of resultant P(DEA-*co*-DMA) mesoglobules formed under different heating rates [139]

should be noted (not straight forward) that intrachain contraction ("folding") can lead to a chain density lower than interchain aggregation (interchain penetration) because no chain is infinitely flexible. This partially explains why the mesoglobules formed in the fast heating have a lower chain density.

Such a lower chain density can also be viewed from the scaling between the scattered light intensity (I) and the scattering vector (q) for resultant stable mesoglobules formed in different heating processes (Fig. 38). It is known that the scaling exponent α in $I \propto q^{-\alpha}$ is the fractal dimension in the scaling between molar mass and size, i.e., $M \propto R^{\alpha}$. The increase of α from 1.8–1.9 to 2.2 indicates that the association changes from a diffusion-limited process to a reaction-limited process [140]. In a reaction-limited process, many collisions only results in a sticking (association), while in a diffusion-limited process, each collision leads to a sticking. Therefore, in a reaction-limited process, each particle or cluster has a much higher chance to penetrate into the "fiords" of the existing aggregates before they stick together [140], which results in a higher chain density. This explains why α is higher for the mesoglobules formed in the slow heating process.

5.2
Effect of Comonomer Distribution

After investigating the effect of comonomer composition on the chain association as well as the effect of comonomer distribution on the chain folding, Siu et al. [141] extended their study to the effect of comonomer distribution on the chain association. They copolymerized NIPAM and vinyl pyrrolidone (VP) at temperatures, respectively, higher and lower than the LCST, which resulted in segmented and random VP distributions on the PNIPAM chain backbone. The synthesis characterization of these PNIPAM-co-VP amphiphilic copolymers with a similar chain length and comonomer composition, but different comonomer distributions, were described in previous sections.

The results showed that after the solution temperature is raised to a temperature higher than the LCST, both M_w and $\langle R_h \rangle$ increase as the time elapses and approach corresponding constants after a certain time, similar to Figs. 35 and 36. Mesoglobules formed in this way were stable for a long time, indicating that the interchain association stopped at a certain stage. It was also found that the average hydrodynamic radius of such resultant stable mesoglobules ($\langle R_h \rangle_{t \to \infty}$) decreases as the aggregation temperature ($T_{\text{aggregation}}$) increases, while their weight average molar mass ($(M_w)_{t \to \infty}$) increases when $T_{\text{aggregation}} < \sim 37\,^\circ\text{C}$, but decreases when $T_{\text{aggregation}}$ is higher than $37\,^\circ\text{C}$. The fact that stable aggregates formed at $36\,^\circ\text{C}$ have a smaller M_w, but a larger $\langle R_h \rangle$, clearly indicates that they have a loose structure. This is because the copolymer chains are only partially collapsed at $36\,^\circ\text{C}$. The stable mesoglobules are narrowly distributed, similar to those in Fig. 37. No precipitation was observed even after a long time, reflecting no change in the scattering intensity.

$\langle R_g \rangle / \langle R_h \rangle$, as shown in Fig. 39, reveals the structural information of these mesoglobules. The date points are scattered due to experimental uncertainties, especially in the measurement of $\langle R_g \rangle$ for large aggregates. However, the decrease of $\langle R_g \rangle / \langle R_h \rangle$ from $\sim 1.5–1.7$ to ~ 0.8 reveals a change from extended random-coil chains to uniform spherical aggregates, i.e., mesoglobules. The mesoglobules made of the NIPAM-co-VP/30/11 chains have the highest ratio of $\langle R_g \rangle / \langle R_h \rangle$. This is because NIPAM-co-VP/30/11 has the highest hydrophobic VP content and a random distribution of VP on the chain backbone so that its contraction is hindered. The formation of such stable mesoglobules is analogous to that described by Timoshenko et al. [136–138].

Figure 40 shows the temperature dependence of the average aggregation number (N_{chain}) of stable mesoglobules made of four different P(NIPAM-

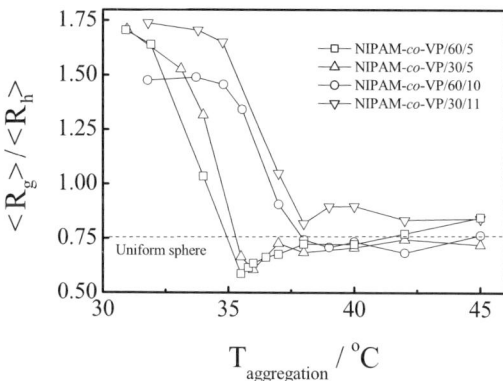

Fig. 39 Aggregation temperature dependence of ratio of average radius of gyration to average hydrodynamic radius ($\langle R_g \rangle / \langle R_h \rangle$) of resultant stable mesoglobules made of different copolymers [141]

co-VP) copolymers, where N_{chain} was obtained from the ratio of the weight average molar masses of the mesoglobules and chains. Note that for the copolymers with 10 mol% VP, N_{chain} reaches its maximum at a higher temperature because they are more hydrophilic. For each pair of copolymers with a similar VP content, the mesoglobules made of the copolymer chains prepared at 60 °C have a higher N_{chain}. On the other hand, for each pair of copolymers prepared at the same temperature, the copolymer with a higher VP content has an expected smaller N_{chain} because the average length of the PNIPAM segments is shorter and the copolymers are more hydrophilic.

It is a hypothesis that the copolymer chains prepared at higher temperatures would have a more segmented structure because it is expected that most of the VP monomers would be copolymerized on the periphery of collapsed PNIPAM segments. In other words, the chains prepared at 60 °C would have a more segmented structure in comparison with those prepared at 30 °C. In this way, for a given VP content, the average length of the PNIPAM segments between two neighboring VP segments should be longer than that of a statistically random copolymer prepared at 30 °C. It is expected that at high temperatures NIPAM-co-VP/60/x copolymers with longer PNIPAM and PVP segments can provide a stronger hydrophobic interaction as well as a stronger hydrophilic stabilization force, which have the opposite effect on the formation of the mesoglobular phase. However, the effect of hydrophobic interaction should be larger because PNIPAM is a major component and the length increase of the PNIPAM segment is much faster when the VP monomers start to group together as short segments. This explains why the copolymer chains prepared at 60 °C result in larger aggregates.

Another feature of Fig. 40 is the initial sharp increase of N_{chain} followed by a gradual decrease. Note that for each copolymer, the temperature at which N_{chain} reaches its maximum roughly corresponds to the temperature at which

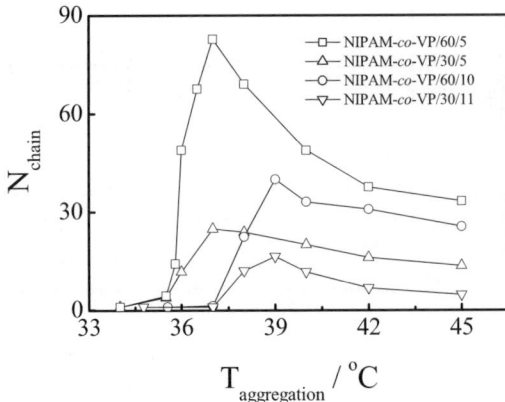

Fig. 40 Aggregation temperature dependence of average aggregation number (N_{chain}) of resultant stable mesoglobules made of different copolymers, where N_{chain} is defined as $M_{w,mesoglobule}/M_{w,chain}$ [141]

individual copolymer chains reach their fully collapsed states. The increase of N_{chain} with the aggregation temperature is understandable because long PNIPAM segments become more and more hydrophobic. Before reaching the collapse temperature, less compact chains can interpenetrate with each other so that the interchain association dominates. At higher aggregation temperatures, intrachain contraction becomes dominant and short hydrophilic VP segments tend to stay on the periphery to minimize the interfacial energy. In this way, the interchain association is retarded, which explains why N_{chain} decreases in the high aggregation temperature range.

Thermodynamically, the formation of stable mesoglobules instead of macroscopic phase separation requires a delicate balance between enthalpic and entropic contributions. In comparison with macroscopic precipitation, the existence of many small mesoglobules must gain in the translational entropy ($\Delta S > 0$) as well as in the interfacial energy ($\Delta H > 0$). For a homopolymer in a poor solvent, the gain of ΔH is larger than $T\Delta S$, i.e., $\Delta G = \Delta H - T\Delta S > 0$. Therefore, the equilibrium moves towards the direction of forming macroscopic precipitation. For amphiphilic copolymers in a selective solvent (often water), a microphase separation can occur, in which the association of hydrophobic segments leads to intrachain contraction and interchain aggregation, but hydrophilic segments tend to stay on the periphery. Under a proper condition, $T\Delta S$ can offset that of ΔH, i.e., $\Delta G = \Delta H - T\Delta S < 0$, so that further interchain aggregation stops. It is expected that more hydrophilic groups on the periphery would lead to smaller mesoglobules. However, due to the chain connectivity, a perfect arrangement to expose all hydrophilic VP components on the periphery is impossible. For a given type of copolymers with a similar composition, longer hydrophobic (shorter hydrophilic) segments should make the arrangement easier. As discussed before, the copolymer synthesized at 60 °C has longer PNIPAM segments than its counterpart prepared at 30 °C for a given comonomer composition. Longer PNIPAM segments provide a stronger hydrophobic attraction at high temperatures so that the copolymer chains prepared at 60 °C have a lower LCST and a higher N_{chain} than its counterpart prepared at 30 °C.

The competition between intrachain contraction and interchain association can be better viewed from the temperature dependence of $\langle R_h \rangle$, as shown in Fig. 41. A comparison of Figs. 40 and 41 shows that such a temperature dependence can be roughly divided into three regions. In the lower temperature range, where N_{chain} remains constant (~ 1), $\langle R_h \rangle$ decreases as the solution temperature increases, reflecting the contraction of individual chains. In the middle temperature range, N_{chain} and $\langle R_h \rangle$ increase before reaching their maximum values, showing that interchain association becomes dominant. In the higher temperature range, both N_{chain} and $\langle R_h \rangle$ decrease as the aggregation temperature increases. It should be noted that the decrease of $\langle R_h \rangle$ in the lower and higher temperature ranges is caused by completely different reasons. In the higher aggregation temperature range, intrachain contraction happens prior to interchain association. The higher the aggregation temperature the faster the contraction rate. Therefore, individual collapsed copolymer chains have much less chance to undergo interchain association. This is why both N_{chain} and $\langle R_h \rangle$ decrease in this region.

Figure 42 shows the temperature dependence of the average chain density ($\langle \rho \rangle$) of stable mesoglobules, where $\langle \rho \rangle$ is defined as $M_w/(4\pi \langle R_h \rangle^3 N_A/3)$. For all the copolymers studied, $\langle \rho \rangle$ always increases with the aggregation temperature. As discussed before, intrachain folding normally results in a lower chain density than interchain penetration because the chains are not infinitely flexible. For the copolymer pair, NIPAM-co-VP/60/10 and NIPAM-

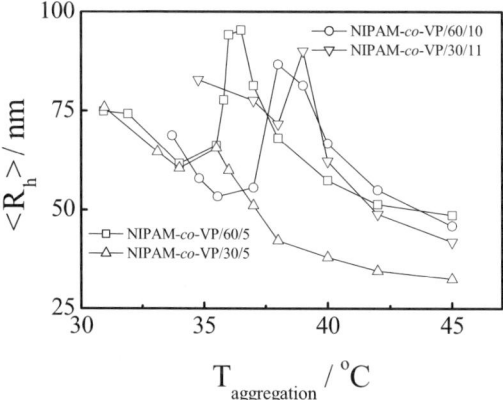

Fig. 41 Aggregation temperature dependence of average hydrodynamic radius ($\langle R_h \rangle$) of resultant stable mesoglobules made of different copolymers [141]

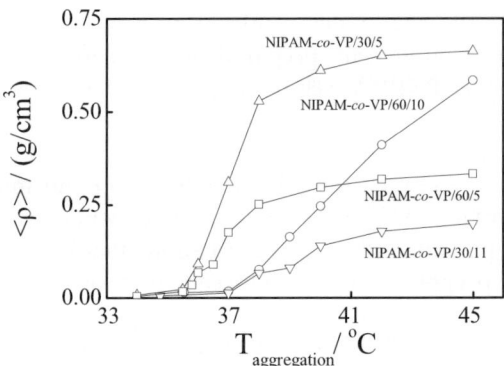

Fig. 42 Aggregation temperature dependence of average chain density ($\langle\rho\rangle$) of resultant stable mesoglobules made of different copolymers, where $\langle\rho\rangle$ is defined as $M_w/(4\pi\langle R_h\rangle^3 N_A/3)$ [141]

co-VP/30/11, the NIPAM-co-VP/60/10 mesoglobules have a higher average chain density because the copolymers prepared at 60 °C have longer PNIPAM segments and tend to form stronger interchain association as discussed earlier. However, for the pair of copolymers with a lower VP content, the NIPAM-co-VP/60/5 mesoglobules have a lower $\langle\rho\rangle$. The study of the chain folding showed that the coil-to-globule transition of individual NIPAM-co-VP/60/5 chains is easier [56]. The lower chain density reflects that NIPAM-co-VP/60/5 mesoglobules consist of many small collapsed single-chain globules, i.e., intrachain contraction is dominant in the formation of the mesoglobular phase.

5.3
Viscoelastic Effect on Formation and Stabilization of Mesoglobules

It is understandable, if not straightforward, that the copolymerization of some hydrophilic comonomers into a hydrophobic chain backbone can prevent macroscopic phase separation (precipitation) in aqueous solutions and result in stable mesoglobules because these water-soluble comonomers have a tendency to stay on the periphery and reduce the surface energy. On the other hand, hydrophobically modified water-soluble chains are known as associating polymers because hydrophobic comonomers in the chain backbone act as "stickers" in water. In a semi-dilute or concentrated solution, such copolymer chains can associate with each other to form insoluble large clusters or even form a hydrogel. Due to such strong interchain association, it would be difficult to imagine that they could form small stable mesoglobules.

Recently, we surprisingly found that even hydrophobically modified PNIPAM-co-MACA comonomer chains can form stable mesoglobules at higher temperatures [142]. Moreover, opposite to our expectation, the PNIPAM copolymer with a higher content of hydrophobic comonomer could form smaller aggregates in water under an identical experimental condition. Note that MACA homopolymer is insoluble in water at room temperature.

Three copolymers, PNIPAM-*co*-1.0-MACA, PNIPAM-*co*-2.9-MACA and PNIPAM-*co*-4.9-MACA were used. Their weight-average molar masses (M_w) are 9.4×10^5, 8.3×10^5 and 6.5×10^5 g/mol, respectively. In order to avoid possible interchain association, we kept each dilute solution in a refrigerator for at least one day to ensure a complete dissolution. The solution was clarified by a 0.45 Millipore (Hydrophlic Millex-LCR, PTFE) filter and then kept in a refrigerator again for a few days before LLS measurements.

Figure 43 shows an expected decrease of the LCST with an increasing hydrophobic MACA content. For PNIPAM-*co*-1.0-MACA, a gradual heating of the solution from 10 °C to 50 °C leads to large mesoglobules with $N_{agg} \sim 4000$, while for PNIPAM-*co*-4.8-MACA, the same heating results in much smaller aggregates only with $N_{agg} \sim 10$, which clearly shows that N_{agg} decreases with an increasing MACA content. On the other hand, the initial decrease of $\langle R_h \rangle$ in the range $T <$ LCST, as shown in Fig. 44, reflects intrachain contraction because $N_{agg} = 1$ in the same temperature range. For PNIPAM-*co*-1.0-MACA, the increases of both N_{agg} and $\langle R_h \rangle$ in the range $T >$ LCST reflect that interchain association becomes dominant. For PNIPAM-*co*-4.8-MACA, Fig. 44 shows a continuous decrease of $\langle R_h \rangle$ before its leveling-off at ~ 32 °C, similar to the coil-to-globule transition of individual single chains [13]. However, the increase of N_{agg} in Fig. 43 reveals that there still exists interchain association even though intrachain contraction is dominant here. For PNIPAM-*co*-2.9-MACA, the transition from intrachain-contraction dominant to interchain-association dominant is fairly clear. The leveling-off of N_{agg} and $\langle R_h \rangle$ for

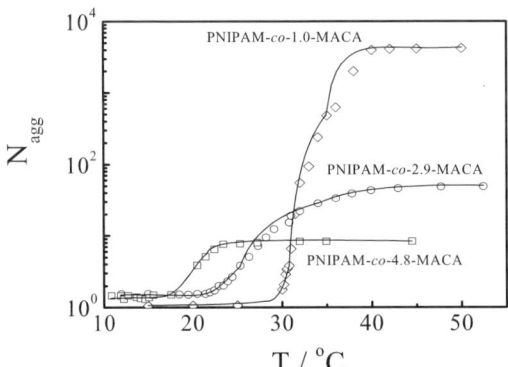

Fig. 43 Temperature dependence of average aggregation number (N_{agg}) of PNIAPM-*co*-MACA mesoglobules formed in a gradual heating process, where $N_{agg} = M_{w,agg}/M_{w,chain}$ with $M_{w,agg}$ and $M_{w,chain}$, the weight-average molar masses of the copolymer aggregates and of the chains, respectively; and copolymer concentration is $\sim 10^{-5}$ g/mL. Note that each data point was obtained after the temperature equilibrium was reached [142]

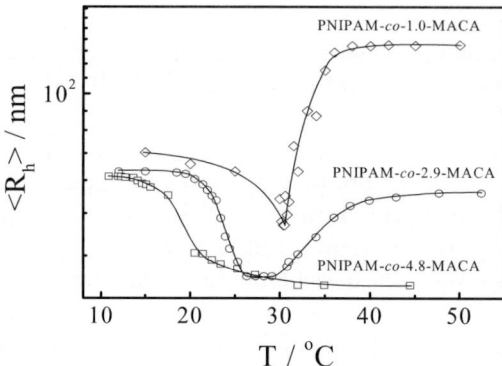

Fig. 44 Temperature dependence of average hydrodynamic radius ($\langle R_h \rangle$) of PNIAPM-co-MACA mesoglobules formed in a gradual heating process, where each data point was obtained after the temperature equilibrium was reached and copolymer concentration is $\sim 10^{-5}$ g/mL [142]

all the samples at higher temperatures indicates the formation of stable mesoglobules.

Now, let us discuss why those chains with a high content of hydrophobic MACA form smaller aggregates. Picarra and Martinho [143] showed that in the phase separation of a thin-layer dilute *homopolymer* solution on the surface, the collision would not be effective as long as the collision (or contact) time (τ_c) is shorter than the time (τ_e) needed to establish a permanent chain entanglement between two approaching aggregates. Quantitatively, Tanaka [144] showed that τ_c and τ_e could be roughly characterized as

$$\frac{l_o}{\langle v \rangle} < \tau_c < \frac{l_o^2}{\langle D \rangle} \quad \text{and} \quad \tau_e \sim \frac{a_m^2 N_m^3 \phi_p^{3/2}}{D_m}, \tag{10}$$

where l_o is the interaction range, $\langle v \rangle$ and $\langle D \rangle$ are the mean thermal velocity and transitional diffusion coefficient of aggregates, respectively, ϕ_p is the average polymer concentration inside chain aggregates, and a_m, N_m and D_m are the length, number and diffusion coefficient of monomer, respectively. When $\tau_c \ll \tau_e$, two colliding aggregates have no time to stick together and they behave just like two tiny elastic nonadhesive "glass" balls. Such an effect is a character of the viscoelasticity of long polymer chains. Our previous study showed that ϕ_p in the collapsed state could be as high as 30 wt % even though the overall concentration was very low ($\sim 10^{-6}$ g/mL) [32]. Therefore, the relaxation of long chains inside each aggregate is very slow. Homopolymer aggregates formed in this way are thermodynamically unstable because there is no hydrophilic stabilization group on the periphery to reduce τ_c.

For a given polymer solution under certain experimental conditions, l_o, a_m, N_m and D_m in Eq. 10 are constants. One can only increase $\langle D \rangle$ and ϕ_p to ensure that $\tau_c \ll \tau_e$ in order to reach a stable mesoglobular phase. Over the years, researchers always tried to make the aggregate's periphery hydrophilic to prevent the aggregation on the basis of thermodynamics, which actually reduces l_o and τ_c, but overlooked the effect of increasing τ_e to keep $\tau_c \ll \tau_e$, i.e., how to use the viscoelasticity to stabilize mesoglobules. As expected, increasing the hydrophobic MACA content promotes both interchain association and intrachain contraction. In dilute solutions, enhancing intrachain contraction increases ϕ_p inside each aggregate even though the overall macroscopic concentration remains. At the same time, τ_c decreases because the strong intrachain contraction reduces the initial size of the aggregates, leading to a higher $\langle D \rangle$. Moreover, hydrophobic association inside each mesoglobule gravely slows down the chain relaxation inside each mesoglobule and increases τ_e. Therefore, in some sense, it is not so crazy to claim that the hydrophobic association plays a partial role in the stabilization.

On the other hand, considering such a competition between intrachain contraction and interchain association as well as the viscoelastic effect, we dilute the copolymer solution to the limit of our LLS detection in order to suppress interchain association. Figure 45 shows that after ~ 10-time dilution, there is no observable change in the time-average scattering intensity $\langle I \rangle$ over the entire temperature range studied. As discussed earlier, $\langle I \rangle$ is extremely sensitive to interchain association; namely, the association of two chains doubles the scattering intensity. The constant value of $\langle I \rangle$ indicates that there is no interchain association. Figure 45 represents a pure intrachain coil-to-globule transition. We must note that due to the help of "hydrophobic stabilization", the experimental realization of the folding of individual MACA modified PNIPAM copolymer chains without interchain association is

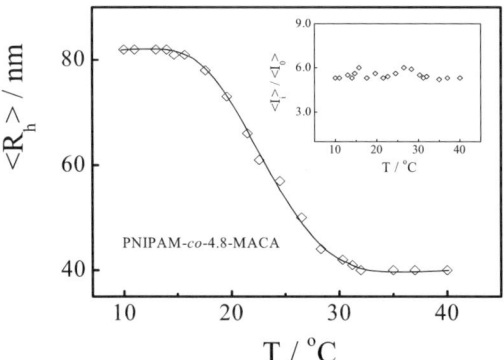

Fig. 45 Temperature dependence of average hydrodynamic radius ($\langle R_h \rangle$) of PNIAPM-co-4.8-MACA mesoglobules formed in a gradual heating process, where each data point was obtained after the temperature equilibrium was reached and copolymer concentration is 10x more dilute than that in Figs. 1 and 2. The *inset* shows the temperature dependence of average scattering intensity ($\langle I \rangle / \langle I \rangle_0$), where $\langle I \rangle_0$ is the reference intensity [142]

much easier in comparison with the formation of single PANPM homopolymer chain globules [32]. We also found that intrachain contraction can be promoted if the copolymer solution is quickly quenched to a desired phase separation condition because of sudden increases of ϕ_p and τ_e.

The above discussion has been schematically summarized in Fig. 46. The formation of the mesoglobular phase of amphiphilic copolymer chains in dilute solution during microphase separation inevitably involves competition between intrachain contraction and interchain association. The resultant mesoglobules are stabilized not only by the well-known concentration of hydrophilic components on the periphery, but also by the overlooked hydrophobic association inside each mesoglobule. This can be attributed to the effect of viscoelasticity even in a dilute solution; namely, when the chain entanglement time (t_e) between two colliding mesoglobules is much longer than their interaction time (t_c), each mesoglobule becomes a tiny nonadhesive "glass" ball. Either increasing the rate of microphase separation or diluting the solution can decrease t_c because intrachain contraction becomes so dominant that smaller initial mesoglobules are formed with a larger diffusion coefficient. On the other hand, increasing either the MACA content or the chain length can increase t_e.

On the basis of Eq. 10, using long chains (larger N) should be a very effective way to increase τ_e. Conventionally, one would think that the association of long chains should lead to larger aggregates. However, long copolymer chains can reach the condition of $\tau_e > \tau_c$ much more easily in the earlier stage of the microphase separation than short chains so that further entan-

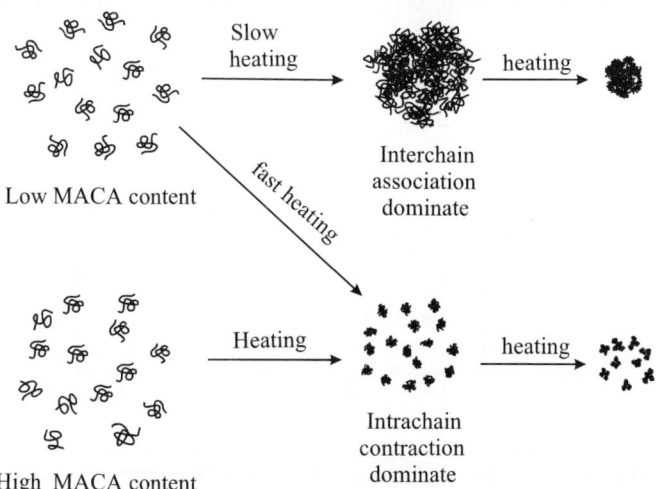

Fig. 46 Schematic of competition between intrachain contraction and interchain association in formation of PNIAPM-*co*-MACA mesoglobules, respectively, made of the chains with low and high hydrophobic MACA contents [142]

glement between longer chains respectively in two different mesoglobules becomes much more difficult, if not impossible, in the experimental time scale. In other words, the viscoelastic effect "overwrites" thermodynamics and leads to a special meta-stable, but very stable, state in dilute solution. Some years ago, Qiu et al. [145] found that longer PNIPAM-g-PEO copolymer chains could form smaller stable mesoglobules if other conditions were kept the same, but they did not attribute this to the effect of viscoelasticity at that time. They used two copolymers, PNIPAM-g-PEO1 ($M_w = 7.29 \times 10^6$ g/mol) and PNIPAM-g-PEO2 ($M_w = 7.85 \times 10^5$ g/mol) copolymers with a similar PEO distribution.

Figures 47 and 48, respectively, show the temperature and concentration dependence of $\langle R_h \rangle$ and $R_{vv}(\theta)/KC$ of PNIPAM-g-PEO1 in water. $R_{vv}(\theta)/KC$ is proportional to the weight average molar mass M_w on the basis of Eq. 1.

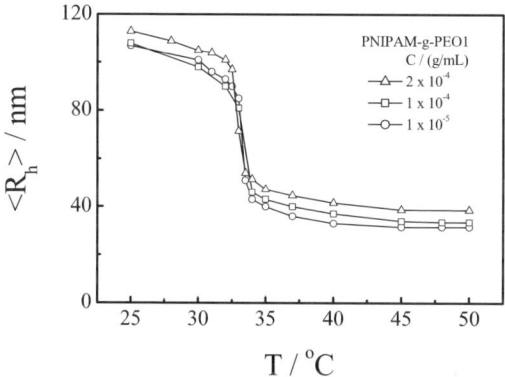

Fig. 47 Temperature and concentration dependence of the average hydrodynamic radius $\langle R_h \rangle$ of PNIPAM-g-PEO1 in deionized water [145]

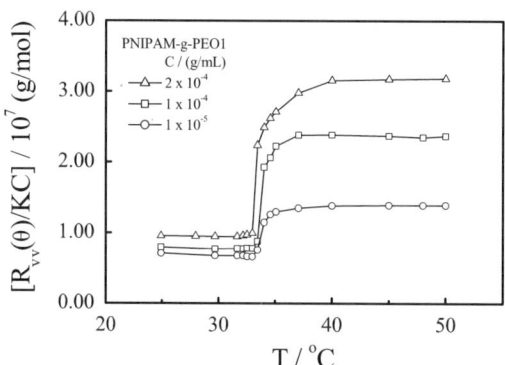

Fig. 48 Temperature and concentration dependence of the excess scattering intensity $R_{vv}(\theta)/KC$ of PNIPAM-g-PEO1 in deionized water [145]

Here, the solution temperature was raised step-by-step from 25 to 50 °C, i.e., it was stopped at each measurement temperature and the LLS measurement was conducted only after the equilibrium was reached. The change of $\langle R_h \rangle$ can be divided into three stages. In the first stage, when the solution temperature increases from 25 to 32 °C, water progressively becomes a poor solvent for the PNIPAM chain backbone, resulting in a slight decrease of $\langle R_h \rangle$. In the second stage ($\sim$ 32–34 °C), the PNIPAM chain backbone undergoes the intrachain "coil-to-globule" transition so that $\langle R_h \rangle$ rapidly decreases. In the last stage ($> \sim$ 34 °C), the PNIPAM chain backbone is already in its fully collapsed state so that further increase of the solution temperature has little effect on $\langle R_h \rangle$.

On the other hand, Fig. 48 shows that in the range $\sim$ 25–32 °C, $R_{vv}(\theta)/KC$ nearly remains a constant ($7 - 8 \times 10^6$ g/mol), very close to the weight average molar mass of individual PNIPAM-g-PEO1 chains, indicating no interchain aggregation. This is expected since both the PNIPAM chain backbone and PEO branches are hydrophilic and soluble in water in this temperature range. An abrupt increase of $R_{vv}(\theta)/KC$ at $\sim$ 33 °C clearly indicates interchain association. Over a narrow temperature range $\sim$ 33–36 °C, each $R_{vv}(\theta)/KC$ curve reaches a higher plateau, implying the formation of stable mesoglobules. From the value of each plateau and the average molar mass of individual PNIPAM-g-PEO1 chains, the average number of the copolymer chains ($N_{\text{PNIPAM-}g\text{-PEO}}$) inside each nanoparticle can be estimated. In Fig. 47, $N_{\text{PNIPAM-}g\text{-PEO}}$ has a value of $\sim$ 2, $\sim$ 3 and $\sim$ 4, respectively. It is easy to understand that the aggregation number increases with the copolymer concentration because the copolymer chains have more chance to undergo interchain association before each of them can collapse into a single-chain globule and is stabilized by the grafted PEO chains.

Figures 49 and 50 show the temperature and concentration dependence of $\langle R_h \rangle$ and $R_{vv}(\theta)/KC$ of PNIPAM-g-PEO2 in water. Note that PNIPAM-g-

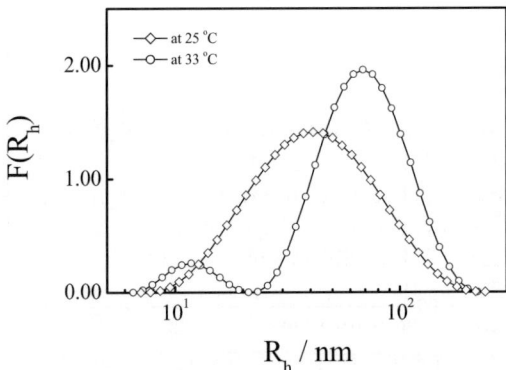

Fig. 49 Temperature and concentration dependence of the average hydrodynamic radius $\langle R_h \rangle$ of PNIPAM-g-PEO2 in deionized water [145]

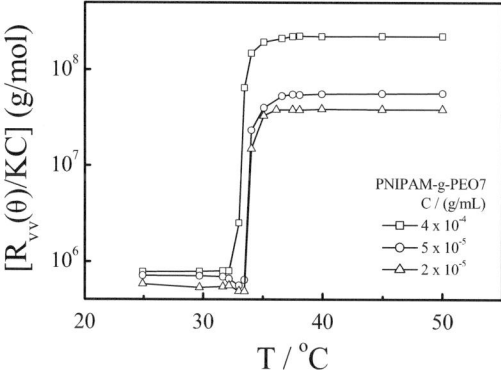

Fig. 50 Temperature and concentration dependence of the excess scattering intensity $R_{vv}(\theta)/KC$ of PNIPAM-g-PEO2 in deionized water [145]

PEO2 is 10 times shorter than PNIPAM-g-PEO1. Figure 49 reveals that short PNIPAM-g-PEO2 chains follow a different path; namely, after reaching a maximum at ∼ 33.5 °C, $\langle R_h \rangle$ starts to decrease, while in the case of long PNIPAM-g-PEO1 chains (Fig. 47) $\langle R_h \rangle$ remains a constant after reaching the maximum value. However, there is no difference in the changing pattern of $R_{vv}(\theta)/KC$ between long and short chains (Figs. 48 and 50). Therefore, the decrease of $\langle R_h \rangle$ in Fig. 49 and the constant values of $R_{vv}(\theta)/KC$ in Fig. 50 lead to only one possibility, i.e., for short PNIPAM-g-PEO2 chains, interchain association is prior to intrachain "coil-to-globule" contraction. The average aggregation numbers of short chains inside each mesoglobule are ∼ 80, ∼ 90, and ∼ 280, respectively, for $C = 2 \times 10^{-5}$, 5×10^{-5}, and 4×10^{-4} g/mL, which also reflect that interchain association is dominant in the case of short chains. A combination of Figs. 47–50 clearly shows that long chains aggregate much less than short chains.

6
Conclusion

In comparison with linear homopolymers, individual amphiphilic copolymer chains in dilute solutions, in which solvents are selectively good for only one of the comonomers, can fold into different single-chain core-shell nanostructures stabilized by a hydrophilic shell. The shell can be made of short hydrophilic chains grafted on the core or small hydrophilic (neutral or ionic) groups or loops (flower-petal) made of hydrophilic segments. The degree of amphiphilicity can strongly influence the folding of individual amphiphilic copolymer chains. For thermally sensitive copolymer chain backbones, the degree of amphiphilicity can be simply alternated by the solution temperature. The folding and unfolding of such thermally sensitive copolymer chains can be switched on and off by a very small temperature change of 1–2 °C. It has been found that the chain folding greatly depends on both comonomer composition and distribution on the chain backbone.

The comonomer distribution can be alternated by controlling the synthesis conditions, such as the copolymerization at different reaction temperatures at which the thermally sensitive chain backbone has different conformations (extended coil or collapsed globule). In this way, hydrophilic comonomers can be incorporated into the thermally sensitive chain backbone in a more random or more segmented (protein-like) fashion. On the other hand, short segments made of hydrophobic comonomers can be inserted into a hydrophilic chain backbone by micelle polymerization. One of the most convenient ways to control and alternate the degree of amphiphilicity of a copolymer chain, i.e., the solubility difference of different comonomers in a selective solvent, is to use a thermally sensitive polymer as the chain backbone, such as poly(N-isopropylacrylamide) (PNIPAM) and Poly(N,N-diethylacrylamide) (PDEA). In this way, the incorporation of a hydrophilic or hydrophobic comonomer into a thermally sensitive chain backbone allows us to adjust the degree of amphiphilicity by a temperature variation.

Using this approach, hydrophilic (neutral or ionic) comonomers, such as end-captured short polyethylene oxide (PEO) chains (macromonomer), 1-vinyl-2-pyrrolidone (VP), acrylic acid (AA) and N,N-dimethylacrylamide (DMA), can be grafted and inserted on the thermally sensitive chain backbone by free radical copolymerization in aqueous solutions at different reaction temperatures higher or lower than its lower critical solution temperature (LCST). When the reaction temperature is higher than the LCST, the chain backbone becomes hydrophobic and collapses into a globular form during the polymerization, which acts as a template so that most of the hydrophilic comonomers are attached on its surface to form a core-shell structure. The dissolution of such a core-shell nanostructure leads to a protein-like heterogeneous distribution of hydrophilic comonomers on the chain backbone.

On the other hand, hydrophobic comonomers, such as $2'$-methacryloylaminoethylene)-$3\alpha,7\alpha,12\alpha$-trihydroxy-5β-cholanoamide (MACA) and styrene (St), can be incorporated into the thermally sensitive chain backbone by copolymerization of different comonomers in a common organic solvent, which results in a random distribution of hydrophobic comonomers on the chain backbone, or by micelle copolymerization in water with the help of surfactant micelles in which hydrophobic comonomer molecules are concentrated, which leads to evenly distributed short hydrophobic segments on the chain backbone.

For copolymers with some protein-like comonomer distributions, individual copolymer chains can "memorize" or "inherit" its parent globular state; namely, their folding back into the core-shell nanostructure is much easier, resulting in a smaller and denser single-chain particle in comparison with their counterparts, randomly distributed comonomers on the chain backbone. On the other hand, for copolymers with evenly distributed short hydrophobic segments, individual random-coil chains can first fold to a novel "ordered" coil state before they collapse into their final single-chain flower-like globules. Conventional wisdom tells us that introducing hydrophobic segments will lead to chain association, resulting in phase separation. Surprisingly, it has been shown that the copolymerization of a few molar percent of hydrophobic comonomers into the hydrophilic chain backbone can actually stabilize the resultant single-chain core-shell globules and make the coil-to-globule transition much easier in extremely dilute solutions.

As the polymer concentration increases, interchain association inevitably occurs, but some amphiphilic chains can undergo a limited interchain association to form a stable mesoglobular phase that exists between microscopic single-chain globules and macroscopic precipitation. As expected, when the solvent quality changes from good to poor, intrachain contraction and interchain association occur simultaneously and there exists a competition between these two processes. Such a competition depends on the comonomer composition and distribution on the chain backbone and also depends on the rate of micro-phase separation. When intrachain contraction happens quickly and prior to interchain association, smaller mesoglobules are formed. A proper adjustment of the rates of intrachain contraction and interchain association can lead to polymeric colloidal particles with different sizes and structures.

Such chain folding and association in solutions can be effectively and accurately studied by a combination of static and dynamic laser light scattering (LLS). This is because LLS can directly measure the weight average molar mass (M_w), the average radius of gyration ($\langle R_g \rangle$) and average hydrodynamic radius ($\langle R_h \rangle$) of individual chains or interchain aggregates. On the basis of these measurable micro-parameters, one can further calculate the average aggregation number (N_{agg}), the average chain density ($\langle \rho \rangle$) and the average surface area occupied per "stabilizer" that can be ionic groups (as $-COO^-$ and $-SO_3^-$) or neutral hydrophilic moieties, such as short grafted hydrophilic chains, small hydrophilic loops and even tiny hydrophilic segments. In particular, the ratio of $\langle R_g \rangle / \langle R_h \rangle$ reflects the chain conformation and the chain density distribution inside interchain aggregates.

Considering the competition between intrachain contraction and interchain association, we have to discuss an overlooked viscoelastic effect in the formation of stable mesoglobules in dilute solutions. Otherwise, it would be difficult to understand why copolymer chains with a high content of hydrophobic comonomers could form smaller interchain aggregates. In the micro-phase separation, copolymer chains in solutions contract and associate. The collision between contracted and associated chains would not be effective if the collision (or contact) time (τ_c) is much shorter than the time (τ_e) needed to establish a permanent chain entanglement between two approaching aggregates. This is because when $\tau_c \ll \tau_e$, two colliding aggregates have no time to stick together and they behave just like two tiny elastic non-adhesive "glass" balls. Such an effect is a character of the viscoelasticity of long polymer chains. It has been shown that the local chain density inside the aggregate in the collapsed state could be as high as 30 wt % even though the overall concentration was very low ($\sim 10^{-6}$ g/mL). Therefore, the relaxation of long chains inside each aggregate is very slow.

Over the years, researchers always tried to make the periphery of aggregates hydrophilic to prevent inter-particle aggregation based on thermodynamics, which actually reduces τ_c, but overlooked that in polymer colloids we can easily increase τ_e to make $\tau_c \ll \tau_e$, i.e., *playing the viscoelastic effect on the stabilization of mesoglobular phase*. Following such a viscoelastic consideration, there are different ways to increase τ_e. For example, using long chains (larger N) should be very effective because long chains can reach the condition of $\tau_e > \tau_c$ much more easily in the initial stage of the micro-phase separation than short chains. As soon as the condition of $\tau_e > \tau_c$ is reached, further inter-particle association via chain entanglements become impossible in the experimental time scale. In other words, the viscoelastic effect "overwrites" thermodynamics and leads to a special metastable, but very stable, state in dilute solutions.

On the other hand, increasing the hydrophobic content promotes both interchain association and intrachain contraction. In dilute solutions, enhancing intrachain contraction increases τ_e because of a higher local chain density, and at the same time, decreases τ_c because strong intrachain contraction reduces the initial size of the aggregates, making the random movement of each particle much faster. In an actual process, during the micro-phase separation, hydrophilic moieties are forced to stay on the periphery, which reduces τ_c, and at the same time, hydrophobic intra- and inter-chain association prevents the relaxation of chains inside each aggregate so that τ_e increases. Such a cooperative effect leads to $\tau_e > \tau_c$.

In some sense, it might not be that crazy to claim that hydrophobic association can make polymer colloidal particles more stable in dispersion. Further, we might have to reconsider the stabilization mechanism of globular proteins. Normally, we consider that hydrophobic amino acid residuals in a protein chain can lead to the chain folding because of hydrophobic association. However, we might have overlooked the stabilization role of such hydrophobic association in preventing further aggregation of resultant protein structures. The current studies of the folding and association of amphiphilic copolymer chains in dilute solutions are only one small step forward in a long journey towards a better understanding of protein folding. It is our opinion that polymer research and polymer industries have to move toward the direction of synthesizing, characterizing and studying more sophisticated and tailored structural copolymers besides the much-studied random copolymers and block copolymers.

Acknowledgements The financial support from the Hong Kong Special Administration Region Earmarked Grants and the National Science Foundation of China over the last ten years is gratefully acknowledged. We are also indebted to many of our students, postdoctorals and collaborators, particularly to Professors Jiang M in Fudan University and Zhou SQ in the City University of New York, to Drs. Qiu XP, Wang XH, Li M, Hu TJ, Peng SF, and to Mr. Li W and Miss Siu MH.

References

1. Stryer L (1988) Biochemistry. Freeman WH, New York
2. Zhou YQ, Karplus M (1999) Nature 401:400
3. Nelson ED, Teneyck LF, Onuchic JN (1997) Phys Rev Lett 79:3534
4. Klimov DK, Thirumalai D (1996) Phys Rev Lett 76:4070
5. Camacho C, Thirumalai D (1993) Phys Rev Lett 71:2505
6. Khokhlov AR, Khalatur PG (1998) Physica A 249:253
7. Zheligovskaya EA, Khalatur PG, Khokhlov AR (1999) Phys Rev E 59:3071
8. Khokhlov AR, Khalatur PG (1999) Phys Rev Lett 82:3456
9. Timoshenko EG, Kuznetsov YA (2000) J Chem Phys 112:8163
10. Schild HG (1992) Prog Polym Sci 17:163
11. Fujishige S, Kubota K, Ando I (1989) J Phys Chem 93:3311
12. Kubota K, Fujishige S, Ando I (1990) J Phys Chem 94:5154
13. Meewes M, Ricka J, Nyffenegger R, Binkert T (1991) Macromolecules 24:5811
14. Zhou SQ, Wu C (1995) Macromolecules 28:5225
15. Tiktopulo EI, Bychkova VE, Ricka J, Ptitsyn OB (1994) Macromolecules 27:2879
16. Virtenen J, Baron C, Tenhu H (2000) Macromolecules 33:336
17. Stockmayer WH (1960) Makromol Chem 35:54
18. Flory PJ (1953) Principles of Polymer Chemistry. Cornell University Press, Ithaca, NY
19. Ptitsyn OB, Kron AK, Eizner YY (1968) J Polym Sci Part C 16:3509
20. Yamakawa H (1971) Modern Theory of Polymer Solutions. Harper & Row, New York, NY
21. Grosberg AY, Kuznetsov DV (1992) Macromolecules 25:1996
22. Yamakawa H (1993) Macromolecules 26:5061
23. Post CB, Zimm BH (1979) Biopolymers 18:1487 and (1982) 21:2123
24. Park IH, Wang QW, Chu B (1987) Macromolecules 20:1965
25. Chu B, Park IH, Wang QW, Wu C (1987) Macromolecules 20:2833
26. Chu B, Yu J, Wang ZL (1993) Prog Colloid Polym Sci 91:142
27. Tanaka F (1985) J Chem Phys 82:4707
28. Tanaka F, Ushiki H (1988) Macromolecules 21:1041
29. Grosberg AY, Kuznetsov DV (1993) Macromolecules 26:4249
30. Yu J, Wang ZL, Chu B (1992) Macromolecules 25:1618
31. Chu B, Ying QC, Grosberg AY (1995) Macromolecules 28:180
32. Wu C, Zhou SQ (1995) Macromolecules 28:5388, 8381
33. Wu C, Zhou SQ (1996) Macromolecules 29:1574
34. Wu C, Zhou SQ (1996) Phys Rev Lett 77:3053
35. Gao J, Wu C (1997) Macromolecules 30:6873
36. Hu TJ, Wu C (1999) Phys Rev Lett 83:4105
37. Wu C, Wang XH (1998) Phys Rev Lett 79:4092
38. Wang XH, Qiu XP, Wu C (1998) Macromolecules 31:2972
39. Wang XH, Wu C (1999) Macromolecules 32:4299

40. Zhang GZ, Wu C (2001) Phys Rev Lett 86:822
41. Zhang GZ, Wu C (2001) J Am Chem Soc 123:1376
42. Zhou SQ, Fan SY, Au-yeung SCF, Wu C (1995) Polymer 36:1341
43. Zimm BH (1948) J Chem Phys 16:1099
44. Wu C, Xia KQ (1994) Rev Sci Instrum 65:587
45. Chu B (1991) Laser Light Scattering. Academic Press, New York, NY
46. Berne B, Pecora R (1976) Dynamic Light Scattering. Plenum Press, New York, NY
47. Stockmayer WH, Schmidt M (1982) Pure Appl Chem 54:407
48. Wu C (1998) Polymer 39:4609
49. Yamamoto I, Iwasaki K, Hirotsu S (1989) J Phys Soc Jpn 58:210
50. de Gennes P-G (1975) J Phys Lett 36:L55
51. Sanchez IC (1979) Macromolecules 12:276
52. Okada Y, Tanaka F (2005) Macromolecules 38:4465
53. Baulin VA, Halperin A (2002) Macromolecules 35:6432
54. Marchetti M, Prager S, Cussler EL (1990) Macromolecules 23:3445
55. Birshtein TM, Pryamitsyn VA (1991) Macromolecules 24:1554
56. Siu MH, Zhang GZ, Wu C (2002) Macromolecules 35:2723
57. Zhang L, Eisenberg A (1995) Science 268:1728 and references therein
58. Antonietti M, Heinz S, Schmidt M (1994) Macromolecules 27:3276
59. Henselwood F, Liu G (1997) Macromolecules 30:488
60. Wu C, Akashi M, Chen MQ (1997) Macromolecules 30:2187
61. Li M, Jiang M, Zhu L, Wu C (1997) Macromolecules 30:2201
62. Berlinova IV, Amzil A, Tsvetkova S, Panayotov IMJ (1994) Polym Sci Polym Chem 32:1523
63. Kawaguchi S, Winnik MA, Ito K (1995) Macromolecules 28:1159
64. Gref R, Minamitake Y, Peracchia MT, Trubetskoy V, Torchilin V, Langer R (1994) Science 263:1600
65. Fessi H, Puisieux F, Devissaguet J Ph, Ammoury N, Benita S (1989) Int J Pharm 55:R1
66. Eckert AR, Webber SE (1996) Macromolecules 29:560
67. Wu C, Qiu XP (1998) Phys Rev Lett 80:620
68. Virtanen J, Tenhu H (2000) Macromolecules 33:5970
69. Chen HW, Li JF, Ding YW, Zhang GZ, Zhang QJ, Wu C (2005) Macromolecules 38:4403
70. Hu TJ, Wu C (2001) Macromolecules 34:6802
71. Ostrovsky B, Bar-Yam Y (1994) Europhys Lett 25:409
72. Cohen-Stua MA, Waajen FHWH, Cosgrove T, Vincent B, Crowley TL (1984) Macromolecules 17:1825
73. Wagner M, Brochard-Wyart F, Hervert H, de Gennes P-G (1993) Colloid Polym Sci 271:621
74. Cook RL, King HE Jr, Peiffer DG (1992) Phys Rev Lett 69:3072
75. Polik WF, Burchard W (1983) Macromolecules 16:978
76. Devanand K, Selser JC (1990) Nature 343:739
77. Tanford C (1980) The Hydrophobic Effect: Formation of Micelles and Biological Membranes. Wiley, New York, NY
78. Bekiranov S, Bruinsma R, Pincus P (1993) Europhys Lett 24:183
79. Martin JI, Wang Z-G (1995) J Phys Chem 99:2833
80. Milner ST (1991) Science 251:905
81. Auroy P, Auvray L, Leger L (1991) Phys Rev Lett 66:719
82. László K, Kosik K, Geissler E (2004) Macromolecules 37:10067
83. Ding YW, Ye XD, Zhang GZ (2005) Macromolecules 38:904

84. Glass JE (1989) Polymers in Aqueous Media: Performance through Association. American Chemical Society, Washington, DC
85. McCormick CL, Bock J, Schulz DN (1989) In: Mark HF, Bikales NM, Overberg CG, Menges G (eds) Encyclopedia of Polymer Science and Engineering. Wiley-Interscience, New York, NY, vol 17
86. Halperin A (1991) Macromolecules 24:1418; Borisov OV, Halperin A (1995) Langmuir 11:2911 and (1996) Macromolecules 29:2612
87. Urakami N, Takasu M (1996) J Phys Soc Jpn 65:2694; Urakami N, Takasu M (1997) Prog Theor Phys Suppl 126:329
88. Semenov AN, Joanny J-F, Khokhlov AR (1995) Macromolecules 28:1066
89. de Gennes P-G (1995) Isr J Chem 35:33
90. Halperin A (2000) In: Ciferri A (ed) Supramolecular Polymers. Marcel Dekker, New York, NY
91. Hu Y, Armentrout RS, McCormick CL (1997) Macromolecules 30:3538
92. Noda T, Morishima Y (1999) Macromolecules 32:4631
93. Kikuchi A, Nose T (1996) Macromolecules 29:6770
94. Zhang GZ, Winnik FM, Wu C (2003) Phys Rev Lett 90:35506
95. Biggs S, Hill A, Selb J, Candau F (1992) J Phys Chem 96:1505
96. Dowling KC, Thomas JK (1990) Macromolecules 23:1059
97. Cui SX, Liu CJ, Zhang WK, Zhang X, Wu C (2003) Macromolecules 36:3779
98. Hugel T, Seitz M (2001) Macromol Rapid Commun 22:989
99. Oesterhelt F, Rief M, Gaub HE (1999) New J Phys 1:61
100. Buchard W (1996) In: Brown W (ed) Light Scattering Principles and Development. Clarendon Press, Oxford
101. Qiu XP, Li M, Kwan CMS, Wu C (1998) J Polym Sci Polym Phys Ed 36:1501
102. Wu C, Chan KK, Woo KF, Qian R, Li X, Chen L, Napper DH, Tan G, Hill AJ (1995) Macromolecules 28:1592
103. Ricka J, Meewes M, Nyffenegger R, Binkert Th (1990) Phys Rev Lett 65:657
104. Meewes M, Ricka J, Nyffenegger R, Binkert Th (1991) Macromolecules 24:5811
105. Wu C (1994) Macromolecules 27:298
106. Antonietti M, Bremser W, Schmidt M (1990) Macromolecules 23:3796
107. Antonietti M, Heinz S, Schmidt M, Rosenauer C (1994) Macromolecules 27:3276
108. Wu C, Akashi M, Chen MQ (1997) Macromolecules 30:2187
109. Li M, Jiang M, Zhu L, Wu C (1997) Macromolecules 30:2201
110. Li M, Jiang M, Zhang YB, Wu C (1999) Macromolecules 31:6841
111. Gao J, Wu C (2000) Macromolecules 33:645
112. Gao J, Wu C (2005) Langmuir 21:782
113. Jordan DS, Green DW, Terry RE, Willhite GP (1982) J Soc Petrol Eng 8:463
114. Flory PJ, Osterheld JE (1954) J Phys Chem 58:653
115. Ohmine I, Tanaka T (1982) J Chem Phys 77:5725
116. Ben Jar PY, Wu YS (1997) Polymer 38:2557
117. Ikegami A, Imai N (1962) J Polym Sci 56:133
118. Dubin P, Bock J, Davies RM, Schulz DN, Thies C (1994) Macromolecular Complexes in Chemistry and Biology. Springer, Berlin Heidelberg New York
119. Narh KA, Keller A (1993) J Polym Sci Polym Phys 31:231
120. Michaeli I (1960) J Polym Sci 48:291
121. Family F, Landau DP (1984) Kinetics of Aggregation and Gelation. North-Holland, Amsterdam
122. Burns JL, Yan YD, Jameson GJ, Biggs S (1997) Langmuir 13:6413
123. Kolb M, Botet R, Jullien R (1983) Phys Rev Lett 51:1123

124. Meakin P (1983) Phys Rev Lett 51:1119
125. von Schulthess GK, Benedek GB, Deblois RW (1980) Macromolecules 13:939
126. Schaefer DW, Martin JE, Wiltzius P, Cannell DS (1984) Phys Rev Lett 52:2371
127. Jullien R, Kolb M (1984) J Phys A 17:L693
128. Brown WD, Ball RC (1985) J Phys A 18:L517
129. Ball RC, Weitz DA, Written TA, Leyvraz F (1987) Phys Rev Lett 58:274
130. Peng SF, Wu C (1999) Macromolecules 32:585
131. Ji HJ, Sun ZW, Zhan WX, Li ZH (1994) Acta Polym Sinica 11:559
132. Tanaka T, Nishio I, Ueno-Nishio ST (1982) Science 218:467
133. Yuko I, Michael B, Manfred S, Klaus H (1998) Macromolecules 31:728
134. Martin TA, Joost HJ, van Opheusden X (1996) Phys Rev E 53:5044
135. Lin MY, Lindsay HM, Weitz DA, Ball RC, Klein R, Meakin P (1990) Phys Rev A 41:2005
136. Timoshenko EG, Kuznetsov YA (2000) J Chem Phys 112:8163
137. Timoshenko EG, Basovsky R, Kuznetsov YA (2001) Coll Surf A 190:129
138. Timoshenko EG, Kuznetsov YA (2001) Europhys Lett 53:322
139. Siu MH, Liu HY, Zhu XX, Wu C (2003) Macromolecules 36:2103
140. Halsey TC (2000) Phys Today 53:36
141. Siu MH, Cheng H, Wu C (2003) Macromolecules 36:6588
142. Wu C, Li W, Zhu XX (2004) Macromolecules 37:4989
143. Picarra S, Martinho JMG (2001) Macromolecules 34:53
144. Tanaka H (1993) Phys Rev Lett 71:3158 (1992) Macromolecules 25:6377
145. Qiu X, Wu C (1997) Macromolecules 30:7921

Water Solutions of Amphiphilic Polymers: Nanostructure Formation and Possibilities for Catalysis

Ivan M. Okhapkin[1,2] · Elena E. Makhaeva[1] · Alexei R. Khokhlov[1,2,3] (✉)

[1]Department of Polymer Science, University of Ulm, Albert-Einstein-Allee 11, 89069 Ulm, Germany
khokhlov@polly.phys.msu.ru

[2]Nesmeyanov Institute of Organoelement Compounds, Vavilov Street 28, 119991 Moscow, Russia
khokhlov@polly.phys.msu.ru

[3]Physics Department, Moscow State University, 119992 Moscow, Russia
khokhlov@polly.phys.msu.ru

1	Introduction	178
2	Amphiphilic Monomers and Polymers: Classification and Conformational Properties	181
3	Nanostructure Formation in Solutions of Thermosensitive Polymers	188
4	Catalytic Properties of Polymer Associates in Aqueous Media	196
5	Conclusion	207
	References	208

Abstract A concept of amphiphilicity in application to monomer units of water-soluble polymers is presented. Molecular simulation and experimental studies of polymers consisting of amphiphilic monomers units are reviewed. Those polymers reveal unusual conformational behavior in aqueous solutions forming nanostructures of nonspherical shape. Self-association of amphiphilic thermosensitive polymers in water solutions is discussed. Possibilities for the use of thermosensitive copolymers as catalysts are described. The sharp water–organic boundaries formed by polymer associates in water solutions are shown to be a prospective medium for catalysis owing to adsorption of interfacially active substrates at the interface.

Keywords Amphiphilic monomers · Interfaces · Polymer catalysts · Water-soluble polymers

Abbreviations
CAC   Critical aggregation concentration
EO   Ethylene oxide
LCST Lower critical solution temperature
MAA   Methacrylic acid
MMA   Methyl methacrylate

NIPA	N-Isopropylacrylamide
NPA	p-Nitrophenyl acetate
NPAlk	p-Nitrophenyl alkanoates
PDEVP	Poly(N-(n-dodecyl)-4-vinylpyridinium-co-N-ethyl-4-vinylpyridinium) bromide
PEO	Poly(ethylene oxide)
PFDA	Perfluorododecanoic acid
PNIPA	Poly(N-isopropylacrylamide)
PS	Polystyrene
PVCL	Poly(N-vinylcaprolactam)
PVim	Poly(1-vinylimidazole)
R_g	Radius of gyration
R_h	Hydrodynamic radius
SAXS	Small-angle X-ray scattering
SMTBA	Styrylmethyl(tributyl)ammonium
SMTMA	Styrylmethyl(trimethyl)ammonium
SMTMAC	Styrylmethyl(trimethyl)ammonium chloride
TEM	Transmission electron microscopy
Tris	Tris(hydroxymethyl)aminomethane
VCL	N-Vinylcaprolactam
Vim	1-Vinylimidazole
VP	N-Vinylpyrrolidone

1
Introduction

Development of systems based on water-soluble polymers is one of the main research tasks of polymer chemistry and physics, since for natural macromolecules, such as proteins and DNA, water is the basic good solvent. In the last few decades, synthetic water-soluble polymers have found wide use in various biomedical applications [1–6] and separation processes [7–12]. Choosing water as a solvent is logical in many areas of research, such as biomedical applications and design of nanosize objects. Normally, nanosized objects are characterized by relatively high interfacial energy owing to their extremely small sizes. Water is an optimal medium for development of those objects. On the one hand, it possesses high interfacial tension at the boundaries with numerous compounds, which makes their molecular dissolution impossible. On the other hand, there are methods (e.g., addition of surfactants) that allow stabilization of the newly formed structures from macrophase separation. Water-soluble polymers are indeed prospective materials for this task. The intrinsic amphiphilic character enables them both to form nanosized structures and to stabilize the systems which are prone to macrophase separation.

Catalysis is a relatively new area for application of water-soluble polymers. Normally, organic solvents are conventional media for catalytic reactions. In recent years significant progress in the design of water-soluble catalysts was achieved [13–15], as water appears a prospective environmentally friendly solvent for "green" chemistry purposes. Also, a number studies were carried out that used biomimetic approaches for construction of synthetic macromolecular enzyme models [16–18]. Water is often used for that task as it is the native medium for enzymes.

Globular proteins (including enzymes) form an important class of water-soluble natural polymers. Their globular structure is believed to be stabilized by hydrophobic interactions. Indeed, a protein globule consists of a hydrophobic core and a hydrophilic shell interacting with the solvent molecules and preventing them from having contact with the core (Fig. 1) [19]. Owing to such structural organization, protein globules are stable against aggregation in solutions. Furthermore, protein folding is also ruled by the balance of hydrophobic and hydrophilic interactions. The approach of treating protein folding and stability in binary terms of hydrophilicity and hydrophobicity is rather simplistic but it has an advantage of employing few parameters and approximations with reasonable results obtained on conformational behavior of proteins. This approach was found to be promising in application to synthetic polymers as well. The initial works in this field were those of Lau and Dill [20, 21], who proposed the subdivision of synthetic monomer units into "hydrophobic" and "polar" classes (HP model). Within this model, an amphiphilic macromolecule is considered as a chain in which hydrophobic and polar units are taken to be pointlike interaction sites distributed along the chain in a linear fashion (Fig. 2a). It is assumed that in water and other polar solvents the hydrophilic groups are in good-solvent conditions, while the hydrophobic groups are in poor-solvent conditions, so that the hydrophobic units attract each other, while the interaction between polar units is effectively repulsive.

However, normally, the groups of both types are present in synthetic and natural monomer units of water-soluble polymers (Scheme 2), which suggests that the units are amphiphilic rather than hydrophobic or hydrophilic. Vasilevskaya et al. [22, 23] reported a dumbbell model of the monomer unit in a chain in which a new representation of monomer units was proposed. In this representation, the amphiphilic character of the monomer units was

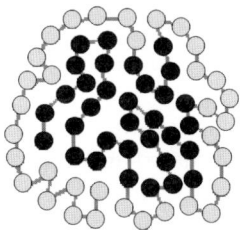

Fig. 1 Schematic representation of spatial structure of globular proteins. *Dark* and *light circles* denote hydrophobic and hydrophilic monomer units, respectively

H-units
P-units
a
H-units
P-units
A-units
b

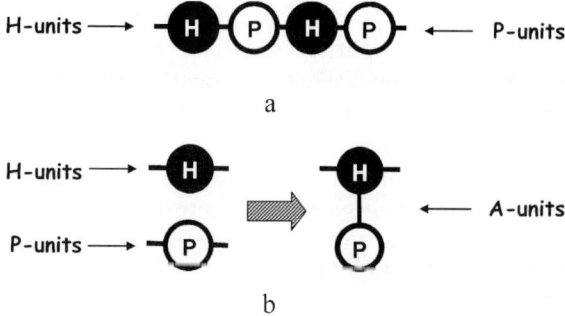

Fig. 2 Schematic representation of **a** the HP model and **b** the dumbbell HA model of an amphiphilic copolymer. *P-units* are hydrophilic (polar), *H-units* are hydrophobic, and *A-units* are amphiphilic. (Adapted from Ref. [25])

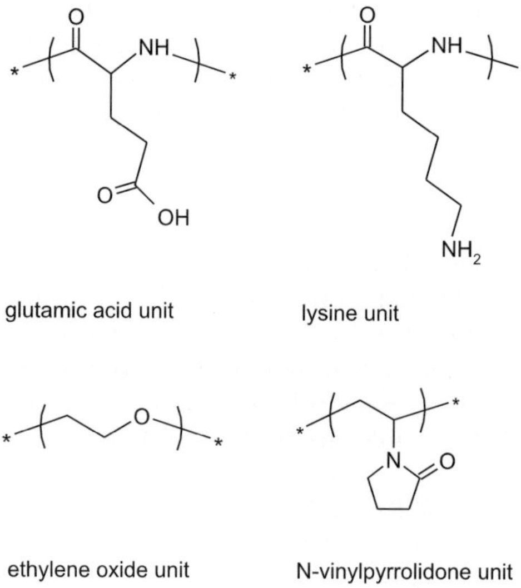

glutamic acid unit
lysine unit
ethylene oxide unit
N-vinylpyrrolidone unit

Scheme 1 Amphiphilic monomer units of synthetic and natural polymers

taken into account: the monomer unit was taken to consist of two parts, namely, a hydrophobic node of backbone and a polar segment attached to the node as a side group. In such a model, conformational properties of polymers, especially in the globular state, were shown to be drastically different from those predicted by the HP model and this prediction was supported by independent experimental results.

The present review has the following structure. In Sect. 2, the properties of amphiphilic monomers are discussed and a special classification of monomer units according to interfacial and partition properties is described. Also, the possibility of nanostructure formation in polymers composed of amphiphilic monomers is touched upon. This topic is more broadly treated in Sect. 3, where conformational properties of a key class of water-soluble polymers are discussed, i.e., thermosensitive polymers. In Sect. 4, catalytic effects associated with the properties of water-soluble polymers are presented. The role of interfaces in catalysis by polymer structures is established and the possibilities of reaction rate increase upon concentrating the substrate in various catalyst areas are shown.

2
Amphiphilic Monomers and Polymers: Classification and Conformational Properties

Amphiphilicity is the key feature which influences the properties of water-soluble polymers in aqueous solutions. Hydrophilic compounds are well soluble in water, but hydrophobic ones are not. Amphiphilic compounds occupy an intermediate position, being able to level off the unprofitable interactions between water and hydrophobic compounds. This applies both to low and high molecular weight compounds. As already indicated, the earliest attempts to formulate the principles of organization in amphiphilic polymer systems led to the development of a simple HP model in which monomers of amphiphilic polymers were divided on a binary alphabet principle into the hydrophilic and hydrophobic [20, 21]. This model was promising from the viewpoint of considering the competition of hydrophobic interactions with the interactions of good solvent type between water molecules and hydrophilic monomer units. The principal disadvantage of the HP model is that it does not take into account the fact that the monomer units, which are considered as hydrophilic, consist actually of hydrophilic and hydrophobic parts, and are actually amphiphilic. This imposes certain restrictions on their behavior: for purely hydrophobic or hydrophilic monomer units, the location only in hydrophobic or hydrophilic media should be suggested; for amphiphilic ones, the location at interfaces is most probable.

In Refs. [24, 25], a two-dimensional thermodynamic classification of nonionic monomers was proposed, which took into account three possible preferential locations of a monomer unit (location either in the hydrophilic or the hydrophobic phase or at the interface between them). The proposed classification incorporates gradations by affinity to polar and nonpolar phases and by interfacial activity (Fig. 3).

Each monomer is ascribed with a two-dimensional coordinate, of which the abscissa dimension corresponds to the affinity to the polar (water) or the nonpolar phase (hexane) and the ordinate dimension corresponds to interfacial activity. The standard free energy of partition between water and hexane is used as a quantitative parameter for the abscissa axis (ΔF_{part}), whereas the standard energy of adsorption at the interface is used for the ordinate axis (ΔF_{ads}). Both parameters are normalized by the kT factor. The normalized values are denoted as Δf_{part} and Δf_{ads}, respectively. Thus,

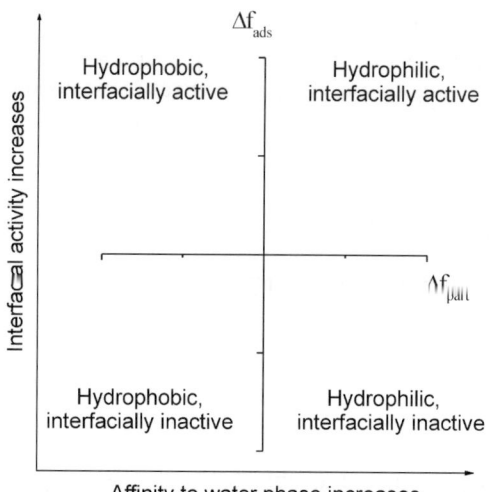

Fig. 3 Two-dimensional diagram of phase affinity and interfacial activity, general view. (Adapted from Ref. [25])

a four-quadrant diagram is constructed. Each quadrant involves a special class of monomers: hydrophobic interfacially active, hydrophobic interfacially inactive, hydrophilic interfacially inactive and hydrophilic interfacially active substances. Each of the two parameters used in the diagram appears as a quantitative tool for "one-dimensional" classification for a series of compounds. Indeed, a scale of n-octanol–water partition coefficients [26] and reverse-phase chromatographic retention parameters [27] are widely used to characterize the hydrophobicity of compounds; the concept of hydrophilic–lipophilic balance [28] was introduced for classification of surfactants according to their emulsifying properties. In the two-dimensional classification, both the potential for localization at the interface and the affinity to bulk phases are taken into account for several monomers, which allowed them to be classified on a broader basis.

The standard free energy of partition was calculated from water–hexane partition coefficients:

$$\Delta f_{part} = \ln P = \ln \frac{c_w^0 - c_h}{c_h}, \qquad (1)$$

where P is the partition coefficient, c_h is the concentration in hexane, and c_w is the concentration in water.

The standard free energy of adsorption from either bulk phase was calculated from interfacial tension isotherms:

$$\Delta f_{ads}^b = \ln \left[1 + \frac{1}{RT\tau} \left(\frac{\gamma^0 - \gamma}{c_b} \bigg|_{c_b \to 0} \right) \right], \qquad (2)$$

where $\Delta f_{\text{ads}}^{h}$ is the standard free energy of adsorption from a bulk phase, c_b is the concentration in the bulk phase, τ is the thickness of the surface layer, and $\gamma^0 - \gamma$ is the interfacial pressure at the water–hexane interface.

When two contacting liquid phases are in equilibrium, the adsorption of surfactant at the interface can take place from either of them. Accordingly, standard Gibbs energies of adsorption can be calculated for each phase using Eq. 2.

Several monomers of water-soluble polymers were analyzed with the help of the proposed two-dimensional diagram (Figs. 4 and 5, Scheme 1). In Fig. 4, the data for four synthetic monomers, N-vinylcaprolactam (VCL), N-vinylpyrrolidone (VP), N-isopropylacrylamide (NIPA), and 1-vinylimidazole (Vim), at different temperatures are presented, and in Fig. 5, the data for eight amino acids are shown.

Three of the four synthetic monomers (NIPA, Vim, and VP) fall into the quadrant for hydrophilic interfacially active compounds; VCL finds itself in that for hydrophobic interfacially active ones and has the highest interfacial activity among the four. All the synthetic monomers considered become more hydrophobic upon temperature increase, as indicated by the decrease of Δf_{part}. VCL and VP become more interfacially active (Δf_{ads} increases) at elevated temperatures, while Vim shows the reverse behavior. Δf_{ads} of NIPA is almost insensitive to the temperature regime. For all the monomers, $|\Delta f_{\text{part}}| < |\Delta f_{\text{ads}}|$ (except NIPA at 10.5 °C) and it may be stated that higher affinity to the interface than to bulk phases is the key factor for those compounds.

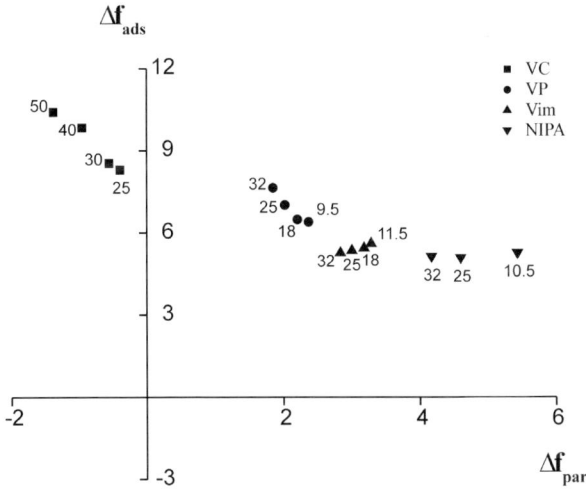

Fig. 4 Two-dimensional diagram for monomers of amphiphilic water-soluble polymers. The *numbers next to the points* are the measurement temperatures in degrees Celsius. (Adapted from Ref. [25])

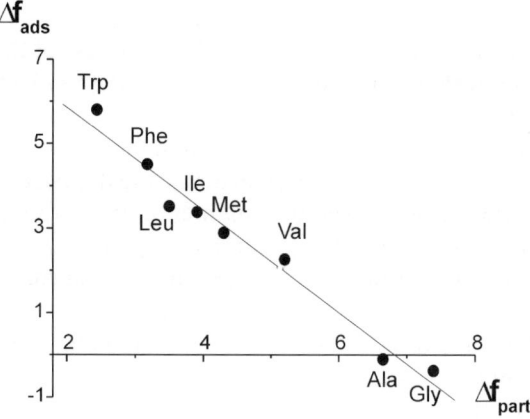

Fig. 5 Two-dimensional classification diagram for natural amino acids at 25 °C. (Adapted from Ref. [24])

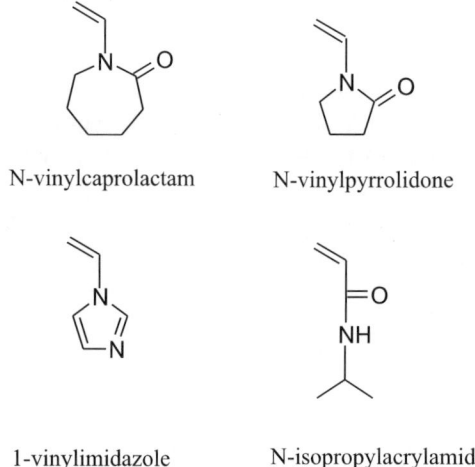

Scheme 2 Monomers characterized with the help of the two-dimensional diagram

The main feature of the amino acid diagram is that Δf_{ads} shows satisfactory linear correlation with Δf_{part}, with a slope of 1.22. Interfacial activity becomes stronger as the hydrophobicity of the amino acid residues increases. Since the amino acids have very hydrophilic amino and carboxyl groups, it may be said that the hydrophobicity increase enhances the amphiphilic character of the amino acids.

Thus, it was shown that many building blocks of natural and synthetic polymers are amphiphilic and interfacially active. For the monomers of several important water-soluble polymers, interfacial activity prevails over affinity to either bulk phase. Accordingly, it appears relevant to try to predict which properties the polymers consisting of interfacially active amphiphilic monomer units may possess.

Vasilevskaya et al. [22, 23] proposed an extended model of the monomer unit in which the amphiphilic character of the latter was taken into account (HA model). Figure 2b illustrates the simplest dumbbell model of an amphiphilic monomer unit constructed from hydrophilic and hydrophobic segments. Within this model, hydrophobic segments represent nodes of a backbone, while polar segments are side groups attached to the nodes. It was shown that in such a model, conformational properties of amphiphilic polymers, especially in the globular state, may differ materially from those predicted by the HP model. In particular, it turned out that the globules formed by macromolecules containing amphiphilic monomer units are not spherical (Fig. 6). It was found that the amphiphilicity and interfacial activity of monomer units are the main reasons for the unusual conformational properties of the corresponding polymers.

The property of the polymers in question to form nonspherical nanostructures was confirmed in experimental studies. Shih et al. [29] synthesized alternating copolymers of 1-alkenes with maleic anhydride. The maleic anhydride units were hydrolyzed to maleic acid units. Fully hydrolyzed macromolecules associated into microstructures of cylindrical and ellipselike shape. The cylindrical shape was characteristic of copolymers with octadecene and hexadecene moieties, while the copolymers with lower alkene copolymers (tetradecene, dodecene, decene, octene) formed ellipsoidal structures. Wataoka et al. [30] investigated the formation of nonspherical helices in a system of maltopentaose-carrying polystyrene (PS). The polymer was synthesized via the homopolymerization of vinylbenzyl maltopentaose amide (Scheme 3).

Its structure was characterized by small-angle X-ray scattering (SAXS) (Fig. 7a). In Fig. 7a, three SAXS profiles are presented. Two of them are calculated theoretically (lines 1, 2) and the third profile (open circles) represents experimental results. Both lines 1 and 2 are obtained using the model of a kinked cylindrical helix as imaged in Fig. 7b. For line 2, the partial aggregation of the helices is taken into account. The theoretical and experimental

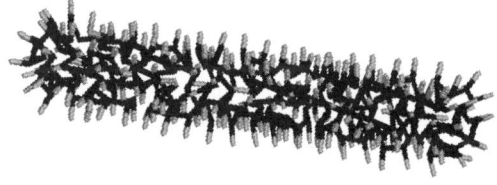

Fig. 6 Typical cylinder-shaped globules of polymers containing amphiphilic monomer units. *Darker spots* denote hydrophobic backbone nodes inside the core, *lighter spots* are hydrophilic side groups comprising the shell of the globule. (Adapted from Ref. [24])

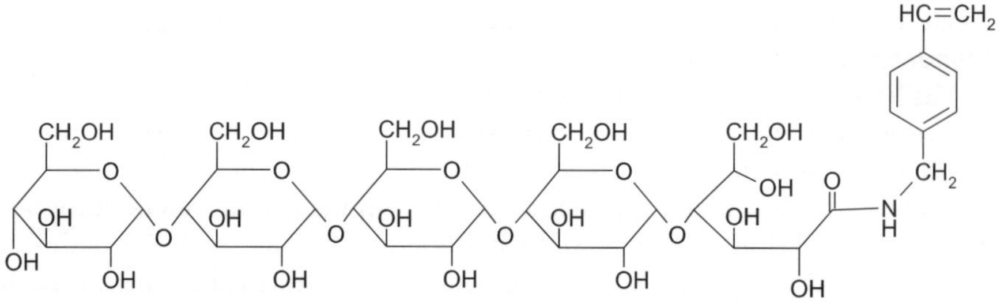

Scheme 3 Maltopentaose-carrying styrene macromonomer

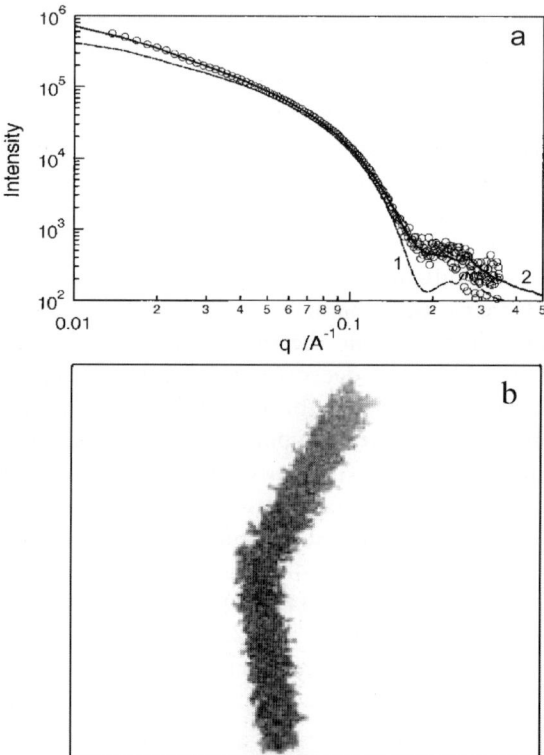

Fig. 7 a Scattering profiles for maltopentaose-carrying polystyrene. *Line 1* represents the calculated profile directly from the molecular model (see **b**) and *line 2* is calculated taking into account the effect of aggregation and backbone-folding. *Open circles* show the observed small-angle X-ray scattering profile of the copolymer in solution. **b** Molecular model of maltopentaose-carrying polystyrene. (Adapted from Ref. [30])

SAXS profiles shown in Fig. 7a coincide at most scattering vector values, which suggests that the proposed model satisfactorily describes the nanostructures formed in solution of maltopentaose-carrying PS as having cylindrical shape.

Theoretical line 2 fits best the experimental results, indicating that the observed cylindrical helices are prone to slight aggregation in water solutions.

The recent study of Thünemann et al. [31] showed the possibility of formation of cylindrical and disk-shaped nanostructures in aqueous solutions of polyampholytic copolymers of styrylmethyl(trimethyl)ammonium chloride (SMTMAC), methacrylic acid (MAA), and methyl methacrylate (MMA) complexed with perfluorododecanoic acid (PFDA) (Scheme 4). Cationic PFDA interacted with anionic SMTMAC to give neutral moieties, which were prone to association, while negatively charged groups of MAA stabilized the macromolecules from macroscopic phase separation. As a result, anisotropic nanostructures were obtained, the formation of which was confirmed by SAXS (Fig. 8). In Fig. 8, experimentally obtained pair distribution functions of the typical structures are plotted along with the theoretical distribution functions for an idealized cylinder with a diameter of 3.0 nm (Fig. 8a) and for a disk with a height of 2.2 nm (Fig. 8b). The intersection points with the abscissa of the distribution functions correspond to the characteristic dimensions of the idealized cylinder or disk. Good consistency was observed between the theoretical and experimental distribution functions, which allowed a conclusion to be made about the formation of nonspherical structures of cylindrical and disklike shape in the systems of amphiphilic copolymers of SMTMAC, MAA, and MMA complexed with PFDA.

Some attention should be also paid to the fact that some copolymers with special sequence distribution do not assume cylindrical shape within the HA model. For example, this is the case for protein-like sequences. Protein-like sequences correspond to a copolymer which forms globules with a hydrophobic core and a hydrophilic shell showing no tendency to aggregation. Protein-like copolymers have been previously studied within the HP model [32–34]. Application of the more realistic HA model showed that the globules formed by protein-like copolymers under worsening solvent quality assume conventional spherical shape and show no tendency to aggregate [23]. The stability for HA model protein-like copolymers is much higher than for those within the HP model.

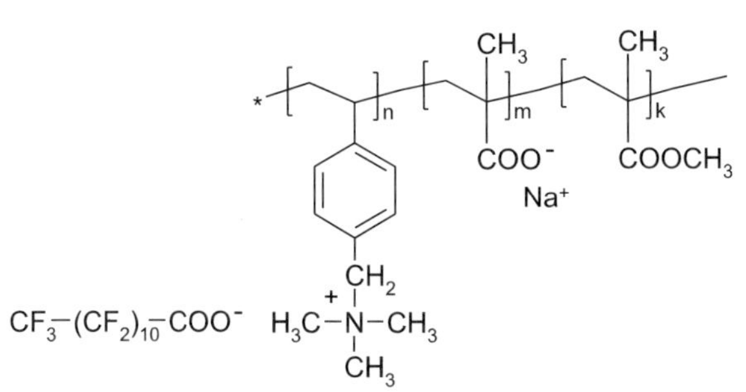

Scheme 4 Polyampholyte complexes with perfluorodecanoate anion

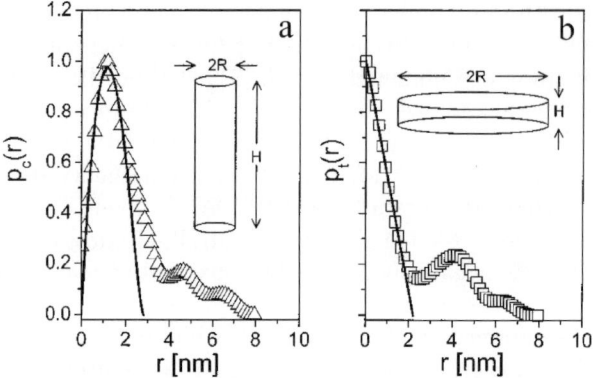

Fig. 8 Pair distribution functions of complexes of **a** cylindrical symmetry (57% styrylmethyl(trimethyl)ammonium, 16% methacrylic acid, 27% methyl methacrylate) and **b** disklike symmetry (79% styrylmethyl(trimethyl)ammonium, 13% methacrylic acid, 8% methyl methacrylate). The curves which were calculated from the scattering data are represented by *triangles* and *squares*. *Solid lines* represent the distribution functions of **a** an idealized cylinder with a diameter of 3.0 nm and of **b** a disk with a height of 2.2 nm. The *insets* depict idealized symmetries of the particles. (Adapted from Ref. [31])

3
Nanostructure Formation in Solutions of Thermosensitive Polymers

The properties associated with the amphiphilic monomer units are strongly exemplified in thermosensitive water-soluble polymers, typical examples of which are shown in Scheme 5. Thermosensitive polymers possess a lower critical solution temperature (LCST) in water solutions. Due to their sharp response to temperature variation, they are widely used in various scientific and technological applications. Drug and gene delivery [1–3], chromatographic [9, 10], membrane technology [11, 12], and catalyst immobiliza-

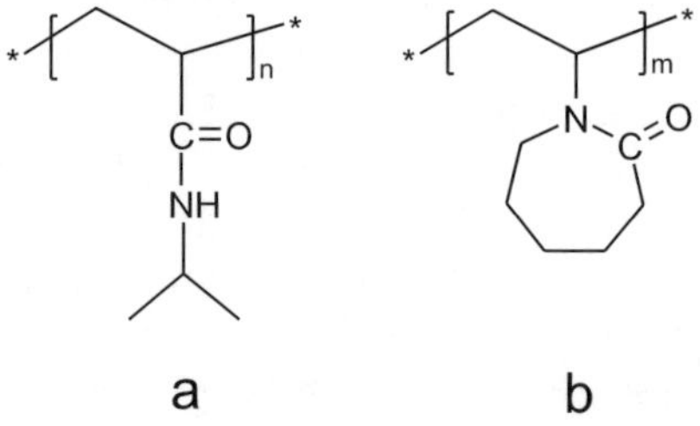

Scheme 5 Typical thermosensitive polymers: **a** poly(*N*-isopropylacrylamide), **b** poly(*N*-vinylcaprolactam)

tion [35] applications have been reported. The property of thermosensitivity is closely connected with the amphiphilicity of monomer units. At low temperature, water is a good solvent for chains of thermosensitive polymers owing to the formation of hydrogen bonds between water molecules and hydrophilic moieties of the macromolecules. Monomer units of the latter also contain hydrophobic CH_2 groups, which are involved in unprofitable hydrophobic interactions. It is well known that temperature increase leads to the intensification of hydrophobic interactions in water [36]. While at low temperature the interactions between water and hydrophilic groups are stronger than the hydrophobic interactions, at higher temperature the latter prevail, which leads to the coil-to-globule transition of macromolecular chains and further aggregation. The coil-to-globule transition may be directly observed in solutions of thermosensitive polymers, both on addition of surfactants [37–40] and in surfactant-free solutions [41–43]. The transition was confirmed by dynamic and static light scattering methods. The collapse of polymer chains was identified by a sharp decrease in the radius of gyration and the hydrodynamic radius of the macromolecules. The formation of globules in surfactant-free solutions of poly(VCL) (PVCL) [41] and poly(NIPA) (PNIPA) [42, 43] can be detected only at extremely low concentrations (below 10^{-3} g/L). At higher concentrations, the collapse of the polymer chains is accompanied by aggregation. However, upon addition of small amounts of surfactants, the aggregation may be suppressed, whereas the intramolecular aggregation is still possible. The corresponding studies were conducted both for PVCL [37, 38] and PNIPA [39, 40] in solutions with relatively high polymer content (approximately 1 g/L). As in the surfactant-free studies, a decrease in hydrodynamic radius was observed upon temperature increase (Figs. 9, 10), which indicated the formation of the globular structures stabilized against aggregation by a surfactant shell.

Globules represent the simplest structures of thermosensitive polymers of the nanometer scale. In the last few years, a series of works were aimed at obtaining nanoscale objects with a complex structure based on PVCL and PNIPA. The amphiphilic character of PVCL and PNIPA chains allowed them to be conjugated both with hydrophilic and with hydrophobic species to yield ordered self-associating systems in solution and at the interface. The main approach for the design of such systems is based on synthesis of block copolymers, which organize in core–shell structures in solution. In particular, block and/or graft copolymers of NIPA and ethylene oxide (PNIPA-b-PEO and PNIPA-g-PEO) [44–47], NIPA and propylene oxide [48], NIPA and styrene (PNIPA-b-PS) [49, 50], NIPA and lactic acid [51], VCL and EO (PVCL-b-PEO) [52, 53] were synthesized. Qiu and Wu [44] obtained PNIPA-g-PEO by radical copolymerization of PNIPA and PEO macromonomers capped with a methacrylate monomer unit. The copolymers underwent a transition accompanied by the formation of a nanostructure with a hydrophobic PNIPA core and a hydrophilic PEO shell. The character of aggregation was

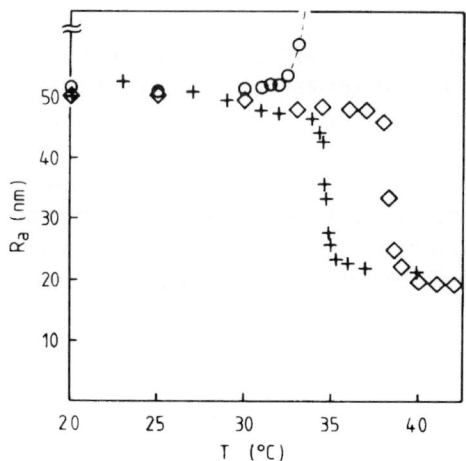

Fig. 9 Temperature dependencies of poly(N-isopropylacrylamide) (*PNIPA*) hydrodynamic radius as a function of surfactant concentration: *circles* 0, *crosses* 1.1, *squares* 1.8 mmol/L. (Adapted from Ref. [39])

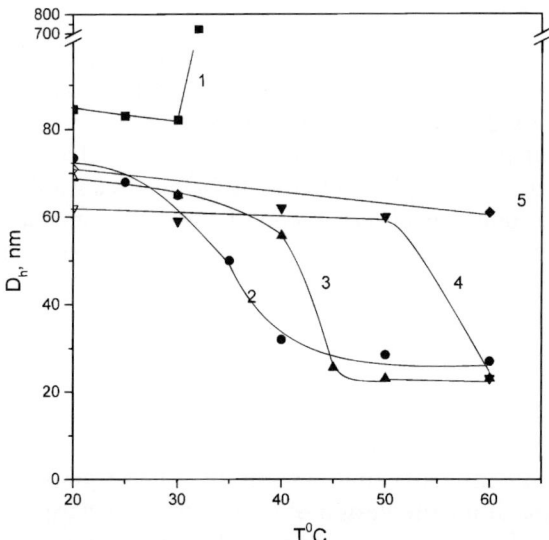

Fig. 10 Temperature dependencies of poly(N-vinylcaprolactam) (PVCL) hydrodynamic diameter as a function of surfactant concentration: *1* 0, *2* 0.25, *3* 0.5, *4* 1, *5* 3 mmol/L sodium dodecyl sulfate. (Reprinted with permission from Ref. [37]. Copyright 1998 American Chemical Society)

molecular-weight-dependent: for shorter chains, intermacromolecular aggregation dominated, as indicated by an increase in R_h at the temperature of the collapse (Fig. 11a), while for longer chains, the coil-to-globule transition of individual macromolecules was observed, as shown by a drop in R_h of polymer particles (Fig. 11b). It should be noted that in the case of longer chains, the intrachain collapse is observed at relatively high concentrations (approximately 0.1 g/L), in contrast to the individual PNIPA chains, for which very low concentrations are needed to detect the coil-to-globule transition.

Diblock copolymers of NIPA and EO showed somewhat different behavior [45]. The copolymers were synthesized using NIPA monomer and PEO-containing macroinitiator. The copolymers aggregated at high temperature with no collapse of individual macromolecules. The essential feature of the polymers in question consisted in the strong sensitivity of the shape of the particles to the polymer concentration and to the molar ratio of EO to NIPA monomer units. At low polymer concentration, the shape of the aggregates was mainly spherical, as indicated by the low R_g/R_h values (Fig. 12). The R_g/R_h ratio is informative of the shape of particles in solution. Anisotropic particles, such as rods and coils, possess high values of the ratio, while for particles of spheroid form, low values of the ratio are observed (for spher-

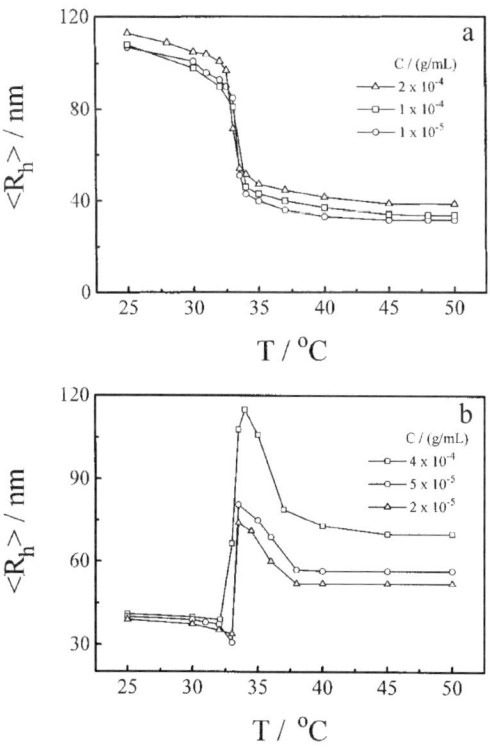

Fig. 11 Temperature dependencies of the hydrodynamic radius of PNIPA-g-PEO **a** high molecular weight and **b** low molecular weight chains in water at different polymer concentrations. *PEO* poly(ethylene oxide). (Adapted from Ref. [44])

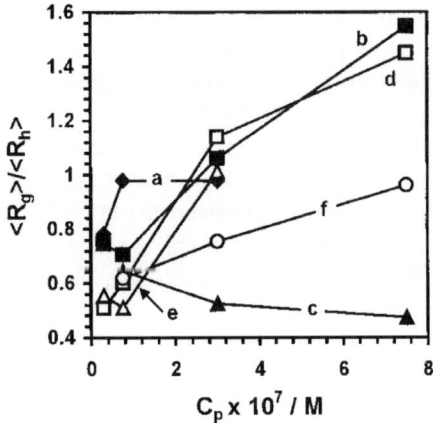

Fig. 12 Ratio of the average radius of gyration to the average hydrodynamic radius of the aggregates as a function of copolymer concentration. Ratios of NIPA-to-EO monomer units: *a* 151, *b* 69, *c* 13, *d* 244, *e* 107, *f* 27. *NIPA* N-isopropylacrylamide, *EO* ethylene oxide. (Reprinted with permission from Ref. [45]. Copyright 2002 American Chemical Society)

ical globules, $R_g/R_h = 0.78$; for rods, $R_g/R_h > 2$; for coils in a θ solvent, $R_g/R_h = 1.78$). Spherical PNIPA-*b*-PEO nanoparticles with high EO content showed in some cases anomalous behavior (Fig. 12, lines c and f): R_g/R_h was lower than the theoretical value, which was explained by the fact that the cores of those particles were much denser than their surface. Upon increase of concentration, the R_g/R_h value of most of the copolymers increased, which indicated that the aggregates assumed anisotropic shape: for the PNIPA-*b*-PEO nanostructures, an ellipsoidal form was suggested (Fig. 12, lines a, b, d, and e).

Besides block copolymers of NIPA and EO, analogous systems based on VCL were reported [52]. PVCL-*g*-PEO was synthesized by copolymerization of VCL and PEO macromonomer end-capped with a methacrylate moiety. The graft copolymers showed unusual conformational behavior depending on temperature (Fig. 13). At temperatures slightly above the LCST, the formation of huge polymer aggregates of 2-μm size was observed. On further heating, the thermosensitive core of the aggregates decorated with PEO grafts shrank readily to give clusters of 300–400 nm in diameter. Increasing the degree of grafting led to larger aggregates at 60 °C owing to a looser PVCL core at a high grafting degree. PVCL-*g*-PEO copolymers were shown to be a prospective material for drug delivery as they were characterized by low cytotoxicity [53].

A new way of obtaining nanostructures in systems of thermosensitive polymers by combining them with hydrophilic species was proposed by Bronstein et al. [54]. The core–shell nanoparticles were obtained by stabilization of the globular conformation of PVCL in aqueous solutions at 45 °C on add-

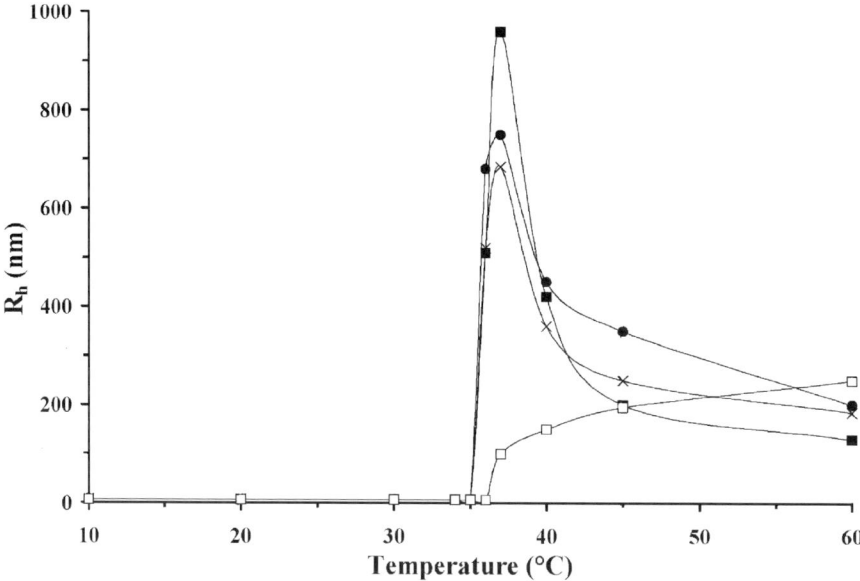

Fig. 13 Temperature dependencies of the hydrodynamic radius of PVCL-g-PEO for copolymers with different degrees of grafting: *squares* 20 wt % PEO, *crosses* 35 wt % PEO, *circles* 50 wt % PEO, at a copolymer concentration of 0.65 mg/mL; *open squares* 20 wt % at a copolymer concentration of 0.065 mg/mL. (Adapted from Ref. [52])

ition of $CoCl_2$. As visualized by transmission electron microscopy (TEM), on cooling PVCL solutions treated with $CoCl_2$ to room temperature, the nanoparticles of 11-nm size were present in the system (Fig. 14a, b). Addition of $CoCl_2$ at room temperature resulted in large aggregates of 200 nm in diameter, as shown in Fig. 14a, inset. A ^{13}C NMR study demonstrated that the spectra of PVCL solutions treated with $CoCl_2$ are identical to those of untreated ones, meaning that VCL units retain their mobility after interaction with Co^{2+} ions. Thus, they do not stay in a compact conformation characteristic of a collapsed state after cooling. To clarify the exact conformation of the particles obtained, the TEM grid was stained with OsO_4, which allowed visualization of the parts of the PVCL macromolecules not coordinated to Co^{2+} ions (Fig. 14c). Staining resulted in loose spherical structures with dark cores. It was suggested that the core consisted of certain areas of PVCL chains cross-linked by multivalent Co ions, while the rest of the molecules protruded through the cross-linked area forming the outer layer visualized in the stained image by loose coronas (Fig. 15).

Some works were reported in which thermosensitive polymers were conjugated with hydrophobic groups. End-capping random poly(NIPA-*co*-dimethylacrylamide) [55] and grafting poly(NIPA-*co*-hydroxymethylacrylamide) [56] with cholesterol moieties led to self-associating polymers with different morphologies. By dissolution of the copolymers in dimethylfor-

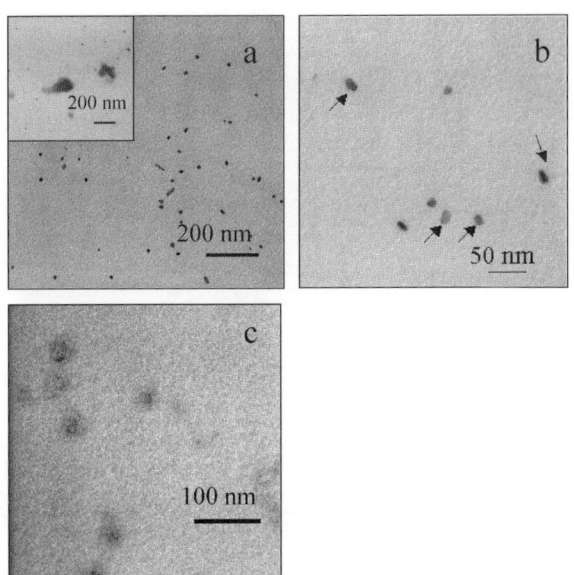

Fig. 14 Transmission electron microscopy (*TEM*) images of PVCL-Co nanoparticles. **a** Nanoparticles obtained upon cooling PVCL solutions treated with CoCl$_2$ at 35 °C (above the lower critical solution temperature). The *inset* shows large aggregates formed upon adding CoCl$_2$ at room temperature. **b** Enlarged image of PVCL nanoparticles. **c** PVCL-Co nanoparticles stained with OsO$_4$. (Reprinted with permission from Ref. [54]. Copyright 2005 American Chemical Society)

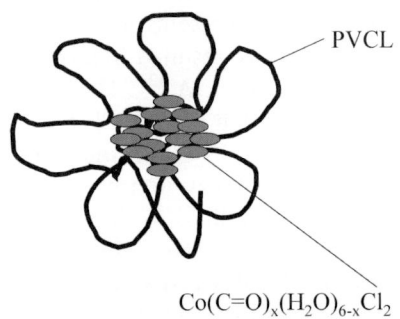

Fig. 15 A core–shell model of PVCL-Co nanoparticles. (Adapted from Ref. [54])

mamide and further dialysis against water, it was possible to obtain copolymer micelles of 30-nm size below the LCST. Heating the dialyzed solutions over 40 °C led to aggregation of the micelles with formation of particles of 200-nm size. Freeze-drying the micellar solutions led to nanoparticles of cuboid, starlike, and spherical shapes depending on the concentration of the initial micellar solution when imaged with TEM (Fig. 16).

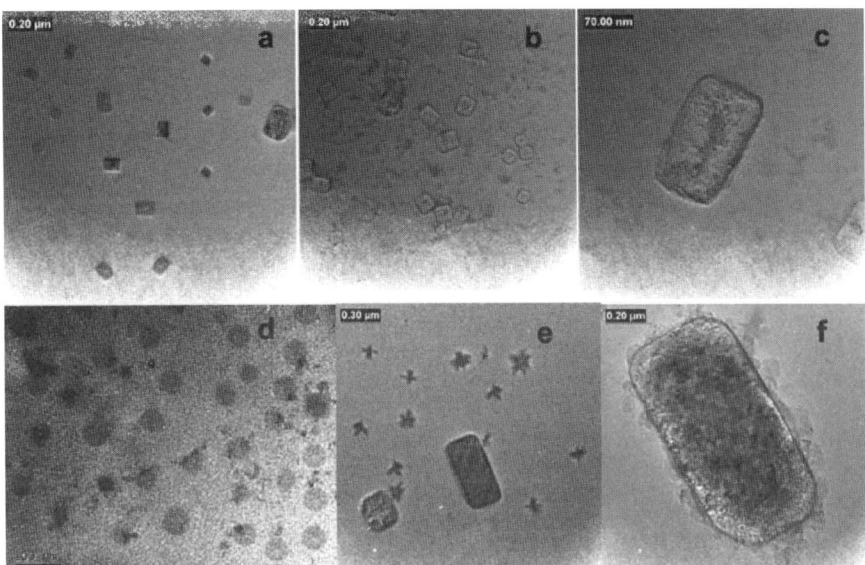

Fig. 16 TEM pictures showing nanoparticles of cholesteryl end-capped poly(NIPA-co-dimethylacrylamide). The nanoparticles were obtained by the dialyzing dimethylformamide solutions of copolymers against water and subsequent freeze-drying. The initial concentrations of the copolymer in dimethylformamide were **a–c** 0.35 wt %, **d** 0.1 wt %, **e, f** 1.2 wt %. **a, c–f** were obtained for the copolymer with $M_w = 3400$; **b** was obtained for the copolymer with $M_w = 8000$. (Reprinted with permission from Ref. [55]. Copyright 2003 Elsevier)

Diblock PNIPA-*b*-PS was synthesized by the reversible addition–fragmentation chain transfer method using PS macroinitiators [49]. The behavior of the copolymers was dependent on the length of the NIPA blocks. Shorter NIPA blocks formed micelles 120 nm in diameter, while the copolymer with longer NIPA blocks associated in huge nonstructured aggregates of 1.2 μm. Upon heating, the aggregates shrank, whereas the size of the micelles did not change considerably. Microcalorimetric studies have shown that the phase transition of the copolymer with longer PNIPA blocks takes place at the same temperature as that of PNIPA homopolymer, meaning that separate PNIPA chains are responsible for shrinking of the aggregates. The collapse of PNIPA-*b*-PS with shorter PNIPA blocks is characterized by lower temperatures of the transition and broad calorimetric peaks, which are characteristic of random PNIPA-PS copolymers (4% of styrene [50]). Indeed, such copolymers are likely to consist of short NIPA blocks separated by styrene monomer units. Thus, the character of the microcalorimetric peaks might be determined by the length of the PNIPA blocks.

As a conclusion to this part, one may state that thermosensitive polymers represent prospective macromolecules for the design of nanoparticles the properties of which are temperature-dependent. Owing to the amphiphilicity of their chains, they can manifest both hydrophilic and hydrophobic properties depending on solvent quality. Therefore, when combined with hydrophobic or hydrophilic moieties, they can self-organize in nanostructures under appropriate conditions.

4
Catalytic Properties of Polymer Associates in Aqueous Media

Catalytic properties of water-soluble synthetic polymers have long been a subject of considerable interest, which was inspired by the investigation of enzyme action mechanisms. As was shown in the works of Overberger et al. [57, 58], Kunitake et al. [59, 60], and Kirsh et al. [61, 62], hydrophobically and electrostatically driven adsorption of substrates at the chains of polymer catalysts is one of the major factors that define the catalytic activity being analogous to the forces that play a leading role in enzyme–substrate interactions. It is worth mentioning here that enzymatic mechanisms of catalysis were also realized in imprinted polymers. Those studies showed that arranging active groups of the catalyst in a stiff spatial configuration with the help of removable template molecules can lead to analogues of enzyme binding sites having a geometrical match with the substrate. Several reviews have already been published on this issue [16, 17], so imprinted polymers will not be considered in this article.

Substrate–catalyst interaction is also essential for micellar catalysis, the principles of which have long been established and consistently described in detail [63–66]. The main feature of micellar catalysis is the ability of reacting species to concentrate inside micelles, which leads to a considerable acceleration of the reaction. The same principle may apply for polymer systems. An interesting way to concentrate the substrate inside polymer catalysts is the use of cross-linked amphiphilic polymer latexes [67–69]. Liu et al. [67] synthesized a histidine-containing resin which was active in hydrolysis of p-nitrophenyl acetate (NPA). The kinetics curve of NPA decomposition in the presence of the resin was of Michaelis–Menten type, indicating that the catalytic act was accompanied by sorption of the substrate. However, no discussion of the possible sorption mechanisms (i.e., sorption by the interfaces or by the core of the resin beads) was presented.

The latexes prepared in the group of Ford [68, 69] consisted of hydrophobic monomers and a cross-linker, i.e., styrene, methacrylate monomers with various substitutes in the ester moiety, divinylbenzene, and hydrophilic monomers bearing charged groups, viz., styrylmethyl(trimethyl)ammonium and styrylmethyl(tributyl)ammonium cations. The latexes were catalytically active in reactions of decarboxylation of 6-nitrobenzioxazole-3-carboxylate (Scheme 6) and p-nitrophenyl hexanoate hydrolysis (Scheme 7, $n = 6$).

The main reason behind the acceleration was suggested to be the concentrating of substrates in a small volume of the latex phase and an increase in intrinsic rate constants, obviously due to the change in environment from the polar medium (water) to the hydrophobic one (latex), since 6-nitrobenzioxazole-3-carboxylate is stabilized in water by hydrogen bonding, while in latex such stabilization does not take place. Hydrolysis of p-nitrophenyl hexanoate in the medium of the latexes was found to have a higher rate with respect to the hydrolysis by OH^- ions in water mainly owing to the concentrating effect of both substrate and catalyst (OH^- ions), as the variation of the intrinsic rate constant was minimal. Data on the catalytic properties of the latexes are summarized in Table 1. The general tendency observed consisted in the fact that the largest rate constants were characteristic for the latexes containing long hydrophobic aliphatic tails in methacrylate and tetraalkylammonium monomer units capable of interacting effectively with hydrophobic substrates.

Interfacial adsorption may also become an effective tool to increase significantly the rate of catalytic reaction since it leads to the concentrating of reactants at the boundaries of immiscible phases, so that the interfacial layer becomes a reactor of nanoscale thickness (*surface nanoreactor*). Recently, Vasilevskaya et al. [70] emphasized that the possibility to accelerate a catalytic reaction exists in surfactant-free miniemulsions, if both catalyst and substrate adsorb at the oil–water interfaces. At emulsion droplet interfaces, a significant concentrating effect may be achieved, since the concentration at interfaces may be several orders of magnitude higher than in bulk phases. Furthermore, the absence of surfactants guarantees that a considerable part of the interface is not occupied by foreign substances, providing greater possibilities for concentrating the reactants. It was shown that at a certain optimum size of miniemulsion droplets (normally around several hundred nanometers), a significant increase in reaction rate occurs compared with the case where the size of the droplets is

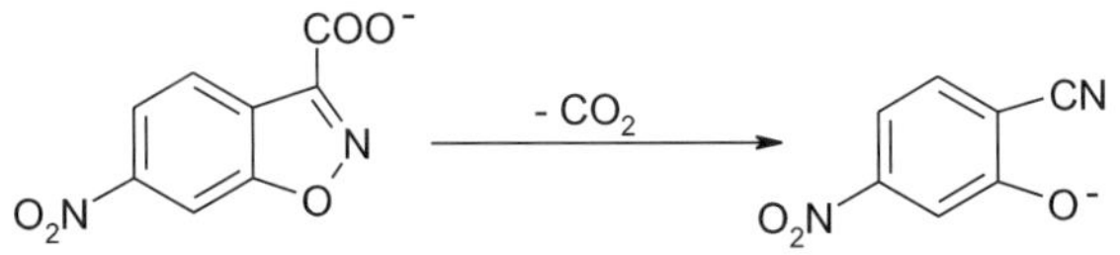

Scheme 6 Decarboxylation of 6-nitrobenzioxazole-3-carboxylate

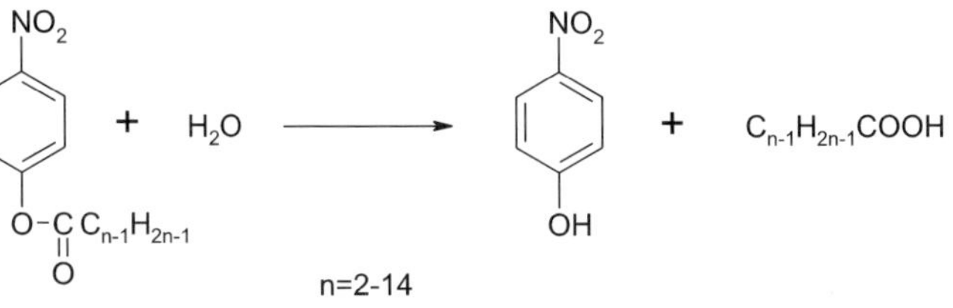

Scheme 7 Hydrolysis of p-nitrophenyl alkanoates

Table 1 Ratios of the rate constant in the presence of latex (k_{lat}) to that in water (k_w) for latexes of different composition. (Data from Ref. [69])

Latex type	Ester substitute in methacrylate monomer	k_{latex}/k_w	
SMTMA latexes	No methacrylate monomer	2.3	900
	n-Butyl	6.1	2900
	n-Hexyl	6.5	
	n-Octyl	9.5	4100
	n-Decyl	10.2	4400
	n-Dodecyl	10.7	
	Isobutyl	6.5	3300
	2-Ethylbutyl	6.1	
	2-Ethylhexyl	12.5	8300
	2-Chloroethyl	5.6	
	Butoxyethoxyethyl	5.8	400
	Ethoxyethoxyethyl	5.5	200
	Tetrahydrofurfuryl	7.1	3200
	Tetrahydropyranyl	5.6	
	Furfuryl	5.8	
SMTBA latexes	No methacrylate monomer	12.0	9600
	n-Butyl	15.7	6600
	2-Ethylhexyl	16.5	10 400

SMTMA styrylmethyl(trimethyl)ammonium, *SMTBA* styrylmethyl(tributyl)ammonium

either too small or too large, as well as with the case of complete phase separation. In Fig. 17, a series of dependencies of reaction rate vs. droplet radius are presented for different adsorption energy at the droplet interface, ε (in kT units). The positions of the maxima were found to be sensitive to ε: a lower energy of adsorption required a lower optimum size of the emulsion droplets. The reason for the presence of the reaction rate maxima in the dependencies is as follows: for small radii of the droplets the overall interfacial area is large, so a high interfacial concentration cannot be achieved, and if the droplets are too large, very few interfacial areas are available, thus reducing the probability that the catalyst and the substrate meet each other in those areas.

The analogous considerations are valid for polymer systems as well. Indeed, amphiphilic monomer units also tend to occupy interfacial areas of macromolecular associates as it is normal for low molecular weight surfactants to adsorb at polymer–poor solvent boundaries. And, if such interfacial groups of the polymer associate catalyze chemical transformation of a compound which tends to adsorb at the associate interfaces, this can result in unusual kinetics effects. Okhapkin et al. [18] studied the influence of temperature-induced aggregation on the catalytic activity of thermosensitive

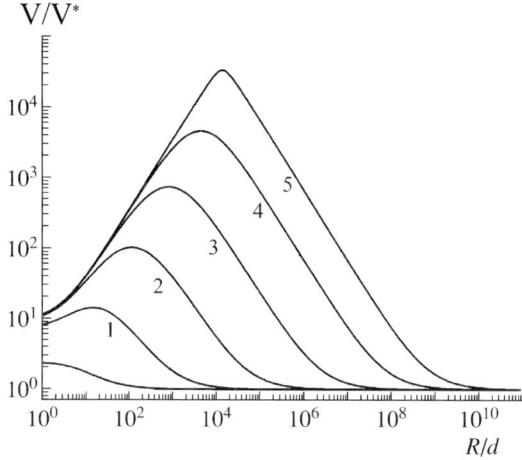

Fig. 17 Reaction rate in an emulsion as a function of emulsion droplet radius and adsorption energy of substrate and catalyst: 1 $\varepsilon = 2$, 2 $\varepsilon = 4$, 3 $\varepsilon = 8$, 4 $\varepsilon = 10$, 5 $\varepsilon = 12$, 6 $\varepsilon = 14$. The reaction rate in the emulsion is normalized by that in the homogeneous phase; the emulsion droplet radius is normalized by the diameter of a substrate molecule. (Adapted from Ref. [70])

copolymers of VCL and Vim (PVCL-Vim) and NIPA and Vim (PNIPA-Vim), which exist in a coiled state in aqueous or 2-propanol–water solutions at room temperature. The copolymers in both coil and aggregate states were tested as catalysts of NPA hydrolysis (Scheme 7, $n = 2$). As the aggregate state of such copolymers is reached by raising the temperature, the effects of aggregation and temperature on the catalytic properties of the copolymers overlapped. Therefore, the correlation of the catalytic properties with the aggregation was investigated using the reaction rate vs. temperature dependencies, which normally give a linear plot on semilogarithmic (Arrhenius) coordinates. The aggregation influenced the reaction rate along with temperature, producing a deviation from the linear law. Figure 18 shows the Arrhenius dependencies for the four copolymer catalysts, 1-methylimidazole and PVim as controls, at identical concentrations of imidazole groups. For 1-methylimidazole and PVim, the dependencies were quite linear, showing that those catalysts followed the Arrhenius-type behavior. For copolymer catalysts, the rate–temperature dependencies were not linear in Arrhenius coordinates. In the temperature range 35–45 °C, the growth law was faster than the linear one. When the temperature was raised further, the opposite effect was observed, viz., the reaction rate slowed down.

The relation between the catalytic and aggregation properties of the copolymers was shown using the dynamic light scattering method. Hydrodynamic radius distributions obtained by processing light scattering data showed that at temperatures below 35 °C the copolymers existed in the form

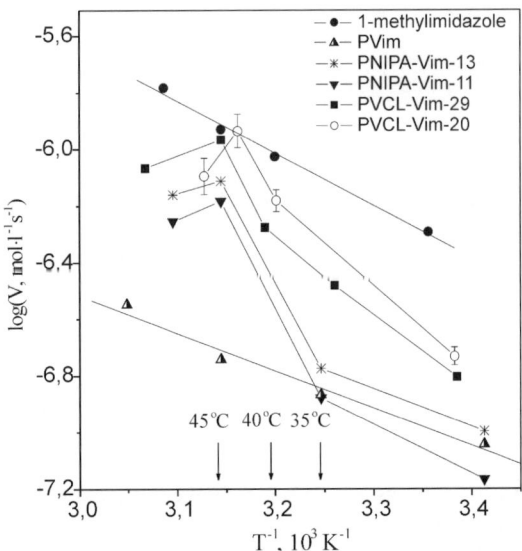

Fig. 18 Reaction rate of hydrolysis of p-nitrophenyl acetate as a function of inverse temperature. Thermosensitive imidazole-containing copolymers (PVCL-Vim, PNIPA-Vim), 1-methylimidazole and poly(1-vinylimidazole) act as catalysts. *Numbers in the copolymer abbreviations* denote the Vim content (in mole percent). *Vim* 1-vinylimidazole. (Adapted from Ref. [18])

of coils, while aggregates were formed upon heating above that temperature (Fig. 19). At low temperature the average radius of the polymer particles did not exceed 10 nm. Upon heating, new peaks at 100–200 nm emerged, indicative of the aggregation. For all the copolymers studied, the temperature intervals of aggregation preceded the temperature interval of rapid growth of the reaction rate in the region 35–45 °C. Thus, the observed acceleration of the reaction was found to be closely connected with the aggregation phenomenon in solutions of the thermosensitive copolymers.

A Michaelis–Menten profile of the catalyzed reaction was observed for the thermosensitive copolymers studied. In enzymatic catalysis, the catalytic act is preceded by the complex formation between catalyst and substrate. Because of the complex formation, enzymatic reactions follow Michaelis–Menten-type kinetics:

$$V = \frac{V_0 [S]}{K_m + [S]}, \qquad (3)$$

where $V_0 = k_{cat}[E_0]$, k_{cat} is the first-order rate constant for breakdown of the substrate–catalyst complex, $[E_0]$ is the concentration of catalyst, $[S]$ is the concentration of substrate, and K_m is the Michaelis constant, which is the dissociation constant of the enzyme–substrate complex. For a PNIPA-Vim copolymer containing 11% of imidazole groups, a kinetics curve in

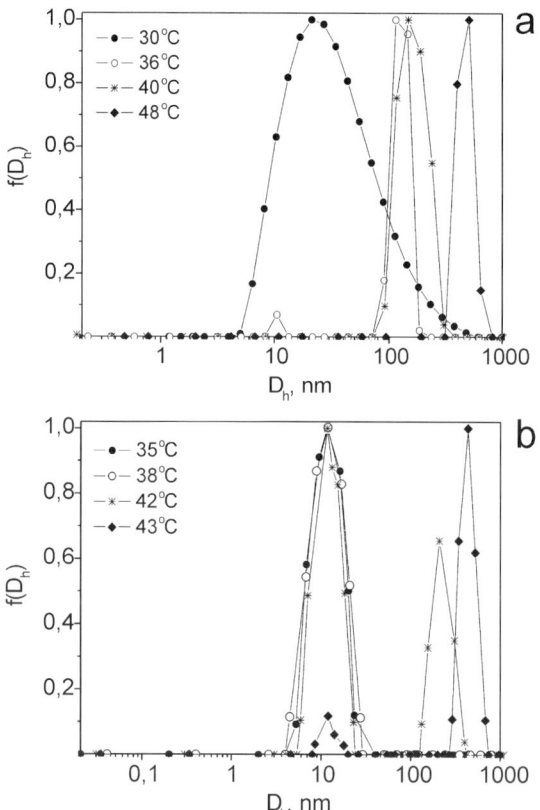

Fig. 19 Distribution functions of the hydrodynamic diameter for the imidazole-containing thermosensitive copolymers in 2-propanol–water solutions at various temperatures: **a** PNIPA-Vim, 11% of imidazole; **b** PVCL-Vim, 29% of imidazole. (Adapted from Ref. [18])

V–$[S]$ coordinates was obtained which could be well fitted with Eq. 3 giving $V_0 = 8.6 \times 10^{-7}$ mol/L s and $K_m = 0.0105$ mol/L. It was possible to explain the phenomenon of enhanced catalytic activity of the copolymer aggregates in the following terms by taking that fact into account.

As PVCL-Vim and PNIPA-Vim form aggregates of submicrometer sizes at elevated temperature, NPA can adsorb at their interfaces forming a kind of "complex" with the outer polymer groups (Fig. 20). Both Vim and NPA are amphiphilic and their affinity to interface is high: when partitioned between hexane and water, their free energies of adsorption to the interface (from water) are 5.8 kT (14 kJ/mol) and 9.7 kT (24 kJ/mol), respectively, whereas the free energies of partition are 3.0 and –2.0 kT. Thus, the location at the phase boundaries is preferential for them, and can stimulate NPA and monomer units of Vim to concentrate at the interfacial areas of the polymer aggregates, which leads to the rapid progress of the reaction in those areas.

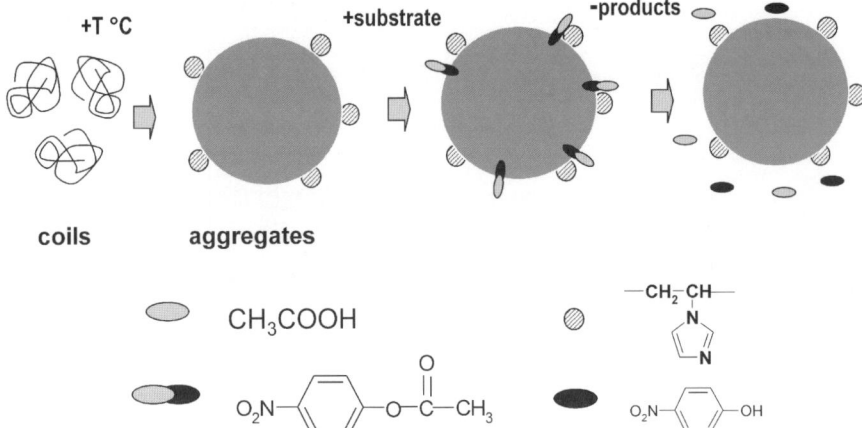

Fig. 20 Interfacial layer of PVCL-Vim and PNIPA-Vim aggregates as a catalytic nanoreactor. (Adapted from Ref. [18])

Furthermore, there can be some additional factors that increase the reaction rate at aggregate interfaces. First, both catalyst and substrate are specifically oriented in the interfacial layer owing to the polymer concentration gradient. This factor can lead to a specific mutual orientation of the interacting species, which is beneficial for an elementary act of the catalytic reaction. Second, the substrate molecules are subjected to a high stress owing to the polymer concentration gradient. The second factor should reduce the activation energy of the reaction as the stress increases the ground-state energy of the substrate molecules.

Binding to the interfaces of polymer aggregates may also result in specific catalytic effects in the case of homologous series of substrates. Lawin et al. [71] studied hydrolysis of *p*-nitrophenyl alkanoates (NPAlk) (Scheme 7) in the presence of hydroxide ions which was mediated by polymer micelles of poly(*N*-(*n*-dodecyl)-4-vinylpyridinium-*co*-*N*-ethyl-4-vinylpyridinium) bromide (PDEVP). PDEVP was reported to form compact micelles with a definite surface, which is mainly covered by charged pyridinium moieties but having as well some hydrophobic areas in the structure. The micelles were supposed to bind the substrates by hydrophobic surface areas and hydroxide ions by positively charged pyridinium ions. The diverse character of the concentration dependencies of the reaction rate was observed for substrates with different chain length (Fig. 21). This was associated with the possibility of aggregation of NPAlk in aqueous solutions suggested in a work of Guthrie [72]. In Fig. 21, one may observe a kind of kink at the concentration dependencies of $n = 8$ and $n = 10$. The kinks of the curves were identified with the critical aggregation concentrations (CACs) of the substrates, meaning that at higher NPAlk concentrations PDEVP associates rather with NPAlk aggregates than

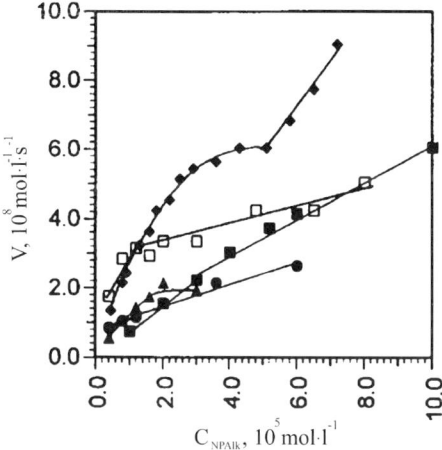

Fig. 21 Reaction rate versus substrate concentration for hydrolysis of *p*-nitrophenyl alkanoates (*NPAlk*) in the presence of poly(*N*-(*n*-dodecyl)-4-vinylpyridinium-*co*-*N*-ethyl-4-vinylpyridinium) bromide (*PDEVP*): *closed squares* $n = 6$, *diamonds* $n = 8$, *open squares* $n = 10$, *circles* $n = 12$, *triangles* $n = 14$. (Adapted from Ref. [71])

with individual molecules (Fig. 22). In particular, the CAC of $n = 8$ is rather high, which ensures the possibility of saturation of the PDEVP surface with individual molecules of the substrates. At the CAC, the newly formed aggregates start to associate with the PDEVP surface contributing to an additional increase of the reaction rate. In the case of $n = 10$, the CAC is rather low, so the saturation by individual molecules is not attained before the kink. For $n = 6$, the kink is not observed at the concentrations studied, obviously because the substrate has a relatively short hydrophobic tail and its critical aggregation concentration is too high to be observed in the concentration range studied. Thus, it was shown that the adsorption of the substrate to the polymeric catalyst may lead to complex catalytic effects which depend on the phase behavior of both substrate and catalyst.

Wang et al. [73–76] performed a study of the catalytic activity of a silicon-based polymer obtained by polycondensation of 4-bis[(3-dimethyl-ethoxy-

 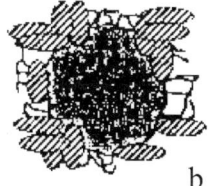

a   b

Fig. 22 Representation of different types of association between PDEVP and NPAlk: **a** adsorption of individual molecules of the substrate, **b** adsorption of aggregates of the substrate. (Adapted from Ref. [71])

silyl)propyl]aminopyridine (Scheme 8). The influence of buffer composition [73, 76], salt content [73], surfactant [74], and polymer concentration [75, 76] was investigated. It was shown that polymer 1 (Scheme 8) revealed specificity to NPAlk of certain length; moreover, the specificity was very sensitive to the solvent composition. The reaction carried out in buffered aqueous solutions was substantially accelerated at a chain length of $n = 6$ in tris(hydroxymethyl)aminomethane (Tris) buffer, whereas in borate and phosphate buffer no specificity was observed (Fig. 23). When the solvent composition was changed, namely, a 1 : 1 (v/v) methanol–water system was used, the substrate specificity in phosphate buffer was recovered for esters with $n = 10$–14 depending on catalyst concentration, whereas in Tris buffer it shifted to $n = 14$ and was constant at all polymer concentrations studied (Table 2).

The selectivity shifts were explained by changes in the morphology of the polymer aggregates, which can adsorb the substrates from methanol–water solution. A transition from spherical particles of 1 to rodlike and vesicle-like particles was suggested for phosphate buffer solutions. It was believed that the substrates with a definite tail length adsorb preferentially at the interfaces of the particles of the corresponding type (Table 2). Tris buffer was shown to increase the solubility of the polymer in the methanol–water mixture, which resulted in no morphological changes and, accordingly, changes in substrate specificity; the polymer aggregates were spherical in Tris buffer at all polymer concentrations studied.

Thus summarizing, the adsorption of substrate at the interfaces of polymer associates or emulsion droplets leads to the unusual effects associated with the progress of catalytic reactions in a nanoscale interfacial layer of the

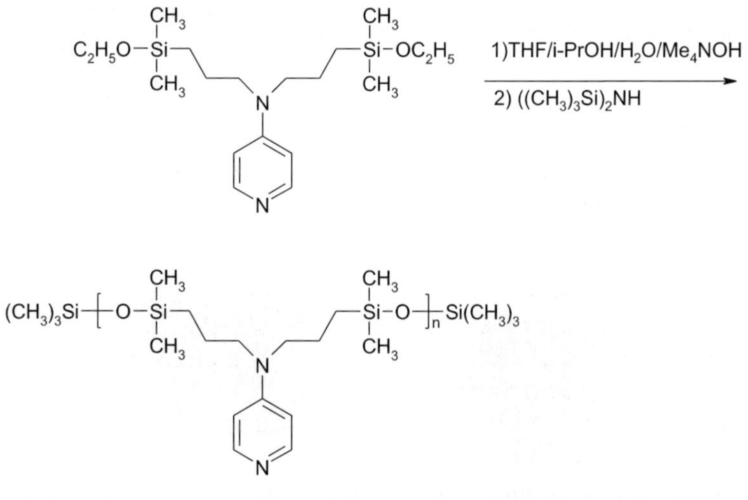

Scheme 8 Polycondensation of 4-[N,N-bis[(3-dimethylethoxysilyl)propyl]]aminopyridine

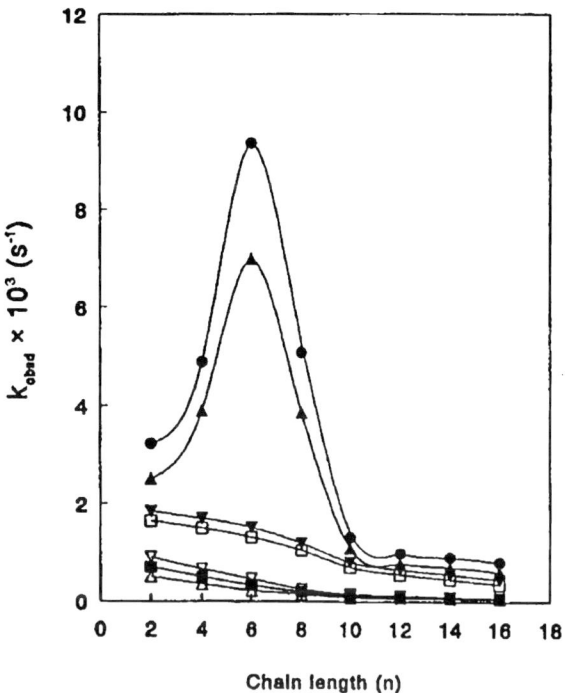

Fig. 23 Pseudo-first-order rate constants for the hydrolysis of NPAlk (n = 2–16) in the absence and in the presence of **1** as a function of alkanoate chain length n, catalyst concentration, and buffer system: *circles* 7.5×10^{-5} mol L^{-1} **1** in Tris(hydroxymethyl)aminomethane (*Tris*) buffer solution; *closed up triangles* 2.5×10^{-5} mol L^{-1} **1** in Tris buffer solution; *closed down triangles* 2.5×10^{-5} mol L^{-1} **1** in phosphate buffer solution; *open squares* 2.5×10^{-5} mol L^{-1} **1** in borate buffer solution; *open down triangles* in Tris buffer solution only; *closed squares* in phosphate buffer solution only; *open up triangles* in borate buffer solution only. (Reprinted with permission from [73]. Copyright 1996 American Chemical Society)

catalyst. It should be pointed out that interfacial reactions are of significant importance not only in catalytic applications. In biological systems, a great variety of interfaces between hydrophobic and hydrophilic areas are available, for example, in liposomes, cell membranes, and mitochondria. However, the role of interfaces is rarely taken into account in biophysical studies. Usually *n*-octanol–water partition coefficients are employed in many environmental and pharmacological studies to evaluate the allocation and fate of chemicals in environmental and biochemical systems. Partition coefficients characterize in a way the overall hydrophobicity of solutes. The latter appears to be one of the most important quantities since many binding sites of enzymes and receptors are quite susceptible to hydrophobic fragments of various low molecular weight compounds. However, as it is possible that reactions may proceed at the interface, the allocation of reacting compounds

Table 2 Substrate specificity [a] for 1-catalyzed hydrolysis of *p*-nitrophenyl alkanoates (*NPAlk*) in 1:1 methanol–aqueous solution in the presence of polymer 1 aggregates. (Data from Ref. [76])

Concentration of 1 (mol/L)	Substrate specificity		Polymer aggregate morphology	
	Phosphate buffer (0.05 mol/L)	Tris buffer (0.05 mol/L)	Phosphate buffer (0.05 mol/L)	Tris buffer (0.05 mol/L)
5.0×10^{-6}	$n = 14$	$n = 14$	Sphere	Sphere
1.0×10^{-5}	$n = 14$	$n = 14$	Sphere	Sphere
2.5×10^{-5}	$n = 12$	$n = 14$	Rod	Sphere
5.0×10^{-5}	$n = 12$	$n = 14$	Rod	Sphere
7.5×10^{-5}	$n = 10$	$n = 14$	Vesicle	Sphere
1.0×10^{-4}	$n = 10$	$n = 14$	Vesicle	Sphere

Tris tris(hydroxymethyl)aminomethane
[a] Substrate specificity is defined by the maxima of plots of pseudo-first-order rate constants for the solvolysis of the series of the substrates as a function of NPAlk chain length *n*, see Fig. 23

in the interfacial areas should be taken into account, which was emphasized in a series of works [24, 25, 77, 78]. Goldar and Sikorav [77] studied the enhancement of renaturation of complementary single-stranded DNA in water–phenol emulsions. It was shown that adsorption of the DNA chains at water–phenol interfaces leads to a dramatic increase in the renaturation rate compared with the case of bulk solution. This was attributed both to the increase of the concentration of DNA strands at the interface and to a special conformation assumed by the strands upon interaction with the interfacial phenol molecules. It was stressed that knowledge of interfacial properties of molecules should be useful in understanding their biological properties. Adam and Delbruck [78] discussed the role of reduction of dimensionality in biological systems. They showed that in diffusion-controlled processes, the transition from three-dimensional to two-dimensional diffusion may result in an increase of the process rate. As an example, the take-up of pheromone molecules from air by developed surfaces of sense organs and further two-dimensional diffusion of the pheromone to target receptors was explored. Finally, along with their role in biological processes, the interfaces were suggested to promote significantly the evolution of living systems. In particular, Oparin [79] advanced a hypothesis that prebiotic reactions were taking place in heterogeneous, coacervated system; according to Onsager [80], oil–brine interfaces of the prebiotic environment might have been the first biochemical reactors (e.g., for interfacial polymerization). Thus, the aforementioned facts allow us to consider that polymer–water and oil–water interfaces are a promising subject of research as media for chemical processes of a broad nature.

5
Conclusion

Recent studies showed that amphiphilic properties have to be taken into account for most water-soluble monomer units when their behavior in water solutions is considered. The amphiphilic properties of monomer units lead to an anisotropic shape of the polymer structures formed under appropriate conditions, which is confirmed both by computer simulation and experimental investigations. The concept of amphiphilicity applied to the monomer units leads to a new classification based on the interfacial and partitioning properties of the monomers. The classification in question opens a broad prospective for predicting properties of polymer systems with developed interfaces (i.e., micelles, polymer globules, fine dispersions of polymer aggregates). The relation between the standard free energy of adsorption and partition makes it possible to estimate semiquantitatively the distribution between the bulk and the interface of monomers and monomer units in complex polymer systems.

Amphiphilicity and surface activity of monomer units have a pronounced effect on self-organization in solutions of thermosensitive polymers. The balance of the hydrophilic and hydrophobic groups is markedly changed with the change of temperature. When the hydrophobic part prevails, a new stable microheterogeneous phase is formed with a definite size of aggregate particles [18, 81]. The possibility of obtaining nanostructures of interesting shape and behavior was shown for amphiphilic thermosensitive polymers when they were combined either with hydrophilic or hydrophobic moieties. Self-organization of thermosensitive polymers was shown to lead to formation of sharp water–polymer interfaces, which can act as catalytic *surface nanoreactors*, where chemical reactions take place with increased reaction rate, compared with the bulk solution. Two factors are responsible for that effect. First, the nanoreactors can adsorb and concentrate interfacially active substrates. Second, both substrate and catalyst moieties might be specifically oriented and subjected to additional high stress owing to the polymer concentration gradient. The adsorption of substrate at polymer–catalyst interfaces results in uncommon catalytic effects not only in the case of thermosensitive catalysts but also in the case of amphiphilic polymers of other types as well.

Finally, reactions in the *surface nanoreactors* were found to be important for biological objects such as DNA and pheromones; some hypotheses were advanced that the analogues of the surface nanoreactors might have played a significant role in biological evolution.

References

1. Vihola H, Laukkanen A, Hirvonen J, Tenhu H (2002) Eur J Pharm Sci 16:69
2. Murthy N, Campbell J, Fausto N, Hoffman AS, Stayton PS (2003) J Controlled Release 89:365
3. Twaites BR, de las Heras Alarcón C, Cunliffe D, Lavigne M, Pennadam S, Smith JR, Górecki DC, Alexander C (2004) J Controlled Release 97:551
4. Lim YB, Han SO, Kong HU (2000) Pharm Res 17:811
5. Lim Y, Choi YH, Park J (1999) J Am Chem Soc 121:5633
6. Bulmus V, Ding Z, Long CJ, Stayton PS, Hoffman AS (2000) Bioconjugate Chem 11:78
7. Hosoya K, Kubo T, Takahashi K, Ikegami T, Tanaka N (2002) J Chromatogr A 979:3
8. Gewehr M, Nakamura K, Ise N, Kitano H (1992) Makromol Chem 193:249
9. Kanazawa H, Yamamoto K, Matsushima Y, Takai N, Kikuchi A, Sakurai Y, Okano T (1996) Anal Chem 68:100
10. Kobayashi J, Kikuchi A, Sakai K, Okano T (2001) Anal Chem 73:2027
11. Kirsh YE, Vorobiev AV, Yanul NA, Fedotov YA, Timashev SF (2001) Sep Purif Technol 22-23:559
12. Hester JF, Olugebefola SC, Mayes AM (2002) J Membr Sci 208:375
13. Verspui GA, ten Brink GJ, Sheldon RA (1999) Chemtracts Org Chem 12:777
14. Franzen R, Xu Y (2005) Can J Chem Rev Can Chim 83:266
15. Cornils B, Herrmann WA (eds) (2004) Aqueous-phase organometallic catalysis: concepts and applications. 2nd edn. Wiley, New York
16. Haupt K, Mosbach K (2000) Chem Rev 100:2495
17. Wulff G (2002) Chem Rev 102:1
18. Okhapkin IM, Bronstein LM, Makhaeva EE, Matveeva VG, Sulman EM, Sulman MG, Khokhlov AR (2004) Macromolecules 37:7879
19. Rose GD, Geselowitz AR, Lesser GJ, Lee RH, Zehfus MH (1985) Science 229:834
20. Lau KF, Dill KA (1989) Macromolecules 22:3986
21. Lau KF, Dill KA (1990) Proc Natl Acad Sci USA 87:6388
22. Vasilevskaya VV, Khalatur PG, Khokhlov AR (2003) Macromolecules 36:10103
23. Vasilevskaya VV, Klochkov AA, Lazutin AA, Khalatur PG, Khokhlov AR (2004) Macromolecules 37:5444
24. Khokhlov AR, Okhapkin IM (2005) In: Adachi K, Sato T (eds) Structure and dynamics in macromolecular systems with specific interactions. Osaka University Press, Osaka, p 57
25. Okhapkin IM, Makhaeva EE, Khokhlov AR (2005) Colloid Polym Sci 284:117
26. Sangster J (1997) Octanol-water partition coefficients: fundamentals and physical chemistry. Wiley, Chichester
27. Kalizan R (1992) Anal Chem 64:619A
28. Griffin WC (1949) J Soc Cosmet Chem 1:311
29. Shih L-B, Mauer DH, Verbrugge CJ, Wu CF, Chang SL, Chen SH (1988) Macromolecules 21:3235
30. Wataoka I, Urakawa H, Kobayashi K, Akaike T, Schmidt M, Kajiwara K (1999) Macromolecules 32:1816
31. Thünemann AF, Wendler U, Jaeger W (2002) Langmuir 18:4500
32. Khokhlov AR, Khalatur PG (1998) Physica A 249:253
33. Khokhlov AR, Khalatur PG (1999) Phys Rev Lett 82:3456

34. Govorun EN, Ivanov VA, Khokhlov AR, Khalatur PG, Borovinsky AL, Grosberg AY (2001) Phys Rev E 64:040903
35. Kokufuta E (1993) Adv Polym Sci 110:157
36. Ben'Naim A (1982) In: Mittal KL, Fendler EJ (eds) Solution behaviour of surfactants: theoretical and applied aspects, vol 1. Plenum, New York, p 27
37. Makhaeva EE, Tenhu H, Khokhlov AR (1998) Macromolecules 31:6112
38. Makhaeva EE, Tenhu H, Khokhlov AR (2000) Polymer 41:9139
39. Rička J, Meewes M, Nyffenegger R, Binkert T (1990) Phys Rev Lett 65:657
40. Walter R, Ricka J, Quellet C, Nyffenegger R, Binkert T (1996) Macromolecules 29:4019
41. Wang X, Qiu X, Wu C (1998) Macromolecules 31:2972
42. Lau ACW, Wu C (1999) Macromolecules 32:58
43. Wang X, Wu C (1999) Macromolecules 32:4299
44. Qiu X, Wu C (1997) Macromolecules 30:7291
45. Virtanen J, Holappa S, Lemmetyinen H, Tenhu H (2002) Macromolecules 35:4763
46. Chen H, Li J, Ding Y, Zhang G, Zhang Q, Wu C (2005) Macromolecules 38:4403
47. Neradovic D, Soga O, Van Nostrum CF, Hennink WE (2004) Biomaterials 25:2619
48. Chen X, Ding X (2004) Colloid Polym Sci 283:452
49. Nuopponen M, Ojala J, Tenhu H (2004) Polymer 45:3643
50. Zhang G, Winnik FM, Wu C (2003) Phys Rev Lett 90:035506
51. Kim IS, Jeong YI, Cho CS, Kim SH (2000) Int J Pharm 211:1
52. Verbrugghe S, Laukkanen A, Aseyev V, Tenhu H, Winnik FM, Du Preza FE (2003) Polymer 44:6807
53. Vihola H, Laukkanen A, Valtola L, Tenhu H, Hirvonen J (2005) Biomaterials 26:3055
54. Bronstein LM, Kostylev M, Tsvetkova I, Tomaszewski J, Stein B, Makhaeva EE, Okhapkin I, Khokhlov AR (2005) Langmuir 21:2652
55. Liu XM, Yang YY, Leong KW (2003) J Colloid Interface Sci 266:295
56. Liu XM, Pramoda KP, Yang YY, Chow SY, He C (2004) Biomaterials 25:2619
57. Overberger CG, Corett R, Salarnone JC, Yaroslavsky S (1968) Macromolecules 1:331
58. Overberger CG, St Pierre T, Vorchheimer N, Lee J, Yaroslavsky S (1965) J Am Chem Soc 87:296
59. Kunitake T, Shimada F, Aso C (1969) J Am Chem Soc 91:2716
60. Kunitake T, Shinkai S (1971) J Am Chem Soc 93:4247
61. Kirsh YE, Kabanov VA, Kargin VA (1967) Dokl Akad Nauk 117:112
62. Starodubtsev SG, Kirsh YE, Kabanov VA (1974) Vysokomolek Soed A 16:2260
63. Yatsimirsky AK, Martinek K, Berezin IV (1971) Tetrahedron 27:2855
64. Berezin IV, Martinek K, Yatsimirsky AK (1973) Uspekhi Khimii 42:1729
65. Bobic C, Anghel FD, Voicu A (1995) Colloids Surf 105:305
66. Vriezema DM, Comellas Aragones M, Elemans JAAW, Cornelissen JJLM, Rowan AE, Nolte RJM (2005) Chem Rev 105:1445
67. Liu CY, Hu CC, Hung WH (1996) J Mol Catal 106:67
68. Ford WT (1997) React Funct Polym 48:3
69. Ford WT (2001) React Funct Polym 33:147
70. Vasilevskaya VV, Aerov AA, Khokhlov AR (2004) Dokl Phys Chem 398(6):1
71. Lawin LR, Fife WK, Tian CX (2000) Langmuir 16:3583
72. Guthrie JP (1973) Can J Chem 51:3494
73. Wang GJ, Ye D, Fife WK (1996) J Am Chem Soc 118:12536
74. Wang GJ, Fife WK (1997) Langmuir 13:3320
75. Wang GJ, Fife WK (1998) J Am Chem Soc 120:883

76. Wang GJ, Fife WK (1999) Macromolecules 32:559
77. Goldar A, Sikorav JL (2004) Eur Phys JE 14:211
78. Adam G, Delbruck M (1968) In: Rich A, Davidson N (eds) Structural chemistry and molecular biology. Freeman, San Francisco, p 198
79. Oparin AI (1965) Adv Enzymol 27:347
80. Onsager L (1974) In: Mintz SL, Widmayer SM (eds) Quantum statistical mechanics in the natural sciences. Plenum, New York, p 1
81. Laukkanen A, Valtola L, Winnik FM, Tenhu H (2004) Macromolecules 37:2268

Author Index Volumes 101–195

Author Index Volumes 1–100 see Volume 100

de Abajo, J. and *de la Campa, J. G.*: Processable Aromatic Polyimides. Vol. 140, pp. 23–60.
Abe, A., Furuya, H., Zhou, Z., Hiejima, T. and *Kobayashi, Y.*: Stepwise Phase Transitions of Chain Molecules: Crystallization/Melting via a Nematic Liquid-Crystalline Phase. Vol. 181, pp. 121–152.
Abetz, V. and *Simon, P. F. W.*: Phase Behaviour and Morphologies of Block Copolymers. Vol. 189, pp. 125–212.
Abetz, V. see Förster, S.: Vol. 166, pp. 173–210.
Adolf, D. B. see Ediger, M. D.: Vol. 116, pp. 73–110.
Aharoni, S. M. and *Edwards, S. F.*: Rigid Polymer Networks. Vol. 118, pp. 1–231.
Alakhov, V. Y. see Kabanov, A. V.: Vol. 193, pp. 173–198.
Albertsson, A.-C. and *Varma, I. K.*: Aliphatic Polyesters: Synthesis, Properties and Applications. Vol. 157, pp. 99–138.
Albertsson, A.-C. see Edlund, U.: Vol. 157, pp. 53–98.
Albertsson, A.-C. see Söderqvist Lindblad, M.: Vol. 157, pp. 139–161.
Albertsson, A.-C. see Stridsberg, K. M.: Vol. 157, pp. 27–51.
Albertsson, A.-C. see Al-Malaika, S.: Vol. 169, pp. 177–199.
Allegra, G. and *Meille, S. V.*: Pre-Crystalline, High-Entropy Aggregates: A Role in Polymer Crystallization? Vol. 191, pp. 87–135.
Allen, S. see Ellis, J. S.: Vol. 193, pp. 123–172.
Al-Malaika, S.: Perspectives in Stabilisation of Polyolefins. Vol. 169, pp. 121–150.
Altstädt, V.: The Influence of Molecular Variables on Fatigue Resistance in Stress Cracking Environments. Vol. 188, pp. 105–152.
Améduri, B., Boutevin, B. and *Gramain, P.*: Synthesis of Block Copolymers by Radical Polymerization and Telomerization. Vol. 127, pp. 87–142.
Améduri, B. and *Boutevin, B.*: Synthesis and Properties of Fluorinated Telechelic Monodispersed Compounds. Vol. 102, pp. 133–170.
Ameduri, B. see Taguet, A.: Vol. 184, pp. 127–211.
Amir, R. J. and *Shabat, D.*: Domino Dendrimers. Vol. 192, pp. 59–94.
Amselem, S. see Domb, A. J.: Vol. 107, pp. 93–142.
Anantawaraskul, S., Soares, J. B. P. and *Wood-Adams, P. M.*: Fractionation of Semicrystalline Polymers by Crystallization Analysis Fractionation and Temperature Rising Elution Fractionation. Vol. 182, pp. 1–54.
Andrady, A. L.: Wavelenght Sensitivity in Polymer Photodegradation. Vol. 128, pp. 47–94.
Andreis, M. and *Koenig, J. L.*: Application of Nitrogen-15 NMR to Polymers. Vol. 124, pp. 191–238.
Angiolini, L. see Carlini, C.: Vol. 123, pp. 127–214.
Anjum, N. see Gupta, B.: Vol. 162, pp. 37–63.
Anseth, K. S., Newman, S. M. and *Bowman, C. N.*: Polymeric Dental Composites: Properties and Reaction Behavior of Multimethacrylate Dental Restorations. Vol. 122, pp. 177–218.

Antonietti, M. see *Cölfen, H.*: Vol. 150, pp. 67–187.
Aoki, H. see *Ito, S.*: Vol. 182, pp. 131–170.
Armitage, B. A. see *O'Brien, D. F.*: Vol. 126, pp. 53–58.
Arnal, M. L. see *Müller, A. J.*: Vol. 190, pp. 1–63.
Arndt, M. see *Kaminski, W.*: Vol. 127, pp. 143–187.
Arnold, A. and *Holm, C.*: Efficient Methods to Compute Long-Range Interactions for Soft Matter Systems. Vol. 185, pp. 59–109.
Arnold Jr., F. E. and *Arnold, F. E.*: Rigid-Rod Polymers and Molecular Composites. Vol. 117, pp. 257–296.
Arora, M. see *Kumar, M. N. V. R.*: Vol. 160, pp. 45–118.
Arshady, R.: Polymer Synthesis via Activated Esters: A New Dimension of Creativity in Macromolecular Chemistry. Vol. 111, pp. 1–42.
Auer, S. and *Frenkel, D.*: Numerical Simulation of Crystal Nucleation in Colloids. Vol. 173, pp. 149–208.
Auriemma, F., de Rosa, C. and *Corradini, P.*: Solid Mesophases in Semicrystalline Polymers: Structural Analysis by Diffraction Techniques. Vol. 181, pp. 1–74.

Bahar, I., Erman, B. and *Monnerie, L.*: Effect of Molecular Structure on Local Chain Dynamics: Analytical Approaches and Computational Methods. Vol. 116, pp. 145–206.
Baietto-Dubourg, M. C. see *Chateauminois, A.*: Vol. 188, pp. 153–193.
Ballauff, M. see *Dingenouts, N.*: Vol. 144, pp. 1–48.
Ballauff, M. see *Holm, C.*: Vol. 166, pp. 1–27.
Ballauff, M. see *Rühe, J.*: Vol. 165, pp. 79–150.
Balsamo, V. see *Müller, A. J.*: Vol. 190, pp. 1–63.
Baltá-Calleja, F. J., González Arche, A., Ezquerra, T. A., Santa Cruz, C., Batallón, F., Frick, B. and *López Cabarcos, E.*: Structure and Properties of Ferroelectric Copolymers of Poly(vinylidene) Fluoride. Vol. 108, pp. 1–48.
Baltussen, J. J. M. see *Northolt, M. G.*: Vol. 178, pp. 1–108.
Barnes, M. D. see *Otaigbe, J. U.*: Vol. 154, pp. 1–86.
Barnes, C. M. see *Satchi-Fainaro, R.*: Vol. 193, pp. 1–65.
Barsett, H. see *Paulsen, S. B.*: Vol. 186, pp. 69–101.
Barshtein, G. R. and *Sabsai, O. Y.*: Compositions with Mineralorganic Fillers. Vol. 101, pp. 1–28.
Barton, J. see *Hunkeler, D.*: Vol. 112, pp. 115–134.
Baschnagel, J., Binder, K., Doruker, P., Gusev, A. A., Hahn, O., Kremer, K., Mattice, W. L., Müller-Plathe, F., Murat, M., Paul, W., Santos, S., Sutter, U. W. and *Tries, V.*: Bridging the Gap Between Atomistic and Coarse-Grained Models of Polymers: Status and Perspectives. Vol. 152, pp. 41–156.
Bassett, D. C.: On the Role of the Hexagonal Phase in the Crystallization of Polyethylene. Vol. 180, pp. 1–16.
Batallán, F. see *Baltá-Calleja, F. J.*: Vol. 108, pp. 1–48.
Batog, A. E., Pet'ko, I. P. and *Penczek, P.*: Aliphatic-Cycloaliphatic Epoxy Compounds and Polymers. Vol. 144, pp. 49–114.
Batrakova, E. V. see *Kabanov, A. V.*: Vol. 193, pp. 173–198.
Baughman, T. W. and *Wagener, K. B.*: Recent Advances in ADMET Polymerization. Vol. 176, pp. 1–42.
Becker, O. and *Simon, G. P.*: Epoxy Layered Silicate Nanocomposites. Vol. 179, pp. 29–82.
Bell, C. L. and *Peppas, N. A.*: Biomedical Membranes from Hydrogels and Interpolymer Complexes. Vol. 122, pp. 125–176.
Bellon-Maurel, A. see *Calmon-Decriaud, A.*: Vol. 135, pp. 207–226.

Bennett, D. E. see *O'Brien, D. F.*: Vol. 126, pp. 53–84.
Berry, G. C.: Static and Dynamic Light Scattering on Moderately Concentraded Solutions: Isotropic Solutions of Flexible and Rodlike Chains and Nematic Solutions of Rodlike Chains. Vol. 114, pp. 233–290.
Bershtein, V. A. and *Ryzhov, V. A.*: Far Infrared Spectroscopy of Polymers. Vol. 114, pp. 43–122.
Bhargava, R., Wang, S.-Q. and *Koenig, J. L*: FTIR Microspectroscopy of Polymeric Systems. Vol. 163, pp. 137–191.
Biesalski, M. see *Rühe, J.*: Vol. 165, pp. 79–150.
Bigg, D. M.: Thermal Conductivity of Heterophase Polymer Compositions. Vol. 119, pp. 1–30.
Binder, K.: Phase Transitions in Polymer Blends and Block Copolymer Melts: Some Recent Developments. Vol. 112, pp. 115–134.
Binder, K.: Phase Transitions of Polymer Blends and Block Copolymer Melts in Thin Films. Vol. 138, pp. 1–90.
Binder, K. see *Baschnagel, J.*: Vol. 152, pp. 41–156.
Binder, K., Müller, M., Virnau, P. and *González MacDowell, L.*: Polymer+Solvent Systems: Phase Diagrams, Interface Free Energies, and Nucleation. Vol. 173, pp. 1–104.
Bird, R. B. see *Curtiss, C. F.*: Vol. 125, pp. 1–102.
Biswas, M. and *Mukherjee, A.*: Synthesis and Evaluation of Metal-Containing Polymers. Vol. 115, pp. 89–124.
Biswas, M. and *Sinha Ray, S.*: Recent Progress in Synthesis and Evaluation of Polymer-Montmorillonite Nanocomposites. Vol. 155, pp. 167–221.
Blankenburg, L. see *Klemm, E.*: Vol. 177, pp. 53–90.
Blumen, A. see *Gurtovenko, A. A.*: Vol. 182, pp. 171–282.
Bogdal, D., Penczek, P., Pielichowski, J. and *Prociak, A.*: Microwave Assisted Synthesis, Crosslinking, and Processing of Polymeric Materials. Vol. 163, pp. 193–263.
Bohrisch, J., Eisenbach, C. D., Jaeger, W., Mori, H., Müller, A. H. E., Rehahn, M., Schaller, C., Traser, S. and *Wittmeyer, P.*: New Polyelectrolyte Architectures. Vol. 165, pp. 1–41.
Bolze, J. see *Dingenouts, N.*: Vol. 144, pp. 1–48.
Bosshard, C.: see *Gubler, U.*: Vol. 158, pp. 123–190.
Boutevin, B. and *Robin, J. J.*: Synthesis and Properties of Fluorinated Diols. Vol. 102, pp. 105–132.
Boutevin, B. see *Améduri, B.*: Vol. 102, pp. 133–170.
Boutevin, B. see *Améduri, B.*: Vol. 127, pp. 87–142.
Boutevin, B. see *Guida-Pietrasanta, F.*: Vol. 179, pp. 1–27.
Boutevin, B. see *Taguet, A.*: Vol. 184, pp. 127–211.
Bowman, C. N. see *Anseth, K. S.*: Vol. 122, pp. 177–218.
Boyd, R. H.: Prediction of Polymer Crystal Structures and Properties. Vol. 116, pp. 1–26.
Bracco, S. see *Sozzani, P.*: Vol. 181, pp. 153–177.
Briber, R. M. see *Hedrick, J. L.*: Vol. 141, pp. 1–44.
Bronnikov, S. V., Vettegren, V. I. and *Frenkel, S. Y.*: Kinetics of Deformation and Relaxation in Highly Oriented Polymers. Vol. 125, pp. 103–146.
Brown, H. R. see *Creton, C.*: Vol. 156, pp. 53–135.
Bruza, K. J. see *Kirchhoff, R. A.*: Vol. 117, pp. 1–66.
Buchmeiser, M. R.: Regioselective Polymerization of 1-Alkynes and Stereoselective Cyclopolymerization of a, w-Heptadiynes. Vol. 176, pp. 89–119.
Budkowski, A.: Interfacial Phenomena in Thin Polymer Films: Phase Coexistence and Segregation. Vol. 148, pp. 1–112.
Bunz, U. H. F.: Synthesis and Structure of PAEs. Vol. 177, pp. 1–52.

Burban, J. H. see Cussler, E. L.: Vol. 110, pp. 67–80.
Burchard, W.: Solution Properties of Branched Macromolecules. Vol. 143, pp. 113–194.
Butté, A. see Schork, F. J.: Vol. 175, pp. 129–255.

Calmon-Decriaud, A., Bellon-Maurel, V., Silvestre, F.: Standard Methods for Testing the Aerobic Biodegradation of Polymeric Materials. Vol. 135, pp. 207–226.
Cameron, N. R. and *Sherrington, D. C.*: High Internal Phase Emulsions (HIPEs)-Structure, Properties and Use in Polymer Preparation. Vol. 126, pp. 163–214.
de la Campa, J. G. see de Abajo, J.: Vol. 140, pp. 23–60.
Candau, F. see Hunkeler, D.: Vol. 112, pp. 115–134.
Canelas, D. A. and *DeSimone, J. M.*: Polymerizations in Liquid and Supercritical Carbon Dioxide. Vol. 133, pp. 103–140.
Canva, M. and *Stegeman, G. I.*: Quadratic Parametric Interactions in Organic Waveguides. Vol. 158, pp. 87–121.
Capek, I.: Kinetics of the Free-Radical Emulsion Polymerization of Vinyl Chloride. Vol. 120, pp. 135–206.
Capek, I.: Radical Polymerization of Polyoxyethylene Macromonomers in Disperse Systems. Vol. 145, pp. 1–56.
Capek, I. and *Chern, C.-S.*: Radical Polymerization in Direct Mini-Emulsion Systems. Vol. 155, pp. 101–166.
Cappella, B. see Munz, M.: Vol. 164, pp. 87–210.
Carlesso, G. see Prokop, A.: Vol. 160, pp. 119–174.
Carlini, C. and *Angiolini, L.*: Polymers as Free Radical Photoinitiators. Vol. 123, pp. 127–214.
Carter, K. R. see Hedrick, J. L.: Vol. 141, pp. 1–44.
Casas-Vazquez, J. see Jou, D.: Vol. 120, pp. 207–266.
Chan, C.-M. and *Li, L.*: Direct Observation of the Growth of Lamellae and Spherulites by AFM. Vol. 188, pp. 1–41.
Chandrasekhar, V.: Polymer Solid Electrolytes: Synthesis and Structure. Vol. 135, pp. 139–206.
Chang, J. Y. see Han, M. J.: Vol. 153, pp. 1–36.
Chang, T.: Recent Advances in Liquid Chromatography Analysis of Synthetic Polymers. Vol. 163, pp. 1–60.
Charleux, B. and *Faust, R.*: Synthesis of Branched Polymers by Cationic Polymerization. Vol. 142, pp. 1–70.
Chateauminois, A. and *Baietto-Dubourg, M. C.*: Fracture of Glassy Polymers Within Sliding Contacts. Vol. 188, pp. 153–193.
Chen, P. see Jaffe, M.: Vol. 117, pp. 297–328.
Chern, C.-S. see Capek, I.: Vol. 155, pp. 101–166.
Chevolot, Y. see Mathieu, H. J.: Vol. 162, pp. 1–35.
Chim, Y. T. A. see Ellis, J. S.: Vol. 193, pp. 123–172.
Choe, E.-W. see Jaffe, M.: Vol. 117, pp. 297–328.
Chow, P. Y. and *Gan, L. M.*: Microemulsion Polymerizations and Reactions. Vol. 175, pp. 257–298.
Chow, T. S.: Glassy State Relaxation and Deformation in Polymers. Vol. 103, pp. 149–190.
Chujo, Y. see Uemura, T.: Vol. 167, pp. 81–106.
Chung, S.-J. see Lin, T.-C.: Vol. 161, pp. 157–193.
Chung, T.-S. see Jaffe, M.: Vol. 117, pp. 297–328.
Clarke, N.: Effect of Shear Flow on Polymer Blends. Vol. 183, pp. 127–173.
Coenjarts, C. see Li, M.: Vol. 190, pp. 183–226.

Cölfen, H. and *Antonietti, M.*: Field-Flow Fractionation Techniques for Polymer and Colloid Analysis. Vol. 150, pp. 67–187.
Colmenero, J. see Richter, D.: Vol. 174, pp. 1–221.
Comanita, B. see Roovers, J.: Vol. 142, pp. 179–228.
Comotti, A. see Sozzani, P.: Vol. 181, pp. 153–177.
Connell, J. W. see Hergenrother, P. M.: Vol. 117, pp. 67–110.
Corradini, P. see Auriemma, F.: Vol. 181, pp. 1–74.
Creton, C., Kramer, E. J., Brown, H. R. and *Hui, C.-Y.*: Adhesion and Fracture of Interfaces Between Immiscible Polymers: From the Molecular to the Continuum Scale. Vol. 156, pp. 53–135.
Criado-Sancho, M. see Jou, D.: Vol. 120, pp. 207–266.
Curro, J. G. see Schweizer, K. S.: Vol. 116, pp. 319–378.
Curtiss, C. F. and *Bird, R. B.*: Statistical Mechanics of Transport Phenomena: Polymeric Liquid Mixtures. Vol. 125, pp. 1–102.
Cussler, E. L., Wang, K. L. and *Burban, J. H.*: Hydrogels as Separation Agents. Vol. 110, pp. 67–80.
Czub, P. see Penczek, P.: Vol. 184, pp. 1–95.

Dalton, L.: Nonlinear Optical Polymeric Materials: From Chromophore Design to Commercial Applications. Vol. 158, pp. 1–86.
Dautzenberg, H. see Holm, C.: Vol. 166, pp. 113–171.
Davidson, J. M. see Prokop, A.: Vol. 160, pp. 119–174.
Davies, M. C. see Ellis, J. S.: Vol. 193, pp. 123–172.
Den Decker, M. G. see Northolt, M. G.: Vol. 178, pp. 1–108.
Desai, S. M. and *Singh, R. P.*: Surface Modification of Polyethylene. Vol. 169, pp. 231–293.
DeSimone, J. M. see Canelas, D. A.: Vol. 133, pp. 103–140.
DeSimone, J. M. see Kennedy, K. A.: Vol. 175, pp. 329–346.
Dhal, P. K., Holmes-Farley, S. R., Huval, C. C. and *Jozefiak, T. H.*: Polymers as Drugs. Vol. 192, pp. 9–58.
DiMari, S. see Prokop, A.: Vol. 136, pp. 1–52.
Dimonie, M. V. see Hunkeler, D.: Vol. 112, pp. 115–134.
Dingenouts, N., Bolze, J., Pötschke, D. and *Ballauf, M.*: Analysis of Polymer Latexes by Small-Angle X-Ray Scattering. Vol. 144, pp. 1–48.
Dodd, L. R. and *Theodorou, D. N.*: Atomistic Monte Carlo Simulation and Continuum Mean Field Theory of the Structure and Equation of State Properties of Alkane and Polymer Melts. Vol. 116, pp. 249–282.
Doelker, E.: Cellulose Derivatives. Vol. 107, pp. 199–266.
Dolden, J. G.: Calculation of a Mesogenic Index with Emphasis Upon LC-Polyimides. Vol. 141, pp. 189–245.
Domb, A. J., Amselem, S., Shah, J. and *Maniar, M.*: Polyanhydrides: Synthesis and Characterization. Vol. 107, pp. 93–142.
Domb, A. J. see Kumar, M. N. V. R.: Vol. 160, pp. 45–118.
Doruker, P. see Baschnagel, J.: Vol. 152, pp. 41–156.
Dubois, P. see Mecerreyes, D.: Vol. 147, pp. 1–60.
Dubrovskii, S. A. see Kazanskii, K. S.: Vol. 104, pp. 97–134.
Dudowicz, J. see Freed, K. F.: Vol. 183, pp. 63–126.
Duncan, R., Ringsdorf, H. and *Satchi-Fainaro, R.*: Polymer Therapeutics: Polymers as Drugs, Drug and Protein Conjugates and Gene Delivery Systems: Past, Present and Future Opportunities. Vol. 192, pp. 1–8.
Duncan, R. see Satchi-Fainaro, R.: Vol. 193, pp. 1–65.

Dunkin, I. R. see Steinke, J.: Vol. 123, pp. 81–126.
Dunson, D. L. see McGrath, J. E.: Vol. 140, pp. 61–106.
Dziezok, P. see Rühe, J.: Vol. 165, pp. 79–150.

Eastmond, G. C.: Poly(e-caprolactone) Blends. Vol. 149, pp. 59–223.
Ebringerová, A., Hromádková, Z. and *Heinze, T.*: Hemicellulose. Vol. 186, pp. 1–67.
Economy, J. and *Goranov, K.*: Thermotropic Liquid Crystalline Polymers for High Performance Applications. Vol. 117, pp. 221–256.
Ediger, M. D. and *Adolf, D. B.*: Brownian Dynamics Simulations of Local Polymer Dynamics. Vol. 116, pp. 73–110.
Edlund, U. and *Albertsson, A.-C.*: Degradable Polymer Microspheres for Controlled Drug Delivery. Vol. 157, pp. 53–98.
Edwards, S. F. see Aharoni, S. M.: Vol. 118, pp. 1–231.
Eisenbach, C. D. see Bohrisch, J.: Vol. 165, pp. 1–41.
Ellis, J. S., Allen, S., Chim, Y. T. A., Roberts, C. J., Tendler, S. J. B. and *Davies, M. C.*: Molecular-Scale Studies on Biopolymers Using Atomic Force Microscopy. Vol. 193, pp. 123–172.
Endo, T. see Yagci, Y.: Vol. 127, pp. 59–86.
Engelhardt, H. and *Grosche, O.*: Capillary Electrophoresis in Polymer Analysis. Vol. 150, pp. 189–217.
Engelhardt, H. and *Martin, H.*: Characterization of Synthetic Polyelectrolytes by Capillary Electrophoretic Methods. Vol. 165, pp. 211–247.
Eriksson, P. see Jacobson, K.: Vol. 169, pp. 151–176.
Erman, B. see Bahar, I.: Vol. 116, pp. 145–206.
Eschner, M. see Spange, S.: Vol. 165, pp. 43–78.
Estel, K. see Spange, S.: Vol. 165, pp. 43–78.
Estevez, R. and *Van der Giessen, E.*: Modeling and Computational Analysis of Fracture of Glassy Polymers. Vol. 188, pp. 195–234.
Ewen, B. and *Richter, D.*: Neutron Spin Echo Investigations on the Segmental Dynamics of Polymers in Melts, Networks and Solutions. Vol. 134, pp. 1–130.
Ezquerra, T. A. see Baltá-Calleja, F. J.: Vol. 108, pp. 1–48.

Fatkullin, N. see Kimmich, R.: Vol. 170, pp. 1–113.
Faust, R. see Charleux, B.: Vol. 142, pp. 1–70.
Faust, R. see Kwon, Y.: Vol. 167, pp. 107–135.
Fekete, E. see Pukánszky, B.: Vol. 139, pp. 109–154.
Fendler, J. H.: Membrane-Mimetic Approach to Advanced Materials. Vol. 113, pp. 1–209.
Fetters, L. J. see Xu, Z.: Vol. 120, pp. 1–50.
Fontenot, K. see Schork, F. J.: Vol. 175, pp. 129–255.
Förster, S., Abetz, V. and *Müller, A. H. E.*: Polyelectrolyte Block Copolymer Micelles. Vol. 166, pp. 173–210.
Förster, S. and *Schmidt, M.*: Polyelectrolytes in Solution. Vol. 120, pp. 51–134.
Freed, K. F. and *Dudowicz, J.*: Influence of Monomer Molecular Structure on the Miscibility of Polymer Blends. Vol. 183, pp. 63–126.
Freire, J. J.: Conformational Properties of Branched Polymers: Theory and Simulations. Vol. 143, pp. 35–112.
Frenkel, D. see Hu, W.: Vol. 191, pp. 1–35.
Frenkel, S. Y. see Bronnikov, S. V.: Vol. 125, pp. 103–146.
Frick, B. see Baltá-Calleja, F. J.: Vol. 108, pp. 1–48.
Fridman, M. L.: see Terent'eva, J. P.: Vol. 101, pp. 29–64.
Fuchs, G. see Trimmel, G.: Vol. 176, pp. 43–87.

Fukui, K. see *Otaigbe, J. U.*: Vol. 154, pp. 1–86.
Funke, W.: Microgels-Intramolecularly Crosslinked Macromolecules with a Globular Structure. Vol. 136, pp. 137–232.
Furusho, Y. see *Takata, T.*: Vol. 171, pp. 1–75.
Furuya, H. see *Abe, A.*: Vol. 181, pp. 121–152.

Galina, H.: Mean-Field Kinetic Modeling of Polymerization: The Smoluchowski Coagulation Equation. Vol. 137, pp. 135–172.
Gan, L. M. see *Chow, P. Y.*: Vol. 175, pp. 257–298.
Ganesh, K. see *Kishore, K.*: Vol. 121, pp. 81–122.
Gaw, K. O. and *Kakimoto, M.*: Polyimide-Epoxy Composites. Vol. 140, pp. 107–136.
Geckeler, K. E. see *Rivas, B.*: Vol. 102, pp. 171–188.
Geckeler, K. E.: Soluble Polymer Supports for Liquid-Phase Synthesis. Vol. 121, pp. 31–80.
Gedde, U. W. and *Mattozzi, A.*: Polyethylene Morphology. Vol. 169, pp. 29–73.
Gehrke, S. H.: Synthesis, Equilibrium Swelling, Kinetics Permeability and Applications of Environmentally Responsive Gels. Vol. 110, pp. 81–144.
Geil, P. H., Yang, J., Williams, R. A., Petersen, K. L., Long, T.-C. and *Xu, P.*: Effect of Molecular Weight and Melt Time and Temperature on the Morphology of Poly(tetrafluorethylene). Vol. 180, pp. 89–159.
de Gennes, P.-G.: Flexible Polymers in Nanopores. Vol. 138, pp. 91–106.
Georgiou, S.: Laser Cleaning Methodologies of Polymer Substrates. Vol. 168, pp. 1–49.
Geuss, M. see *Munz, M.*: Vol. 164, pp. 87–210.
Giannelis, E. P., Krishnamoorti, R. and *Manias, E.*: Polymer-Silicate Nanocomposites: Model Systems for Confined Polymers and Polymer Brushes. Vol. 138, pp. 107–148.
Van der Giessen, E. see *Estevez, R.*: Vol. 188, pp. 195–234.
Godovsky, D. Y.: Device Applications of Polymer-Nanocomposites. Vol. 153, pp. 163–205.
Godovsky, D. Y.: Electron Behavior and Magnetic Properties Polymer-Nanocomposites. Vol. 119, pp. 79–122.
Gohy, J.-F.: Block Copolymer Micelles. Vol. 190, pp. 65–136.
González Arche, A. see *Baltá-Calleja, F. J.*: Vol. 108, pp. 1–48.
Goranov, K. see *Economy, J.*: Vol. 117, pp. 221–256.
Gramain, P. see *Améduri, B.*: Vol. 127, pp. 87–142.
Grein, C.: Toughness of Neat, Rubber Modified and Filled β-Nucleated Polypropylene: From Fundamentals to Applications. Vol. 188, pp. 43–104.
Greish, K. see *Maeda, H.*: Vol. 193, pp. 103–121.
Grest, G. S.: Normal and Shear Forces Between Polymer Brushes. Vol. 138, pp. 149–184.
Grigorescu, G. and *Kulicke, W.-M.*: Prediction of Viscoelastic Properties and Shear Stability of Polymers in Solution. Vol. 152, p. 1–40.
Gröhn, F. see *Rühe, J.*: Vol. 165, pp. 79–150.
Grosberg, A. and *Nechaev, S.*: Polymer Topology. Vol. 106, pp. 1–30.
Grosche, O. see *Engelhardt, H.*: Vol. 150, pp. 189–217.
Grubbs, R., Risse, W. and *Novac, B.*: The Development of Well-defined Catalysts for Ring-Opening Olefin Metathesis. Vol. 102, pp. 47–72.
Gubler, U. and *Bosshard, C.*: Molecular Design for Third-Order Nonlinear Optics. Vol. 158, pp. 123–190.
Guida-Pietrasanta, F. and *Boutevin, B.*: Polysilalkylene or Silarylene Siloxanes Said Hybrid Silicones. Vol. 179, pp. 1–27.
van Gunsteren, W. F. see *Gusev, A. A.*: Vol. 116, pp. 207–248.
Gupta, B. and *Anjum, N.*: Plasma and Radiation-Induced Graft Modification of Polymers for Biomedical Applications. Vol. 162, pp. 37–63.

Gurtovenko, A. A. and *Blumen, A.*: Generalized Gaussian Structures: Models for Polymer Systems with Complex Topologies. Vol. 182, pp. 171–282.
Gusev, A. A., Müller-Plathe, F., van Gunsteren, W. F. and *Suter, U. W.*: Dynamics of Small Molecules in Bulk Polymers. Vol. 116, pp. 207–248.
Gusev, A. A. see Baschnagel, J.: Vol. 152, pp. 41–156.
Guillot, J. see Hunkeler, D.: Vol. 112, pp. 115–134.
Guyot, A. and *Tauer, K.*: Reactive Surfactants in Emulsion Polymerization. Vol. 111, pp. 43–66.

Hadjichristidis, N., Pispas, S., Pitsikalis, M., Iatrou, H. and *Vlahos, C.*: Asymmetric Star Polymers Synthesis and Properties. Vol. 142, pp. 71–128.
Hadjichristidis, N., Pitsikalis, M. and *Iatrou, H.*: Synthesis of Block Copolymers. Vol. 189, pp. 1–124.
Hadjichristidis, N. see Xu, Z.: Vol. 120, pp. 1–50.
Hadjichristidis, N. see Pitsikalis, M.: Vol. 135, pp. 1–138.
Hahn, O. see Baschnagel, J.: Vol. 152, pp. 41–156.
Hakkarainen, M.: Aliphatic Polyesters: Abiotic and Biotic Degradation and Degradation Products. Vol. 157, pp. 1–26.
Hakkarainen, M. and *Albertsson, A.-C.*: Environmental Degradation of Polyethylene. Vol. 169, pp. 177–199.
Halary, J. L. see Monnerie, L.: Vol. 187, pp. 35–213.
Halary, J. L. see Monnerie, L.: Vol. 187, pp. 215–364.
Hall, H. K. see Penelle, J.: Vol. 102, pp. 73–104.
Hamley, I. W.: Crystallization in Block Copolymers. Vol. 148, pp. 113–138.
Hammouda, B.: SANS from Homogeneous Polymer Mixtures: A Unified Overview. Vol. 106, pp. 87–134.
Han, M. J. and *Chang, J. Y.*: Polynucleotide Analogues. Vol. 153, pp. 1–36.
Harada, A.: Design and Construction of Supramolecular Architectures Consisting of Cyclodextrins and Polymers. Vol. 133, pp. 141–192.
Haralson, M. A. see Prokop, A.: Vol. 136, pp. 1–52.
Harding, S. E.: Analysis of Polysaccharides by Ultracentrifugation. Size, Conformation and Interactions in Solution. Vol. 186, pp. 211–254.
Hasegawa, N. see Usuki, A.: Vol. 179, pp. 135–195.
Hassan, C. M. and *Peppas, N. A.*: Structure and Applications of Poly(vinyl alcohol) Hydrogels Produced by Conventional Crosslinking or by Freezing/Thawing Methods. Vol. 153, pp. 37–65.
Hawker, C. J.: Dentritic and Hyperbranched Macromolecules Precisely Controlled Macromolecular Architectures. Vol. 147, pp. 113–160.
Hawker, C. J. see Hedrick, J. L.: Vol. 141, pp. 1–44.
He, G. S. see Lin, T.-C.: Vol. 161, pp. 157–193.
Hedrick, J. L., Carter, K. R., Labadie, J. W., Miller, R. D., Volksen, W., Hawker, C. J., Yoon, D. Y., Russell, T. P., McGrath, J. E. and *Briber, R. M.*: Nanoporous Polyimides. Vol. 141, pp. 1–44.
Hedrick, J. L., Labadie, J. W., Volksen, W. and *Hilborn, J. G.*: Nanoscopically Engineered Polyimides. Vol. 147, pp. 61–112.
Hedrick, J. L. see Hergenrother, P. M.: Vol. 117, pp. 67–110.
Hedrick, J. L. see Kiefer, J.: Vol. 147, pp. 161–247.
Hedrick, J. L. see McGrath, J. E.: Vol. 140, pp. 61–106.
Heine, D. R., Grest, G. S. and *Curro, J. G.*: Structure of Polymer Melts and Blends: Comparison of Integral Equation theory and Computer Sumulation. Vol. 173, pp. 209–249.
Heinrich, G. and *Klüppel, M.*: Recent Advances in the Theory of Filler Networking in Elastomers. Vol. 160, pp. 1–44.

Heinze, T. see Ebringerová, A.: Vol. 186, pp. 1–67.
Heinze, T. see El Seoud, O. A.: Vol. 186, pp. 103–149.
Heller, J.: Poly (Ortho Esters). Vol. 107, pp. 41–92.
Helm, C. A. see Möhwald, H.: Vol. 165, pp. 151–175.
Hemielec, A. A. see Hunkeler, D.: Vol. 112, pp. 115–134.
Hergenrother, P. M., Connell, J. W., Labadie, J. W. and *Hedrick, J. L.*: Poly(arylene ether)s Containing Heterocyclic Units. Vol. 117, pp. 67–110.
Hernández-Barajas, J. see Wandrey, C.: Vol. 145, pp. 123–182.
Hervet, H. see Léger, L.: Vol. 138, pp. 185–226.
Hiejima, T. see Abe, A.: Vol. 181, pp. 121–152.
Hikosaka, M., Watanabe, K., Okada, K. and *Yamazaki, S.*: Topological Mechanism of Polymer Nucleation and Growth – The Role of Chain Sliding Diffusion and Entanglement. Vol. 191, pp. 137–186.
Hilborn, J. G. see Hedrick, J. L.: Vol. 147, pp. 61–112.
Hilborn, J. G. see Kiefer, J.: Vol. 147, pp. 161–247.
Hillborg, H. see Vancso, G. J.: Vol. 182, pp. 55–129.
Hillmyer, M. A.: Nanoporous Materials from Block Copolymer Precursors. Vol. 190, pp. 137–181.
Hiramatsu, N. see Matsushige, M.: Vol. 125, pp. 147–186.
Hirasa, O. see Suzuki, M.: Vol. 110, pp. 241–262.
Hirotsu, S.: Coexistence of Phases and the Nature of First-Order Transition in Poly-N-isopropylacrylamide Gels. Vol. 110, pp. 1–26.
Höcker, H. see Klee, D.: Vol. 149, pp. 1–57.
Holm, C. see Arnold, A.: Vol. 185, pp. 59–109.
Holm, C., Hofmann, T., Joanny, J. F., Kremer, K., Netz, R. R., Reineker, P., Seidel, C., Vilgis, T. A. and *Winkler, R. G.*: Polyelectrolyte Theory. Vol. 166, pp. 67–111.
Holm, C., Rehahn, M., Oppermann, W. and *Ballauff, M.*: Stiff-Chain Polyelectrolytes. Vol. 166, pp. 1–27.
Holmes-Farley, S. R. see Dhal, P. K.: Vol. 192, pp. 9–58.
Hornsby, P.: Rheology, Compounding and Processing of Filled Thermoplastics. Vol. 139, pp. 155–216.
Houbenov, N. see Rühe, J.: Vol. 165, pp. 79–150.
Hromádková, Z. see Ebringerová, A.: Vol. 186, pp. 1–67.
Hu, W. and *Frenkel, D.*: Polymer Crystallization Driven by Anisotropic Interactions. Vol. 191, pp. 1–35.
Huber, K. see Volk, N.: Vol. 166, pp. 29–65.
Hugenberg, N. see Rühe, J.: Vol. 165, pp. 79–150.
Hui, C.-Y. see Creton, C.: Vol. 156, pp. 53–135.
Hult, A., Johansson, M. and *Malmström, E.*: Hyperbranched Polymers. Vol. 143, pp. 1–34.
Hünenberger, P. H.: Thermostat Algorithms for Molecular-Dynamics Simulations. Vol. 173, pp. 105–147.
Hunkeler, D., Candau, F., Pichot, C., Hemielec, A. E., Xie, T. Y., Barton, J., Vaskova, V., Guillot, J., Dimonie, M. V. and *Reichert, K. H.*: Heterophase Polymerization: A Physical and Kinetic Comparision and Categorization. Vol. 112, pp. 115–134.
Hunkeler, D. see Macko, T.: Vol. 163, pp. 61–136.
Hunkeler, D. see Prokop, A.: Vol. 136, pp. 1–52; 53–74.
Hunkeler, D. see Wandrey, C.: Vol. 145, pp. 123–182.
Huval, C. C. see Dhal, P. K.: Vol. 192, pp. 9–58.

Iatrou, H. see Hadjichristidis, N.: Vol. 142, pp. 71–128.

Iatrou, H. see Hadjichristidis, N.: Vol. 189, pp. 1–124.
Ichikawa, T. see Yoshida, H.: Vol. 105, pp. 3–36.
Ihara, E. see Yasuda, H.: Vol. 133, pp. 53–102.
Ikada, Y. see Uyama, Y.: Vol. 137, pp. 1–40.
Ikehara, T. see Jinnuai, H.: Vol. 170, pp. 115–167.
Ilavsky, M.: Effect on Phase Transition on Swelling and Mechanical Behavior of Synthetic Hydrogels. Vol. 109, pp. 173–206.
Imai, M. see Kaji, K.: Vol. 191, pp. 187–240.
Imai, Y.: Rapid Synthesis of Polyimides from Nylon-Salt Monomers. Vol. 140, pp. 1–23.
Imanishi, Y. see Saito, S.: Vol. 160, pp. 107–131.
Inoue, S. see Sugimoto, H.: Vol. 146, pp. 39–120.
Irie, M.: Stimuli-Responsive Poly(N-isopropylacrylamide), Photo- and Chemical-Induced Phase Transitions. Vol. 110, pp. 49–66.
Ise, N. see Matsuoka, H.: Vol. 114, pp. 187–232.
Ishikawa, T.: Advances in Inorganic Fibers. Vol. 178, pp. 109–144.
Ito, H.: Chemical Amplification Resists for Microlithography. Vol. 172, pp. 37–245.
Ito, K. and *Kawaguchi, S.*: Poly(macronomers), Homo- and Copolymerization. Vol. 142, pp. 129–178.
Ito, K. see Kawaguchi, S.: Vol. 175, pp. 299–328.
Ito, S. and *Aoki, H.*: Nano-Imaging of Polymers by Optical Microscopy. Vol. 182, pp. 131–170.
Ito, Y. see Suginome, M.: Vol. 171, pp. 77–136.
Ivanov, A. E. see Zubov, V. P.: Vol. 104, pp. 135–176.

Jacob, S. and *Kennedy, J.*: Synthesis, Characterization and Properties of OCTA-ARM Polyisobutylene-Based Star Polymers. Vol. 146, pp. 1–38.
Jacobson, K., Eriksson, P., Reitberger, T. and *Stenberg, B.*: Chemiluminescence as a Tool for Polyolefin. Vol. 169, pp. 151–176.
Jaeger, W. see Bohrisch, J.: Vol. 165, pp. 1–41.
Jaffe, M., Chen, P., Choe, E.-W., Chung, T.-S. and *Makhija, S.*: High Performance Polymer Blends. Vol. 117, pp. 297–328.
Jancar, J.: Structure-Property Relationships in Thermoplastic Matrices. Vol. 139, pp. 1–66.
Jen, A. K.-Y. see Kajzar, F.: Vol. 161, pp. 1–85.
Jerome, R. see Mecerreyes, D.: Vol. 147, pp. 1–60.
de Jeu, W. H. see Li, L.: Vol. 181, pp. 75–120.
Jiang, M., Li, M., Xiang, M. and *Zhou, H.*: Interpolymer Complexation and Miscibility and Enhancement by Hydrogen Bonding. Vol. 146, pp. 121–194.
Jin, J. see Shim, H.-K.: Vol. 158, pp. 191–241.
Jinnai, H., Nishikawa, Y., Ikehara, T. and *Nishi, T.*: Emerging Technologies for the 3D Analysis of Polymer Structures. Vol. 170, pp. 115–167.
Jo, W. H. and *Yang, J. S.*: Molecular Simulation Approaches for Multiphase Polymer Systems. Vol. 156, pp. 1–52.
Joanny, J.-F. see Holm, C.: Vol. 166, pp. 67–111.
Joanny, J.-F. see Thünemann, A. F.: Vol. 166, pp. 113–171.
Johannsmann, D. see Rühe, J.: Vol. 165, pp. 79–150.
Johansson, M. see Hult, A.: Vol. 143, pp. 1–34.
Joos-Müller, B. see Funke, W.: Vol. 136, pp. 137–232.
Jou, D., Casas-Vazquez, J. and *Criado-Sancho, M.*: Thermodynamics of Polymer Solutions under Flow: Phase Separation and Polymer Degradation. Vol. 120, pp. 207–266.
Jozefiak, T. H. see Dhal, P. K.: Vol. 192, pp. 9–58.

Kabanov, A. V., Batrakova, E. V., Sherman, S. and *Alakhov, V. Y.*: Polymer Genomics. Vol. 193, pp. 173–198.
Kaetsu, I.: Radiation Synthesis of Polymeric Materials for Biomedical and Biochemical Applications. Vol. 105, pp. 81–98.
Kaji, K., Nishida, K., Kanaya, T., Matsuba, G., Konishi, T. and *Imai, M.*: Spinodal Crystallization of Polymers: Crystallization from the Unstable Melt. Vol. 191, pp. 187–240.
Kaji, K. see Kanaya, T.: Vol. 154, pp. 87–141.
Kajzar, F., Lee, K.-S. and *Jen, A. K.-Y.*: Polymeric Materials and their Orientation Techniques for Second-Order Nonlinear Optics. Vol. 161, pp. 1–85.
Kakimoto, M. see Gaw, K. O.: Vol. 140, pp. 107–136.
Kaminski, W. and *Arndt, M.*: Metallocenes for Polymer Catalysis. Vol. 127, pp. 143–187.
Kammer, H. W., Kressler, H. and *Kummerloewe, C.*: Phase Behavior of Polymer Blends – Effects of Thermodynamics and Rheology. Vol. 106, pp. 31–86.
Kanaya, T. and *Kaji, K.*: Dynamcis in the Glassy State and Near the Glass Transition of Amorphous Polymers as Studied by Neutron Scattering. Vol. 154, pp. 87–141.
Kanaya, T. see Kaji, K.: Vol. 191, pp. 187–240.
Kandyrin, L. B. and *Kuleznev, V. N.*: The Dependence of Viscosity on the Composition of Concentrated Dispersions and the Free Volume Concept of Disperse Systems. Vol. 103, pp. 103–148.
Kaneko, M. see Ramaraj, R.: Vol. 123, pp. 215–242.
Kang, E. T., Neoh, K. G. and *Tan, K. L.*: X-Ray Photoelectron Spectroscopic Studies of Electroactive Polymers. Vol. 106, pp. 135–190.
Kaplan, D. L. see Singh, A.: Vol. 194, pp. 211–224.
Kaplan, D. L. see Xu, P.: Vol. 194, pp. 69–94.
Karlsson, S. see Söderqvist Lindblad, M.: Vol. 157, pp. 139–161.
Karlsson, S.: Recycled Polyolefins. Material Properties and Means for Quality Determination. Vol. 169, pp. 201–229.
Kataoka, K. see Nishiyama, N.: Vol. 193, pp. 67–101.
Kato, K. see Uyama, Y.: Vol. 137, pp. 1–40.
Kato, M. see Usuki, A.: Vol. 179, pp. 135–195.
Kausch, H.-H. and *Michler, G. H.*: The Effect of Time on Crazing and Fracture. Vol. 187, pp. 1–33.
Kausch, H.-H. see Monnerie, L. Vol. 187, pp. 215–364.
Kautek, W. see Krüger, J.: Vol. 168, pp. 247–290.
Kawaguchi, S. see Ito, K.: Vol. 142, pp. 129–178.
Kawaguchi, S. and *Ito, K.*: Dispersion Polymerization. Vol. 175, pp. 299–328.
Kawata, S. see Sun, H.-B.: Vol. 170, pp. 169–273.
Kazanskii, K. S. and *Dubrovskii, S. A.*: Chemistry and Physics of Agricultural Hydrogels. Vol. 104, pp. 97–134.
Kennedy, J. P. see Jacob, S.: Vol. 146, pp. 1–38.
Kennedy, J. P. see Majoros, I.: Vol. 112, pp. 1–113.
Kennedy, K. A., Roberts, G. W. and *DeSimone, J. M.*: Heterogeneous Polymerization of Fluoroolefins in Supercritical Carbon Dioxide. Vol. 175, pp. 329–346.
Khalatur, P. G. and *Khokhlov, A. R.*: Computer-Aided Conformation-Dependent Design of Copolymer Sequences. Vol. 195, pp. 1–100.
Khokhlov, A., Starodybtzev, S. and *Vasilevskaya, V.*: Conformational Transitions of Polymer Gels: Theory and Experiment. Vol. 109, pp. 121–172.
Khokhlov, A. R. see Khalatur, P. G.: Vol. 195, pp. 1–100.
Khokhlov, A. R. see Okhapkin, I. M.: Vol. 195, pp. 177–210.

Kiefer, J., Hedrick, J. L. and *Hiborn, J. G.*: Macroporous Thermosets by Chemically Induced Phase Separation. Vol. 147, pp. 161–247.
Kihara, N. see Takata, T.: Vol. 171, pp. 1–75.
Kilian, H. G. and *Pieper, T.*: Packing of Chain Segments. A Method for Describing X-Ray Patterns of Crystalline, Liquid Crystalline and Non-Crystalline Polymers. Vol. 108, pp. 49–90.
Kim, J. see Quirk, R. P.: Vol. 153, pp. 67–162.
Kim, K.-S. see Lin, T.-C.: Vol. 161, pp. 157–193.
Kimmich, R. and *Fatkullin, N.*: Polymer Chain Dynamics and NMR. Vol. 170, pp. 1–113.
Kippelen, D. and *Peyghambarian, N.*: Photorefractive Polymers and their Applications Vol. 161, pp. 87–156.
Kirchhoff, R. A. and *Bruza, K. J.*: Polymers from Benzocyclobutenes. Vol. 117, pp. 1–66.
Kishore, K. and *Ganesh, K.*: Polymers Containing Disulfide, Tetrasulfide, Diselenide and Ditelluride Linkages in the Main Chain. Vol. 121, pp. 81–122.
Kitamaru, R.: Phase Structure of Polyethylene and Other Crystalline Polymers by Solid-State 13C/MNR. Vol. 137, pp. 41–102.
Klapper, M. see Rusanov, A. L.: Vol. 179, pp. 83–134.
Klee, D. and *Höcker, H.*: Polymers for Biomedical Applications: Improvement of the Interface Compatibility. Vol. 149, pp. 1–57.
Klemm, E., Pautzsch, T. and *Blankenburg, L.*: Organometallic PAEs. Vol. 177, pp. 53–90.
Klier, J. see Scranton, A. B.: Vol. 122, pp. 1–54.
v. Klitzing, R. and *Tieke, B.*: Polyelectrolyte Membranes. Vol. 165, pp. 177–210.
Kloeckner, J. see Wagner, E.: Vol. 192, pp. 135–173.
Klüppel, M.: The Role of Disorder in Filler Reinforcement of Elastomers on Various Length Scales. Vol. 164, pp. 1–86.
Klüppel, M. see Heinrich, G.: Vol. 160, pp. 1–44.
Knuuttila, H., Lehtinen, A. and *Nummila-Pakarinen, A.*: Advanced Polyethylene Technologies – Controlled Material Properties. Vol. 169, pp. 13–27.
Kobayashi, S. and *Ohmae, M.*: Enzymatic Polymerization to Polysaccharides. Vol. 194, pp. 159–210.
Kobayashi, S. see Uyama, H.: Vol. 194, pp. 51–67.
Kobayashi, S. see Uyama, H.: Vol. 194, pp. 133–158.
Kobayashi, S., Shoda, S. and *Uyama, H.*: Enzymatic Polymerization and Oligomerization. Vol. 121, pp. 1–30.
Kobayashi, T. see Abe, A.: Vol. 181, pp. 121–152.
Köhler, W. and *Schäfer, R.*: Polymer Analysis by Thermal-Diffusion Forced Rayleigh Scattering. Vol. 151, pp. 1–59.
Koenig, J. L. see Bhargava, R.: Vol. 163, pp. 137–191.
Koenig, J. L. see Andreis, M.: Vol. 124, pp. 191–238.
Koike, T.: Viscoelastic Behavior of Epoxy Resins Before Crosslinking. Vol. 148, pp. 139–188.
Kokko, E. see Löfgren, B.: Vol. 169, pp. 1–12.
Kokufuta, E.: Novel Applications for Stimulus-Sensitive Polymer Gels in the Preparation of Functional Immobilized Biocatalysts. Vol. 110, pp. 157–178.
Konishi, T. see Kaji, K.: Vol. 191, pp. 187–240.
Konno, M. see Saito, S.: Vol. 109, pp. 207–232.
Konradi, R. see Rühe, J.: Vol. 165, pp. 79–150.
Kopecek, J. see Putnam, D.: Vol. 122, pp. 55–124.
Koßmehl, G. see Schopf, G.: Vol. 129, pp. 1–145.
Kostoglodov, P. V. see Rusanov, A. L.: Vol. 179, pp. 83–134.
Kozlov, E. see Prokop, A.: Vol. 160, pp. 119–174.

Kramer, E. J. see Creton, C.: Vol. 156, pp. 53–135.
Kremer, K. see Baschnagel, J.: Vol. 152, pp. 41–156.
Kremer, K. see Holm, C.: Vol. 166, pp. 67–111.
Kressler, J. see Kammer, H. W.: Vol. 106, pp. 31–86.
Kricheldorf, H. R.: Liquid-Cristalline Polyimides. Vol. 141, pp. 83–188.
Krishnamoorti, R. see Giannelis, E. P.: Vol. 138, pp. 107–148.
Krüger, J. and *Kautek, W.*: Ultrashort Pulse Laser Interaction with Dielectrics and Polymers, Vol. 168, pp. 247–290.
Kuchanov, S. I.: Modern Aspects of Quantitative Theory of Free-Radical Copolymerization. Vol. 103, pp. 1–102.
Kuchanov, S. I.: Principles of Quantitive Description of Chemical Structure of Synthetic Polymers. Vol. 152, pp. 157–202.
Kudaibergennow, S. E.: Recent Advances in Studying of Synthetic Polyampholytes in Solutions. Vol. 144, pp. 115–198.
Kuleznev, V. N. see Kandyrin, L. B.: Vol. 103, pp. 103–148.
Kulichkhin, S. G. see Malkin, A. Y.: Vol. 101, pp. 217–258.
Kulicke, W.-M. see Grigorescu, G.: Vol. 152, pp. 1–40.
Kumar, M. N. V. R., Kumar, N., Domb, A. J. and *Arora, M.*: Pharmaceutical Polymeric Controlled Drug Delivery Systems. Vol. 160, pp. 45–118.
Kumar, N. see Kumar, M. N. V. R.: Vol. 160, pp. 45–118.
Kummerloewe, C. see Kammer, H. W.: Vol. 106, pp. 31–86.
Kuznetsova, N. P. see Samsonov, G. V.: Vol. 104, pp. 1–50.
Kwon, Y. and *Faust, R.*: Synthesis of Polyisobutylene-Based Block Copolymers with Precisely Controlled Architecture by Living Cationic Polymerization. Vol. 167, pp. 107–135.

Labadie, J. W. see Hergenrother, P. M.: Vol. 117, pp. 67–110.
Labadie, J. W. see Hedrick, J. L.: Vol. 141, pp. 1–44.
Labadie, J. W. see Hedrick, J. L.: Vol. 147, pp. 61–112.
Lamparski, H. G. see O'Brien, D. F.: Vol. 126, pp. 53–84.
Laschewsky, A.: Molecular Concepts, Self-Organisation and Properties of Polysoaps. Vol. 124, pp. 1–86.
Laso, M. see Leontidis, E.: Vol. 116, pp. 283–318.
Lauprêtre, F. see Monnerie, L.: Vol. 187, pp. 35–213.
Lazár, M. and *Rychl, R.*: Oxidation of Hydrocarbon Polymers. Vol. 102, pp. 189–222.
Lechowicz, J. see Galina, H.: Vol. 137, pp. 135–172.
Léger, L., Raphaël, E. and *Hervet, H.*: Surface-Anchored Polymer Chains: Their Role in Adhesion and Friction. Vol. 138, pp. 185–226.
Lenz, R. W.: Biodegradable Polymers. Vol. 107, pp. 1–40.
Leontidis, E., de Pablo, J. J., Laso, M. and *Suter, U. W.*: A Critical Evaluation of Novel Algorithms for the Off-Lattice Monte Carlo Simulation of Condensed Polymer Phases. Vol. 116, pp. 283–318.
Lee, B. see Quirk, R. P.: Vol. 153, pp. 67–162.
Lee, K.-S. see Kajzar, F.: Vol. 161, pp. 1–85.
Lee, Y. see Quirk, R. P.: Vol. 153, pp. 67–162.
Lehtinen, A. see Knuuttila, H.: Vol. 169, pp. 13–27.
Leónard, D. see Mathieu, H. J.: Vol. 162, pp. 1–35.
Lesec, J. see Viovy, J.-L.: Vol. 114, pp. 1–42.
Levesque, D. see Weis, J.-J.: Vol. 185, pp. 163–225.
Li, L. and *de Jeu, W. H.*: Flow-induced mesophases in crystallizable polymers. Vol. 181, pp. 75–120.

Li, L. see *Chan, C.-M.*: Vol. 188, pp. 1–41.
Li, M., Coenjarts, C. and *Ober, C. K.*: Patternable Block Copolymers. Vol. 190, pp. 183–226.
Li, M. see *Jiang, M.*: Vol. 146, pp. 121–194.
Liang, G. L. see *Sumpter, B. G.*: Vol. 116, pp. 27–72.
Lienert, K.-W.: Poly(ester-imide)s for Industrial Use. Vol. 141, pp. 45–82.
Likhatchev, D. see *Rusanov, A. L.*: Vol. 179, pp. 83–134.
Lin, J. and *Sherrington, D. C.*: Recent Developments in the Synthesis, Thermostability and Liquid Crystal Properties of Aromatic Polyamides. Vol. 111, pp. 177–220.
Lin, T.-C., Chung, S.-J., Kim, K.-S., Wang, X., He, G. S., Swiatkiewicz, J., Pudavar, H. E. and *Prasad, P. N.*: Organics and Polymers with High Two-Photon Activities and their Applications. Vol. 161, pp. 157–193.
Linse, P.: Simulation of Charged Colloids in Solution. Vol. 185, pp. 111–162.
Lippert, T.: Laser Application of Polymers. Vol. 168, pp. 51–246.
Liu, Y. see *Söderqvist Lindblad, M.*: Vol. 157, pp. 139–161.
Long, T.-C. see *Geil, P. H.*: Vol. 180, pp. 89–159.
López Cabarcos, E. see *Baltá-Calleja, F. J.*: Vol. 108, pp. 1–48.
Lotz, B.: Analysis and Observation of Polymer Crystal Structures at the Individual Stem Level. Vol. 180, pp. 17–44.
Löfgren, B., Kokko, E. and *Seppälä, J.*: Specific Structures Enabled by Metallocene Catalysis in Polyethenes. Vol. 169, pp. 1–12.
Löwen, H. see *Thünemann, A. F.*: Vol. 166, pp. 113–171.
Luo, Y. see *Schork, F. J.*: Vol. 175, pp. 129–255.

Macko, T. and *Hunkeler, D.*: Liquid Chromatography under Critical and Limiting Conditions: A Survey of Experimental Systems for Synthetic Polymers. Vol. 163, pp. 61–136.
Maeda, H., Greish, K. and *Fang, J.*: The EPR Effect and Polymeric Drugs: A Paradigm Shift for Cancer Chemotherapy in the 21st Century. Vol. 193, pp. 103–121.
Majoros, I., Nagy, A. and *Kennedy, J. P.*: Conventional and Living Carbocationic Polymerizations United. I. A Comprehensive Model and New Diagnostic Method to Probe the Mechanism of Homopolymerizations. Vol. 112, pp. 1–113.
Makhaeva, E. E. see *Okhapkin, I. M.*: Vol. 195, pp. 177–210.
Makhija, S. see *Jaffe, M.*: Vol. 117, pp. 297–328.
Malmström, E. see *Hult, A.*: Vol. 143, pp. 1–34.
Malkin, A. Y. and *Kulichkhin, S. G.*: Rheokinetics of Curing. Vol. 101, pp. 217–258.
Maniar, M. see *Domb, A. J.*: Vol. 107, pp. 93–142.
Manias, E. see *Giannelis, E. P.*: Vol. 138, pp. 107–148.
Martin, H. see *Engelhardt, H.*: Vol. 165, pp. 211–247.
Marty, J. D. and *Mauzac, M.*: Molecular Imprinting: State of the Art and Perspectives. Vol. 172, pp. 1–35.
Mashima, K., Nakayama, Y. and *Nakamura, A.*: Recent Trends in Polymerization of a-Olefins Catalyzed by Organometallic Complexes of Early Transition Metals. Vol. 133, pp. 1–52.
Mathew, D. see *Reghunadhan Nair, C. P.*: Vol. 155, pp. 1–99.
Mathieu, H. J., Chevolot, Y., Ruiz-Taylor, L. and *Leónard, D.*: Engineering and Characterization of Polymer Surfaces for Biomedical Applications. Vol. 162, pp. 1–35.
Matsuba, G. see *Kaji, K.*: Vol. 191, pp. 187–240.
Matsumura S.: Enzymatic Synthesis of Polyesters via Ring-Opening Polymerization. Vol. 194, pp. 95–132.
Matsumoto, A.: Free-Radical Crosslinking Polymerization and Copolymerization of Multivinyl Compounds. Vol. 123, pp. 41–80.
Matsumoto, A. see *Otsu, T.*: Vol. 136, pp. 75–138.

Matsuoka, H. and *Ise, N.*: Small-Angle and Ultra-Small Angle Scattering Study of the Ordered Structure in Polyelectrolyte Solutions and Colloidal Dispersions. Vol. 114, pp. 187–232.
Matsushige, K., Hiramatsu, N. and *Okabe, H.*: Ultrasonic Spectroscopy for Polymeric Materials. Vol. 125, pp. 147–186.
Mattice, W. L. see Rehahn, M.: Vol. 131/132, pp. 1–475.
Mattice, W. L. see Baschnagel, J.: Vol. 152, pp. 41–156.
Mattozzi, A. see Gedde, U. W.: Vol. 169, pp. 29–73.
Mauzac, M. see Marty, J. D.: Vol. 172, pp. 1–35.
Mays, W. see Xu, Z.: Vol. 120, pp. 1–50.
Mays, J. W. see Pitsikalis, M.: Vol. 135, pp. 1–138.
McGrath, J. E. see Hedrick, J. L.: Vol. 141, pp. 1–44.
McGrath, J. E., Dunson, D. L. and *Hedrick, J. L.*: Synthesis and Characterization of Segmented Polyimide-Polyorganosiloxane Copolymers. Vol. 140, pp. 61–106.
McLeish, T. C. B. and *Milner, S. T.*: Entangled Dynamics and Melt Flow of Branched Polymers. Vol. 143, pp. 195–256.
Mecerreyes, D., Dubois, P. and *Jerome, R.*: Novel Macromolecular Architectures Based on Aliphatic Polyesters: Relevance of the Coordination-Insertion Ring-Opening Polymerization. Vol. 147, pp. 1–60.
Mecham, S. J. see McGrath, J. E.: Vol. 140, pp. 61–106.
Meille, S. V. see Allegra, G.: Vol. 191, pp. 87–135.
Menzel, H. see Möhwald, H.: Vol. 165, pp. 151–175.
Meyer, T. see Spange, S.: Vol. 165, pp. 43–78.
Michler, G. H. see Kausch, H.-H.: Vol. 187, pp. 1–33.
Mikos, A. G. see Thomson, R. C.: Vol. 122, pp. 245–274.
Milner, S. T. see McLeish, T. C. B.: Vol. 143, pp. 195–256.
Mison, P. and *Sillion, B.*: Thermosetting Oligomers Containing Maleimides and Nadiimides End-Groups. Vol. 140, pp. 137–180.
Miyasaka, K.: PVA-Iodine Complexes: Formation, Structure and Properties. Vol. 108, pp. 91–130.
Miller, R. D. see Hedrick, J. L.: Vol. 141, pp. 1–44.
Minko, S. see Rühe, J.: Vol. 165, pp. 79–150.
Möhwald, H., Menzel, H., Helm, C. A. and *Stamm, M.*: Lipid and Polyampholyte Monolayers to Study Polyelectrolyte Interactions and Structure at Interfaces. Vol. 165, pp. 151–175.
Monkenbusch, M. see Richter, D.: Vol. 174, pp. 1–221.
Monnerie, L., Halary, J. L. and *Kausch, H.-H.*: Deformation, Yield and Fracture of Amorphous Polymers: Relation to the Secondary Transitions. Vol. 187, pp. 215–364.
Monnerie, L., Lauprêtre, F. and *Halary, J. L.*: Investigation of Solid-State Transitions in Linear and Crosslinked Amorphous Polymers. Vol. 187, pp. 35–213.
Monnerie, L. see Bahar, I.: Vol. 116, pp. 145–206.
Moore, J. S. see Ray, C. R.: Vol. 177, pp. 99–149.
Mori, H. see Bohrisch, J.: Vol. 165, pp. 1–41.
Morishima, Y.: Photoinduced Electron Transfer in Amphiphilic Polyelectrolyte Systems. Vol. 104, pp. 51–96.
Morton, M. see Quirk, R. P.: Vol. 153, pp. 67–162.
Motornov, M. see Rühe, J.: Vol. 165, pp. 79–150.
Mours, M. see Winter, H. H.: Vol. 134, pp. 165–234.
Müllen, K. see Scherf, U.: Vol. 123, pp. 1–40.
Müller, A. H. E. see Bohrisch, J.: Vol. 165, pp. 1–41.
Müller, A. H. E. see Förster, S.: Vol. 166, pp. 173–210.

Müller, A. J., Balsamo, V. and *Arnal, M. L.*: Nucleation and Crystallization in Diblock and Triblock Copolymers. Vol. 190, pp. 1–63.
Müller, M. and *Schmid, F.*: Incorporating Fluctuations and Dynamics in Self-Consistent Field Theories for Polymer Blends. Vol. 185, pp. 1–58.
Müller, M. see Thünemann, A. F.: Vol. 166, pp. 113–171.
Müller-Plathe, F. see Gusev, A. A.: Vol. 116, pp. 207–248.
Müller-Plathe, F. see Baschnagel, J.: Vol. 152, p. 41–156.
Mukerherjee, A. see Biswas, M.: Vol. 115, pp. 89–124.
Munz, M., Cappella, B., Sturm, H., Geuss, M. and *Schulz, E.*: Materials Contrasts and Nanolithography Techniques in Scanning Force Microscopy (SFM) and their Application to Polymers and Polymer Composites. Vol. 164, pp. 87–210.
Murat, M. see Baschnagel, J.: Vol. 152, p. 41–156.
Muthukumar, M.: Modeling Polymer Crystallization. Vol. 191, pp. 241–274.
Muzzarelli, C. see Muzzarelli, R. A. A.: Vol. 186, pp. 151–209.
Muzzarelli, R. A. A. and *Muzzarelli, C.*: Chitosan Chemistry: Relevance to the Biomedical Sciences. Vol. 186, pp. 151–209.
Mylnikov, V.: Photoconducting Polymers. Vol. 115, pp. 1–88.

Nagy, A. see Majoros, I.: Vol. 112, pp. 1–11.
Naka, K. see Uemura, T.: Vol. 167, pp. 81–106.
Nakamura, A. see Mashima, K.: Vol. 133, pp. 1–52.
Nakayama, Y. see Mashima, K.: Vol. 133, pp. 1–52.
Narasinham, B. and *Peppas, N. A.*: The Physics of Polymer Dissolution: Modeling Approaches and Experimental Behavior. Vol. 128, pp. 157–208.
Nechaev, S. see Grosberg, A.: Vol. 106, pp. 1–30.
Neoh, K. G. see Kang, E. T.: Vol. 106, pp. 135–190.
Netz, R. R. see Holm, C.: Vol. 166, pp. 67–111.
Netz, R. R. see Rühe, J.: Vol. 165, pp. 79–150.
Newman, S. M. see Anseth, K. S.: Vol. 122, pp. 177–218.
Nijenhuis, K. te: Thermoreversible Networks. Vol. 130, pp. 1–252.
Ninan, K. N. see Reghunadhan Nair, C. P.: Vol. 155, pp. 1–99.
Nishi, T. see Jinnai, H.: Vol. 170, pp. 115–167.
Nishida, K. see Kaji, K.: Vol. 191, pp. 187–240.
Nishikawa, Y. see Jinnai, H.: Vol. 170, pp. 115–167.
Nishiyama, N. and *Kataoka, K.*: Nanostructured Devices Based on Block Copolymer Assemblies for Drug Delivery: Designing Structures for Enhanced Drug Function. Vol. 193, pp. 67–101.
Noid, D. W. see Otaigbe, J. U.: Vol. 154, pp. 1–86.
Noid, D. W. see Sumpter, B. G.: Vol. 116, pp. 27–72.
Nomura, M., Tobita, H. and *Suzuki, K.*: Emulsion Polymerization: Kinetic and Mechanistic Aspects. Vol. 175, pp. 1–128.
Northolt, M. G., Picken, S. J., Den Decker, M. G., Baltussen, J. J. M. and *Schlatmann, R.*: The Tensile Strength of Polymer Fibres. Vol. 178, pp. 1–108.
Novac, B. see Grubbs, R.: Vol. 102, pp. 47–72.
Novikov, V. V. see Privalko, V. P.: Vol. 119, pp. 31–78.
Nummila-Pakarinen, A. see Knuuttila, H.: Vol. 169, pp. 13–27.

Ober, C. K. see Li, M.: Vol. 190, pp. 183–226.
O'Brien, D. F., Armitage, B. A., Bennett, D. E. and *Lamparski, H. G.*: Polymerization and Domain Formation in Lipid Assemblies. Vol. 126, pp. 53–84.

Ogasawara, M.: Application of Pulse Radiolysis to the Study of Polymers and Polymerizations. Vol.105, pp. 37–80.
Ohmae, M. see Kobayashi, S.: Vol. 194, pp. 159–210.
Okabe, H. see Matsushige, K.: Vol. 125, pp. 147–186.
Okada, M.: Ring-Opening Polymerization of Bicyclic and Spiro Compounds. Reactivities and Polymerization Mechanisms. Vol. 102, pp. 1–46.
Okada, K. see Hikosaka, M.: Vol. 191, pp. 137–186.
Okano, T.: Molecular Design of Temperature-Responsive Polymers as Intelligent Materials. Vol. 110, pp. 179–198.
Okay, O. see Funke, W.: Vol. 136, pp. 137–232.
Okhapkin, I. M., Makhaeva, E. E. and *Khokhlov, A. R.*: Water Solutions of Amphiphilic Polymers: Nanostructure Formation and Possibilities for Catalysis. Vol. 195, pp. 177–210.
Onuki, A.: Theory of Phase Transition in Polymer Gels. Vol. 109, pp. 63–120.
Oppermann, W. see Holm, C.: Vol. 166, pp. 1–27.
Oppermann, W. see Volk, N.: Vol. 166, pp. 29–65.
Osad'ko, I. S.: Selective Spectroscopy of Chromophore Doped Polymers and Glasses. Vol. 114, pp. 123–186.
Osakada, K. and *Takeuchi, D.*: Coordination Polymerization of Dienes, Allenes, and Methylenecycloalkanes. Vol. 171, pp. 137–194.
Otaigbe, J. U., Barnes, M. D., Fukui, K., Sumpter, B. G. and *Noid, D. W.*: Generation, Characterization, and Modeling of Polymer Micro- and Nano-Particles. Vol. 154, pp. 1–86.
Otsu, T. and *Matsumoto, A.*: Controlled Synthesis of Polymers Using the Iniferter Technique: Developments in Living Radical Polymerization. Vol. 136, pp. 75–138.

de Pablo, J. J. see Leontidis, E.: Vol. 116, pp. 283–318.
Padias, A. B. see Penelle, J.: Vol. 102, pp. 73–104.
Pascault, J.-P. see Williams, R. J. J.: Vol. 128, pp. 95–156.
Pasch, H.: Analysis of Complex Polymers by Interaction Chromatography. Vol. 128, pp. 1–46.
Pasch, H.: Hyphenated Techniques in Liquid Chromatography of Polymers. Vol. 150, pp. 1–66.
Pasut, G. and *Veronese, F. M.*: PEGylation of Proteins as Tailored Chemistry for Optimized Bioconjugates. Vol. 192, pp. 95–134.
Paul, W. see Baschnagel, J.: Vol. 152, pp. 41–156.
Paulsen, S. B. and *Barsett, H.*: Bioactive Pectic Polysaccharides. Vol. 186, pp. 69–101.
Pautzsch, T. see Klemm, E.: Vol. 177, pp. 53–90.
Penczek, P., Czub, P. and *Pielichowski, J.*: Unsaturated Polyester Resins: Chemistry and Technology. Vol. 184, pp. 1–95.
Penczek, P. see Batog, A. E.: Vol. 144, pp. 49–114.
Penczek, P. see Bogdal, D.: Vol. 163, pp. 193–263.
Penelle, J., Hall, H. K., Padias, A. B. and *Tanaka, H.*: Captodative Olefins in Polymer Chemistry. Vol. 102, pp. 73–104.
Peppas, N. A. see Bell, C. L.: Vol. 122, pp. 125–176.
Peppas, N. A. see Hassan, C. M.: Vol. 153, pp. 37–65.
Peppas, N. A. see Narasimhan, B.: Vol. 128, pp. 157–208.
Petersen, K. L. see Geil, P. H.: Vol. 180, pp. 89–159.
Pet'ko, I. P. see Batog, A. E.: Vol. 144, pp. 49–114.
Pheyghambarian, N. see Kippelen, B.: Vol. 161, pp. 87–156.
Pichot, C. see Hunkeler, D.: Vol. 112, pp. 115–134.
Picken, S. J. see Northolt, M. G.: Vol. 178, pp. 1–108.
Pielichowski, J. see Bogdal, D.: Vol. 163, pp. 193–263.

Pielichowski, J. see Penczek, P.: Vol. 184, pp. 1–95.
Pieper, T. see Kilian, H. G.: Vol. 108, pp. 49–90.
Pispas, S. see Pitsikalis, M.: Vol. 135, pp. 1–138.
Pispas, S. see Hadjichristidis, N.: Vol. 142, pp. 71–128.
Pitsikalis, M., Pispas, S., Mays, J. W. and *Hadjichristidis, N.*: Nonlinear Block Copolymer Architectures. Vol. 135, pp. 1–138.
Pitsikalis, M. see Hadjichristidis, N.: Vol. 142, pp. 71–128.
Pitsikalis, M. see Hadjichristidis, N.: Vol. 189, pp. 1–124.
Pleul, D. see Spange, S.: Vol. 165, pp. 43–78.
Plummer, C. J. G.: Microdeformation and Fracture in Bulk Polyolefins. Vol. 169, pp. 75–119.
Pötschke, D. see Dingenouts, N.: Vol. 144, pp. 1–48.
Pokrovskii, V. N.: The Mesoscopic Theory of the Slow Relaxation of Linear Macromolecules. Vol. 154, pp. 143–219.
Pospíšil, J.: Functionalized Oligomers and Polymers as Stabilizers for Conventional Polymers. Vol. 101, pp. 65–168.
Pospíšil, J.: Aromatic and Heterocyclic Amines in Polymer Stabilization. Vol. 124, pp. 87–190.
Powers, A. C. see Prokop, A.: Vol. 136, pp. 53–74.
Prasad, P. N. see Lin, T.-C.: Vol. 161, pp. 157–193.
Priddy, D. B.: Recent Advances in Styrene Polymerization. Vol. 111, pp. 67–114.
Priddy, D. B.: Thermal Discoloration Chemistry of Styrene-co-Acrylonitrile. Vol. 121, pp. 123–154.
Privalko, V. P. and *Novikov, V. V.*: Model Treatments of the Heat Conductivity of Heterogeneous Polymers. Vol. 119, pp. 31–78.
Prociak, A. see Bogdal, D.: Vol. 163, pp. 193–263.
Prokop, A., Hunkeler, D., DiMari, S., Haralson, M. A. and *Wang, T. G.*: Water Soluble Polymers for Immunoisolation I: Complex Coacervation and Cytotoxicity. Vol. 136, pp. 1–52.
Prokop, A., Hunkeler, D., Powers, A. C., Whitesell, R. R. and *Wang, T. G.*: Water Soluble Polymers for Immunoisolation II: Evaluation of Multicomponent Microencapsulation Systems. Vol. 136, pp. 53–74.
Prokop, A., Kozlov, E., Carlesso, G. and *Davidsen, J. M.*: Hydrogel-Based Colloidal Polymeric System for Protein and Drug Delivery: Physical and Chemical Characterization, Permeability Control and Applications. Vol. 160, pp. 119–174.
Pruitt, L. A.: The Effects of Radiation on the Structural and Mechanical Properties of Medical Polymers. Vol. 162, pp. 65–95.
Pudavar, H. E. see Lin, T.-C.: Vol. 161, pp. 157–193.
Pukánszky, B. and *Fekete, E.*: Adhesion and Surface Modification. Vol. 139, pp. 109–154.
Putnam, D. and *Kopecek, J.*: Polymer Conjugates with Anticancer Acitivity. Vol. 122, pp. 55–124.
Putra, E. G. R. see Ungar, G.: Vol. 180, pp. 45–87.

Quirk, R. P., Yoo, T., Lee, Y., M., Kim, J. and *Lee, B.*: Applications of 1,1-Diphenylethylene Chemistry in Anionic Synthesis of Polymers with Controlled Structures. Vol. 153, pp. 67–162.

Ramaraj, R. and *Kaneko, M.*: Metal Complex in Polymer Membrane as a Model for Photosynthetic Oxygen Evolving Center. Vol. 123, pp. 215–242.
Rangarajan, B. see Scranton, A. B.: Vol. 122, pp. 1–54.
Ranucci, E. see Söderqvist Lindblad, M.: Vol. 157, pp. 139–161.
Raphaël, E. see Léger, L.: Vol. 138, pp. 185–226.

Rastogi, S. and *Terry, A. E.*: Morphological implications of the interphase bridging crystalline and amorphous regions in semi-crystalline polymers. Vol. 180, pp. 161–194.
Ray, C. R. and *Moore, J. S.*: Supramolecular Organization of Foldable Phenylene Ethynylene Oligomers. Vol. 177, pp. 99–149.
Reddinger, J. L. and *Reynolds, J. R.*: Molecular Engineering of p-Conjugated Polymers. Vol. 145, pp. 57–122.
Reghunadhan Nair, C. P., Mathew, D. and *Ninan, K. N.*: Cyanate Ester Resins, Recent Developments. Vol. 155, pp. 1–99.
Reichert, K. H. see Hunkeler, D.: Vol. 112, pp. 115–134.
Reihmann, M. and *Ritter, H.*: Synthesis of Phenol Polymers Using Peroxidases. Vol. 194, pp. 1–49.
Rehahn, M., Mattice, W. L. and *Suter, U. W.*: Rotational Isomeric State Models in Macromolecular Systems. Vol. 131/132, pp. 1–475.
Rehahn, M. see Bohrisch, J.: Vol. 165, pp. 1–41.
Rehahn, M. see Holm, C.: Vol. 166, pp. 1–27.
Reineker, P. see Holm, C.: Vol. 166, pp. 67–111.
Reitberger, T. see Jacobson, K.: Vol. 169, pp. 151–176.
Ritter, H. see Reihmann, M.: Vol. 194, pp. 1–49.
Reynolds, J. R. see Reddinger, J. L.: Vol. 145, pp. 57–122.
Richter, D. see Ewen, B.: Vol. 134, pp. 1–130.
Richter, D., Monkenbusch, M. and *Colmenero, J.*: Neutron Spin Echo in Polymer Systems. Vol. 174, pp. 1–221.
Riegler, S. see Trimmel, G.: Vol. 176, pp. 43–87.
Ringsdorf, H. see Duncan, R.: Vol. 192, pp. 1–8.
Risse, W. see Grubbs, R.: Vol. 102, pp. 47–72.
Rivas, B. L. and *Geckeler, K. E.*: Synthesis and Metal Complexation of Poly(ethyleneimine) and Derivatives. Vol. 102, pp. 171–188.
Roberts, C. J. see Ellis, J. S.: Vol. 193, pp. 123–172.
Roberts, G. W. see Kennedy, K. A.: Vol. 175, pp. 329–346.
Robin, J. J.: The Use of Ozone in the Synthesis of New Polymers and the Modification of Polymers. Vol. 167, pp. 35–79.
Robin, J. J. see Boutevin, B.: Vol. 102, pp. 105–132.
Rodríguez-Pérez, M. A.: Crosslinked Polyolefin Foams: Production, Structure, Properties, and Applications. Vol. 184, pp. 97–126.
Roe, R.-J.: MD Simulation Study of Glass Transition and Short Time Dynamics in Polymer Liquids. Vol. 116, pp. 111–114.
Roovers, J. and *Comanita, B.*: Dendrimers and Dendrimer-Polymer Hybrids. Vol. 142, pp. 179–228.
Rothon, R. N.: Mineral Fillers in Thermoplastics: Filler Manufacture and Characterisation. Vol. 139, pp. 67–108.
de Rosa, C. see Auriemma, F.: Vol. 181, pp. 1–74.
Rozenberg, B. A. see Williams, R. J. J.: Vol. 128, pp. 95–156.
Rühe, J., Ballauff, M., Biesalski, M., Dziezok, P., Gröhn, F., Johannsmann, D., Houbenov, N., Hugenberg, N., Konradi, R., Minko, S., Motornov, M., Netz, R. R., Schmidt, M., Seidel, C., Stamm, M., Stephan, T., Usov, D. and *Zhang, H.*: Polyelectrolyte Brushes. Vol. 165, pp. 79–150.
Ruckenstein, E.: Concentrated Emulsion Polymerization. Vol. 127, pp. 1–58.
Ruiz-Taylor, L. see Mathieu, H. J.: Vol. 162, pp. 1–35.
Rusanov, A. L.: Novel Bis (Naphtalic Anhydrides) and Their Polyheteroarylenes with Improved Processability. Vol. 111, pp. 115–176.

Rusanov, A. L., Likhatchev, D., Kostoglodov, P. V., Müllen, K. and *Klapper, M.*: Proton-Exchanging Electrolyte Membranes Based on Aromatic Condensation Polymers. Vol. 179, pp. 83–134.
Russel, T. P. see Hedrick, J. L.: Vol. 141, pp. 1–44.
Russum, J. P. see Schork, F. J.: Vol. 175, pp. 129–255.
Rychly, J. see Lazár, M.: Vol. 102, pp. 189–222.
Ryner, M. see Stridsberg, K. M.: Vol. 157, pp. 27–51.
Ryzhov, V. A. see Bershtein, V. A.: Vol. 114, pp. 43–122.

Sabsai, O. Y. see Barshtein, G. R.: Vol. 101, pp. 1–28.
Saburov, V. V. see Zubov, V. P.: Vol. 104, pp. 135–176.
Saito, S., Konno, M. and *Inomata, H.*: Volume Phase Transition of N-Alkylacrylamide Gels. Vol. 109, pp. 207–232.
Samsonov, G. V. and *Kuznetsova, N. P.*: Crosslinked Polyelectrolytes in Biology. Vol. 104, pp. 1–50.
Santa Cruz, C. see Baltá-Calleja, F. J.: Vol. 108, pp. 1–48.
Santos, S. see Baschnagel, J.: Vol. 152, p. 41–156.
Satchi-Fainaro, R., Duncan, R. and *Barnes, C. M.*: Polymer Therapeutics for Cancer: Current Status and Future Challenges. Vol. 193, pp. 1–65.
Satchi-Fainaro, R. see Duncan, R.: Vol. 192, pp. 1–8.
Sato, T. and *Teramoto, A.*: Concentrated Solutions of Liquid-Christalline Polymers. Vol. 126, pp. 85–162.
Schaller, C. see Bohrisch, J.: Vol. 165, pp. 1–41.
Schäfer, R. see Köhler, W.: Vol. 151, pp. 1–59.
Scherf, U. and *Müllen, K.*: The Synthesis of Ladder Polymers. Vol. 123, pp. 1–40.
Sherman, S. see Kabanov, A. V.: Vol. 193, pp. 173–198.
Schlatmann, R. see Northolt, M. G.: Vol. 178, pp. 1–108.
Schmid, F. see Müller, M.: Vol. 185, pp. 1–58.
Schmidt, M. see Förster, S.: Vol. 120, pp. 51–134.
Schmidt, M. see Rühe, J.: Vol. 165, pp. 79–150.
Schmidt, M. see Volk, N.: Vol. 166, pp. 29–65.
Scholz, M.: Effects of Ion Radiation on Cells and Tissues. Vol. 162, pp. 97–158.
Schönherr, H. see Vancso, G. J.: Vol. 182, pp. 55–129.
Schopf, G. and *Koßmehl, G.*: Polythiophenes – Electrically Conductive Polymers. Vol. 129, pp. 1–145.
Schork, F. J., Luo, Y., Smulders, W., Russum, J. P., Butté, A. and *Fontenot, K.*: Miniemulsion Polymerization. Vol. 175, pp. 127–255.
Schulz, E. see Munz, M.: Vol. 164, pp. 97–210.
Schwahn, D.: Critical to Mean Field Crossover in Polymer Blends. Vol. 183, pp. 1–61.
Seppälä, J. see Löfgren, B.: Vol. 169, pp. 1–12.
Sturm, H. see Munz, M.: Vol. 164, pp. 87–210.
Schweizer, K. S.: Prism Theory of the Structure, Thermodynamics, and Phase Transitions of Polymer Liquids and Alloys. Vol. 116, pp. 319–378.
Scranton, A. B., Rangarajan, B. and *Klier, J.*: Biomedical Applications of Polyelectrolytes. Vol. 122, pp. 1–54.
Sefton, M. V. and *Stevenson, W. T. K.*: Microencapsulation of Live Animal Cells Using Polycrylates. Vol. 107, pp. 143–198.
Seidel, C. see Holm, C.: Vol. 166, pp. 67–111.
Seidel, C. see Rühe, J.: Vol. 165, pp. 79–150.

El Seoud, O. A. and *Heinze, T.*: Organic Esters of Cellulose: New Perspectives for Old Polymers. Vol. 186, pp. 103–149.
Shabat, D. see Amir, R. J.: Vol. 192, pp. 59–94.
Shamanin, V. V.: Bases of the Axiomatic Theory of Addition Polymerization. Vol. 112, pp. 135–180.
Shcherbina, M. A. see Ungar, G.: Vol. 180, pp. 45–87.
Sheiko, S. S.: Imaging of Polymers Using Scanning Force Microscopy: From Superstructures to Individual Molecules. Vol. 151, pp. 61–174.
Sherrington, D. C. see Cameron, N. R.: Vol. 126, pp. 163–214.
Sherrington, D. C. see Lin, J.: Vol. 111, pp. 177–220.
Sherrington, D. C. see Steinke, J.: Vol. 123, pp. 81–126.
Shibayama, M. see Tanaka, T.: Vol. 109, pp. 1–62.
Shiga, T.: Deformation and Viscoelastic Behavior of Polymer Gels in Electric Fields. Vol. 134, pp. 131–164.
Shim, H.-K. and *Jin, J.*: Light-Emitting Characteristics of Conjugated Polymers. Vol. 158, pp. 191–241.
Shoda, S. see Kobayashi, S.: Vol. 121, pp. 1–30.
Siegel, R. A.: Hydrophobic Weak Polyelectrolyte Gels: Studies of Swelling Equilibria and Kinetics. Vol. 109, pp. 233–268.
de Silva, D. S. M. see Ungar, G.: Vol. 180, pp. 45–87.
Silvestre, F. see Calmon-Decriaud, A.: Vol. 207, pp. 207–226.
Sillion, B. see Mison, P.: Vol. 140, pp. 137–180.
Simon, F. see Spange, S.: Vol. 165, pp. 43–78.
Simon, G. P. see Becker, O.: Vol. 179, pp. 29–82.
Simon, P. F. W. see Abetz, V.: Vol. 189, pp. 125–212.
Simonutti, R. see Sozzani, P.: Vol. 181, pp. 153–177.
Singh, A. and *Kaplan, D. L.*: In Vitro Enzyme-Induced Vinyl Polymerization. Vol. 194, pp. 211–224.
Singh, A. see Xu, P.: Vol. 194, pp. 69–94.
Singh, R. P. see Sivaram, S.: Vol. 101, pp. 169–216.
Singh, R. P. see Desai, S. M.: Vol. 169, pp. 231–293.
Sinha Ray, S. see Biswas, M.: Vol. 155, pp. 167–221.
Sivaram, S. and *Singh, R. P.*: Degradation and Stabilization of Ethylene-Propylene Copolymers and Their Blends: A Critical Review. Vol. 101, pp. 169–216.
Slugovc, C. see Trimmel, G.: Vol. 176, pp. 43–87.
Smulders, W. see Schork, F. J.: Vol. 175, pp. 129–255.
Soares, J. B. P. see Anantawaraskul, S.: Vol. 182, pp. 1–54.
Sozzani, P., Bracco, S., Comotti, A. and *Simonutti, R.*: Motional Phase Disorder of Polymer Chains as Crystallized to Hexagonal Lattices. Vol. 181, pp. 153–177.
Söderqvist Lindblad, M., Liu, Y., Albertsson, A.-C., Ranucci, E. and *Karlsson, S.*: Polymer from Renewable Resources. Vol. 157, pp. 139–161.
Spange, S., Meyer, T., Voigt, I., Eschner, M., Estel, K., Pleul, D. and *Simon, F.*: Poly(Vinylformamide-co-Vinylamine)/Inorganic Oxid Hybrid Materials. Vol. 165, pp. 43–78.
Stamm, M. see Möhwald, H.: Vol. 165, pp. 151–175.
Stamm, M. see Rühe, J.: Vol. 165, pp. 79–150.
Starodybtzev, S. see Khokhlov, A.: Vol. 109, pp. 121–172.
Stegeman, G. I. see Canva, M.: Vol. 158, pp. 87–121.
Steinke, J., Sherrington, D. C. and *Dunkin, I. R.*: Imprinting of Synthetic Polymers Using Molecular Templates. Vol. 123, pp. 81–126.
Stelzer, F. see Trimmel, G.: Vol. 176, pp. 43–87.

Stenberg, B. see Jacobson, K.: Vol. 169, pp. 151–176.
Stenzenberger, H. D.: Addition Polyimides. Vol. 117, pp. 165–220.
Stephan, T. see Rühe, J.: Vol. 165, pp. 79–150.
Stevenson, W. T. K. see Sefton, M. V.: Vol. 107, pp. 143–198.
Stridsberg, K. M., Ryner, M. and *Albertsson, A.-C.*: Controlled Ring-Opening Polymerization: Polymers with Designed Macromoleculars Architecture. Vol. 157, pp. 27–51.
Sturm, H. see Munz, M.: Vol. 164, pp. 87–210.
Suematsu, K.: Recent Progress of Gel Theory: Ring, Excluded Volume, and Dimension. Vol. 156, pp. 136–214.
Sugimoto, H. and *Inoue, S.*: Polymerization by Metalloporphyrin and Related Complexes. Vol. 146, pp. 39–120.
Suginome, M. and *Ito, Y.*: Transition Metal-Mediated Polymerization of Isocyanides. Vol. 171, pp. 77–136.
Sumpter, B. G., Noid, D. W., Liang, G. L. and *Wunderlich, B.*: Atomistic Dynamics of Macromolecular Crystals. Vol. 116, pp. 27–72.
Sumpter, B. G. see Otaigbe, J. U.: Vol. 154, pp. 1–86.
Sun, H.-B. and *Kawata, S.*: Two-Photon Photopolymerization and 3D Lithographic Microfabrication. Vol. 170, pp. 169–273.
Suter, U. W. see Gusev, A. A.: Vol. 116, pp. 207–248.
Suter, U. W. see Leontidis, E.: Vol. 116, pp. 283–318.
Suter, U. W. see Rehahn, M.: Vol. 131/132, pp. 1–475.
Suter, U. W. see Baschnagel, J.: Vol. 152, pp. 41–156.
Suzuki, A.: Phase Transition in Gels of Sub-Millimeter Size Induced by Interaction with Stimuli. Vol. 110, pp. 199–240.
Suzuki, A. and *Hirasa, O.*: An Approach to Artifical Muscle by Polymer Gels due to Micro-Phase Separation. Vol. 110, pp. 241–262.
Suzuki, K. see Nomura, M.: Vol. 175, pp. 1–128.
Swiatkiewicz, J. see Lin, T.-C.: Vol. 161, pp. 157–193.

Tagawa, S.: Radiation Effects on Ion Beams on Polymers. Vol. 105, pp. 99–116.
Taguet, A., Ameduri, B. and *Boutevin, B.*: Crosslinking of Vinylidene Fluoride-Containing Fluoropolymers. Vol. 184, pp. 127–211.
Takata, T., Kihara, N. and *Furusho, Y.*: Polyrotaxanes and Polycatenanes: Recent Advances in Syntheses and Applications of Polymers Comprising of Interlocked Structures. Vol. 171, pp. 1–75.
Takeuchi, D. see Osakada, K.: Vol. 171, pp. 137–194.
Tan, K. L. see Kang, E. T.: Vol. 106, pp. 135–190.
Tanaka, H. and *Shibayama, M.*: Phase Transition and Related Phenomena of Polymer Gels. Vol. 109, pp. 1–62.
Tanaka, T. see Penelle, J.: Vol. 102, pp. 73–104.
Tauer, K. see Guyot, A.: Vol. 111, pp. 43–66.
Tendler, S. J. B. see Ellis, J. S.: Vol. 193, pp. 123–172.
Teramoto, A. see Sato, T.: Vol. 126, pp. 85–162.
Terent'eva, J. P. and *Fridman, M. L.*: Compositions Based on Aminoresins. Vol. 101, pp. 29–64.
Terry, A. E. see Rastogi, S.: Vol. 180, pp. 161–194.
Theodorou, D. N. see Dodd, L. R.: Vol. 116, pp. 249–282.
Thomson, R. C., Wake, M. C., Yaszemski, M. J. and *Mikos, A. G.*: Biodegradable Polymer Scaffolds to Regenerate Organs. Vol. 122, pp. 245–274.
Thünemann, A. F., Müller, M., Dautzenberg, H., Joanny, J.-F. and *Löwen, H.*: Polyelectrolyte complexes. Vol. 166, pp. 113–171.

Tieke, B. see v. Klitzing, R.: Vol. 165, pp. 177–210.
Tobita, H. see Nomura, M.: Vol. 175, pp. 1–128.
Tokita, M.: Friction Between Polymer Networks of Gels and Solvent. Vol. 110, pp. 27–48.
Traser, S. see Bohrisch, J.: Vol. 165, pp. 1–41.
Tries, V. see Baschnagel, J.: Vol. 152, p. 41–156.
Trimmel, G., Riegler, S., Fuchs, G., Slugovc, C. and *Stelzer, F.*: Liquid Crystalline Polymers by Metathesis Polymerization. Vol. 176, pp. 43–87.
Tsuruta, T.: Contemporary Topics in Polymeric Materials for Biomedical Applications. Vol. 126, pp. 1–52.

Uemura, T., Naka, K. and *Chujo, Y.*: Functional Macromolecules with Electron-Donating Dithiafulvene Unit. Vol. 167, pp. 81–106.
Ungar, G., Putra, E. G. R., de Silva, D. S. M., Shcherbina, M. A. and *Waddon, A. J.*: The Effect of Self-Poisoning on Crystal Morphology and Growth Rates. Vol. 180, pp. 45–87.
Usov, D. see Rühe, J.: Vol. 165, pp. 79–150.
Usuki, A., Hasegawa, N. and *Kato, M.*: Polymer-Clay Nanocomposites. Vol. 179, pp. 135–195.
Uyama, H. and *Kobayashi, S.*: Enzymatic Synthesis and Properties of Polymers from Polyphenols. Vol. 194, pp. 51–67.
Uyama, H. and *Kobayashi, S.*: Enzymatic Synthesis of Polyesters via Polycondensation. Vol. 194, pp. 133–158.
Uyama, H. see Kobayashi, S.: Vol. 121, pp. 1–30.
Uyama, Y.: Surface Modification of Polymers by Grafting. Vol. 137, pp. 1–40.

Vancso, G. J., Hillborg, H. and *Schönherr, H.*: Chemical Composition of Polymer Surfaces Imaged by Atomic Force Microscopy and Complementary Approaches. Vol. 182, pp. 55–129.
Varma, I. K. see Albertsson, A.-C.: Vol. 157, pp. 99–138.
Vasilevskaya, V. see Khokhlov, A.: Vol. 109, pp. 121–172.
Vaskova, V. see Hunkeler, D.: Vol. 112, pp. 115–134.
Verdugo, P.: Polymer Gel Phase Transition in Condensation-Decondensation of Secretory Products. Vol. 110, pp. 145–156.
Veronese, F. M. see Pasut, G.: Vol. 192, pp. 95–134.
Vettegren, V. I. see Bronnikov, S. V.: Vol. 125, pp. 103–146.
Vilgis, T. A. see Holm, C.: Vol. 166, pp. 67–111.
Viovy, J.-L. and *Lesec, J.*: Separation of Macromolecules in Gels: Permeation Chromatography and Electrophoresis. Vol. 114, pp. 1–42.
Vlahos, C. see Hadjichristidis, N.: Vol. 142, pp. 71–128.
Voigt, I. see Spange, S.: Vol. 165, pp. 43–78.
Volk, N., Vollmer, D., Schmidt, M., Oppermann, W. and *Huber, K.*: Conformation and Phase Diagrams of Flexible Polyelectrolytes. Vol. 166, pp. 29–65.
Volksen, W.: Condensation Polyimides: Synthesis, Solution Behavior, and Imidization Characteristics. Vol. 117, pp. 111–164.
Volksen, W. see Hedrick, J. L.: Vol. 141, pp. 1–44.
Volksen, W. see Hedrick, J. L.: Vol. 147, pp. 61–112.
Vollmer, D. see Volk, N.: Vol. 166, pp. 29–65.
Voskerician, G. and *Weder, C.*: Electronic Properties of PAEs. Vol. 177, pp. 209–248.

Waddon, A. J. see Ungar, G.: Vol. 180, pp. 45–87.
Wagener, K. B. see Baughman, T. W.: Vol. 176, pp. 1–42.

Wagner, E. and *Kloeckner, J.*: Gene Delivery Using Polymer Therapeutics. Vol. 192, pp. 135–173.
Wake, M. C. see Thomson, R. C.: Vol. 122, pp. 245–274.
Wandrey, C., Hernández-Barajas, J. and *Hunkeler, D.*: Diallyldimethylammonium Chloride and its Polymers. Vol. 145, pp. 123–182.
Wang, K. L. see Cussler, E. L.: Vol. 110, pp. 67–80.
Wang, S.-Q.: Molecular Transitions and Dynamics at Polymer/Wall Interfaces: Origins of Flow Instabilities and Wall Slip. Vol. 138, pp. 227–276.
Wang, S.-Q. see Bhargava, R.: Vol. 163, pp. 137–191.
Wang, T. G. see Prokop, A.: Vol. 136, pp. 1–52; 53–74.
Wang, X. see Lin, T.-C.: Vol. 161, pp. 157–193.
Watanabe, K. see Hikosaka, M.: Vol. 191, pp. 137–186.
Webster, O. W.: Group Transfer Polymerization: Mechanism and Comparison with Other Methods of Controlled Polymerization of Acrylic Monomers. Vol. 167, pp. 1–34.
Weder, C. see Voskerician, G.: Vol. 177, pp. 209–248.
Weis, J.-J. and *Levesque, D.*: Simple Dipolar Fluids as Generic Models for Soft Matter. Vol. 185, pp. 163–225.
Whitesell, R. R. see Prokop, A.: Vol. 136, pp. 53–74.
Williams, R. A. see Geil, P. H.: Vol. 180, pp. 89–159.
Williams, R. J. J., Rozenberg, B. A. and *Pascault, J.-P.*: Reaction Induced Phase Separation in Modified Thermosetting Polymers. Vol. 128, pp. 95–156.
Winkler, R. G. see Holm, C.: Vol. 166, pp. 67–111.
Winter, H. H. and *Mours, M.*: Rheology of Polymers Near Liquid-Solid Transitions. Vol. 134, pp. 165–234.
Wittmeyer, P. see Bohrisch, J.: Vol. 165, pp. 1–41.
Wood-Adams, P. M. see Anantawaraskul, S.: Vol. 182, pp. 1–54.
Wu, C.: Laser Light Scattering Characterization of Special Intractable Macromolecules in Solution. Vol. 137, pp. 103–134.
Wu, C. see Zhang, G.: Vol. 195, pp. 101–176.
Wunderlich, B. see Sumpter, B. G.: Vol. 116, pp. 27–72.

Xiang, M. see Jiang, M.: Vol. 146, pp. 121–194.
Xie, T. Y. see Hunkeler, D.: Vol. 112, pp. 115–134.
Xu, P., Singh, A. and *Kaplan, D. L.*: Enzymatic Catalysis in the Synthesis of Polyanilines and Derivatives of Polyanilines. Vol. 194, pp. 69–94.
Xu, P. see Geil, P. H.: Vol. 180, pp. 89–159.
Xu, Z., Hadjichristidis, N., Fetters, L. J. and *Mays, J. W.*: Structure/Chain-Flexibility Relationships of Polymers. Vol. 120, pp. 1–50.

Yagci, Y. and *Endo, T.*: N-Benzyl and N-Alkoxy Pyridium Salts as Thermal and Photochemical Initiators for Cationic Polymerization. Vol. 127, pp. 59–86.
Yamaguchi, I. see Yamamoto, T.: Vol. 177, pp. 181–208.
Yamamoto, T.: Molecular Dynamics Modeling of the Crystal-Melt Interfaces and the Growth of Chain Folded Lamellae. Vol. 191, pp. 37–85.
Yamamoto, T., Yamaguchi, I. and *Yasuda, T.*: PAEs with Heteroaromatic Rings. Vol. 177, pp. 181–208.
Yamaoka, H.: Polymer Materials for Fusion Reactors. Vol. 105, pp. 117–144.
Yamazaki, S. see Hikosaka, M.: Vol. 191, pp. 137–186.
Yannas, I. V.: Tissue Regeneration Templates Based on Collagen-Glycosaminoglycan Copolymers. Vol. 122, pp. 219–244.

Yang, J. see Geil, P. H.: Vol. 180, pp. 89–159.
Yang, J. S. see Jo, W. H.: Vol. 156, pp. 1–52.
Yasuda, H. and *Ihara, E.*: Rare Earth Metal-Initiated Living Polymerizations of Polar and Nonpolar Monomers. Vol. 133, pp. 53–102.
Yasuda, T. see Yamamoto, T.: Vol. 177, pp. 181–208.
Yaszemski, M. J. see Thomson, R. C.: Vol. 122, pp. 245–274.
Yoo, T. see Quirk, R. P.: Vol. 153, pp. 67–162.
Yoon, D. Y. see Hedrick, J. L.: Vol. 141, pp. 1–44.
Yoshida, H. and *Ichikawa, T.*: Electron Spin Studies of Free Radicals in Irradiated Polymers. Vol. 105, pp. 3–36.

Zhang, G. and *Wu, C.*: Folding and Formation of Mesoglobules in Dilute Copolymer Solutions. Vol. 195, pp. 101–176.
Zhang, H. see Rühe, J.: Vol. 165, pp. 79–150.
Zhang, Y.: Synchrotron Radiation Direct Photo Etching of Polymers. Vol. 168, pp. 291–340.
Zheng, J. and *Swager, T. M.*: Poly(arylene ethynylene)s in Chemosensing and Biosensing. Vol. 177, pp. 151–177.
Zhou, H. see Jiang, M.: Vol. 146, pp. 121–194.
Zhou, Z. see Abe, A.: Vol. 181, pp. 121–152.
Zubov, V. P., Ivanov, A. E. and *Saburov, V. V.*: Polymer-Coated Adsorbents for the Separation of Biopolymers and Particles. Vol. 104, pp. 135–176.

Subject Index

Acrylic acid   *I* 112
Adsorption selectivity   *I* 90
Adsorption-tuned copolymers   *I* 23, 90
AIBN   *I* 108
Alzheimer's disease   *I* 80
Amino acids   *I* 10
Amphiphilic monomers   *I* 177
Amphiphilic polymers   *II* 5
Amyloid fibrils   *I* 80
ATRP   *I* 8

Biopolymers   *II* 189
Blockiness, heterogeneous   *II* 87

Catalysis, micellar   *I* 196
Cholic acid   *I* 113
Cluster aggregation, diffusion-limited/reaction-limited   *I* 151
Coil-globule transition   *I* 79, 86, 102, 105, 118; *II* 189
Collapse transition   *I* 54
Colloidal stability   *II* 1
Configurational statistics   *II* 133
Conformation-dependent sequence design (CDSD)   *I* 1, 8
Controlled radical polymerization   *I* 8
Critical aggregation concentration   *I* 202

Degenerative chain transfer technique   *I* 8
Density functional theory   *I* 57
Derjaguin-Landau-Verwey-Overbeek theory (DLVO)   *II* 4
Designability of conformation   *II* 205
Diethylacrylamide (DEA)   *I* 155
Dimethylacrylamide (DMA)   *I* 155
DNA   *I* 15; *II* 189
–, heteropolymer, melting   *II* 203
DNA collapse   *II* 194

Emulsion copolymerization   *I* 36

Gels, superabsorbing   *II* 189
Gibbs energy   *II* 3
Globular proteins   *I* 179
Globules   *II* 1, 189
Glycidyl methacrylate   *II* 92

HA copolymer   *I* 84
Heteropolymers, charged   *I* 1
HP model   *I* 9, 179
Hydrophobic effect   *II* 5
Hydrophobization   *II* 94

Immobilized metal chelate chromatography (IMCC)   *II* 117
Interfaces   *I* 177
Ionomers   *I* 145
N-Isopropylacrylamide   *I* 108, 183; *II* 14, 92

Laser light scattering   *I* 106, 114
LCST   *I* 188; *II* 12
Lifshitz, I.M.   *II* 189
Long-range correlations   *I* 14

Markovian copolymers   *II* 135
Mesoglobular phase   *I* 154
Mesoglobules   *II* 1
–, colloidal stability   *II* 63
Methyl methacrylate   *I* 187
N-Methyldiethanolamine   *II* 121
Molecular dispenser   *I* 24, 91; *II* 202
Molecular structure distribution   *II* 132

Neutral amphiphilic polymers   *II* 5
NIPAM   *I* 108, 183; *II* 14, 92
p-Nitrophenyl acetate   *I* 196
Nitrophenyl alkanoates   *I* 202
Nonradiative energy transfer   *II* 54

Oil-water boundary   *II* 168

P(DEA-co-DMA)   *I* 113
Pair correlation functions   *I* 59
PAM-co-NaAA   *I* 151
PDEVP   *I* 202
PEO   *I* 110, 189; *II* 4, 192
PEO-methacrylates   *II* 46
Phase behavior   *I* 57
Phenolfluorene   *II* 122
PMMA   *I* 85
PNIPAM   *I* 102, 105, 188; *II* 1, 12, 15
PNIPAM-co-PEO   *II* 29
Poly(acrylamide)   *I* 112
Poly(*N*-alkylacrylamide)   *II* 15
Poly(*N*,*N*-diethylacrylamide)   *I* 113, 170
Poly(ethylene oxide) PEO   *I* 110, 189; *II* 4, 192
Poly(*N*-isopropylacrylamide) PNIPAM   *I* 102, 105, 188; *II* 1, 12, 15
Poly(*N*-vinylamides)   *II* 36
Poly(*N*-vinylcaprolactam)   *I* 189; *II* 1, 12, 36
Poly(vinyl methyl ether)   *II* 1, 12, 59
Polyacrylamide, hydrophobized   *II* 95
Polyamphiphiles   *I* 1
Polyelectrolyte gels   *II* 196
Polyelectrolytes, hydrophobic   *I* 70
Polymer catalysts   *I* 196
Polymeranalogous reactions   *II* 130
Polystyrene, maltopentaose   *I* 186
Protein globules   *I* 79
Protein-like copolymers   *I* 10; *II* 27, 74, 87

PVCL   *I* 189; *II* 1, 12, 36
PVME   *II* 1, 12, 59

Random energy model   *II* 196
Random phase approximation   *I* 57
Repulsive forces   *II* 4
RISM theory   *I* 57, 58, 63

SAXS   *I* 185
Scattering   *II* 152
Self-consistent field methods (SCF)   *I* 63
Self-consistent mean field (SCMF)   *I* 57
Sequence assembly   *I* 57
Sequence design   *II* 197, 199, 202
Sequence evolution   *I* 26
Size-composition distribution   *II* 132
Solution properties   *I* 1
Solvent-accessible surface-area   *I* 21
Stable free-radical polymerization   *I* 8
Styrene, maltopentaose   *I* 186
Superabsorbing gels   *II* 189
Surface nanoreactor   *I* 198

Terephthaloyl dichloride   *II* 122

N-Vinylcaprolactam   *I* 35, 183; *II* 98
N-Vinylimidazole   *I* 35; *II* 98
N-Vinylpyrrolidone   *I* 109, 158, 183; *II* 14, 119

Zipping transition   *I* 84

Printing: Krips bv, Meppel
Binding: Stürtz, Würzburg